Textbook of
Fluid
Mechanics

Textbook of
Fluid
Mechanics

Suparna Mukhopadhyay MTech PhD
Faculty
Power Management Institute
National Thermal Power Corporation (NTPC)
NOIDA, UP
former
Faculty/Scientist, Engineering Mechanics Unit, JN Centre for Advanced Scientific Research
Bangalore, Karnataka
Assistant Professor, Mechanical Engineering, Kalyani Engineering College, Nadia, WB
Assistant Professor, Mechanical Engineering, Lingaya's Institute of Technology and Management, Faridabad, Haryana
Assistant Professor, Mechanical Engineering, Institute of Technology and Management
Gurgaon, Haryana

CBS Publishers & Distributors Pvt Ltd

New Delhi • Bengaluru • Chennai • Kochi • Mumbai • Pune
Hyderabad • Kolkata • Nagpur • Patna • Vijayawada

Textbook of
Fluid Mechanics

ISBN: 978-81-239-2340-6

First Edition: 2014

Reprint:2017

Published by Satish Kumar Jain for

CBS Publishers & Distributors Pvt Ltd

4819/XI Prahlad Street, 24 Ansari Road, Daryaganj, New Delhi 110 002, India.

Ph: 23289259, 23266861, 23266867   Fax: 011-23243014   Website: www.cbspd.com
e-mail: delhi@cbspd.com; cbspubs@airtelmail.in

Corporate Office: 204 FIE, Industrial Area, Patparganj, Delhi 110 092

Ph: 4934 4934   Fax: 4934 4935   e-mail: publishing@cbspd.com; publicity@cbspd.com

Branches

- **Bengaluru:** Seema House 2975, 17th Cross, K.R. Road,
  Banasankari 2nd Stage, Bengaluru 560 070, Karnataka
  Ph: +91-80-26771678/79   Fax: +91-80-26771680   e-mail: bangalore@cbspd.com
- **Chennai:** 20, West Park Road, Shenoy Nagar, Chennai 600 030, Tamil Nadu
  Ph: +91-44-26260666, 26208620   Fax: +91-44-42032115   e-mail: chennai@cbspd.com
- **Kochi:** 36/14 Kalluvilakam, Lissie Hospital Road, Kochi 682 018, Kerala
  Ph: +91-484-4059061-65   Fax: +91-484-4059065   e-mail: kochi@cbspd.com
- **Mumbai:** 83-C, Dr E Moses Road, Worli, Mumbai-400018, Maharashtra
  Ph: +91-9833017933   e-mail: mumbai@cbspd.com
- **Pune:** Bhuruk Prestige, Sr. No. 52/12/2+1+3/2 Narhe, Haveli
  (Near Katraj-Dehu Road Bypass), Pune 411 041, Maharashtra
  Ph: +91-20-64704058, 64704059, 32392277 Fax: +91-20-24300160   e-mail: pune@cbspd.com

Representatives

- **Hyderabad**   0-9885175004
- **Nagpur**   0-9021734563
- **Kolkata**   0-9831437309, 0-9051152362
- **Patna**   0-9334159340
- **Vijayawada**   0-9000660880

Printed at India Binding House, Noida, UP

to

my father
Mr Tarapada Mukhopadhyay

and

my students

Foreword

Emphasis on the enhanced generation by exploiting the potential of renewable energy, particularly hydroelectricity, has gained importance in the light of worldwide environmental concerns. In this context, knowledge of the subject of fluid mechanics, which forms the foundation for the development of the hydroelectrical power plants, is crucial for the engineering students and practising engineers.

Dr Suparna Mukhopadhyay has been involved in teaching fluid mechanics and other mechanical engineering subjects for nearly 15 years. She has a strong academic, research and industrial background to her credit. This book *Textbook of Fluid Mechanics,* written by her, aims to bridge the knowledge-gap in this field. I am happy to note that the book has been able to capture the practical need of fluid mechanics with respect to pipe flow, open channel flow and instrumentation needed for fluid mechanics and its contemporary subject hydraulics.

The language of the book is lucid and is modeled on the approach of 'telling you'. The book, therefore, will not only be immensely useful to mechanical engineering students but also to the students of other disciplines and those who wish to acquire the knowledge on fluid mechanics. In addition, readers with a quantitative mindset, eager to build rigorous problem-solving technique, will also be benefitted by the contents.

Chapters on properties of fluid, dimensional analysis and fluid kinematics, key tool for design of blades of turbines used in hydroelectric power plants, would particularly be useful to power professionals.

There is always need for books which do not entirely depend on complex presentation but are aimed to enhance the reader's understanding. The coverage of topics in this book is very comprehensive and the content and style, easy and handy. I hope that the book would be very useful and serve its intended purpose. I extend my heartiest congratulation to Dr Suparna on this endeavour and effort of hers.

Best wishes

Avinash Chandra Chaturvedi

Executive Director
National Thermal Power Corporation (NTPC)

From Mentor's desk

Writing a technical book for undergraduate students is always a difficult task because on one hand it requires dealing the subject lucidly to get it understood by young knowledge-seeker, and on the other hand, the content is updated with the latest information.

I am glad to see that *Textbook of Fluid Mechanics* by Dr Suparna Mukhopadhyay has covered the whole gamut of fluid mechanics in a simpler way to help the undergraduate students to assimilate the subject.

Solving numerical problems is another difficult area for the students who require skills to catch the problem from the right perspective. I would like to appreciate Dr Mukhopadhyay for taking all the pains to add last eight years question papers with solutions so that the students can acquire problem-solving skills for solving problems.

I convey my best wishes for this endeavour.

Pradip Chanda
Additional General Manager
National Thermal Power Corporation (PMI)

Preface

Concepts of fluid mechanics are important for all engineering disciplines. I have been teaching this subject for the past 20 years and basically this textbook is based on my lecture notes, and prompted by the requests from groups of my students in Bangalore, Haryana, Delhi and West Bengal, in different phases of my teaching life, to write this book. Acknowledgement is due and is hereby made to all the authors and students along with my family members and NTPC. Any suggestions from the students and my fellow teachers will be acknowledged.

I believe that the chapter-wise style of writing the text with solved problems will not only help the undergraduate level students but also help the candidates preparing for AMIE examinations, and other professionals.

Finally, I wish to thank the publisher, CBS Publishers & Distributors, New Delhi, for bringing out the book in a pleasing format.

Dr Suparna Mukhopadhyay

Contents

Part A

Properties of Fluids

Introduction, properties of fluids, viscosity, surface tension and capillarity, vapour pressure and cavitation, thermodynamic properties.

1. INTRODUCTION

1.1 Concept of Fluid and Flow

A matter exists in either the solid state or the fluid state. The fluid state is further divided into liquid and gaseous state.

1.1.1 Continuum concept of a fluid

Fluid is considered as a continuum—a hypothetical continuous substance. A continuous and homogenous fluid medium is called continuum. Quantities such as velocity and pressure can then be considered to be constant at any point.

1.2 What is Fluid?

A fluid is a substance which deforms continuously under the action of shearing forces, however small they may be. Example: Liquids and Gases.

1.3 Differences between Solids and Fluids

(a) For a solid, the strain is a function of the applied stress, providing that the elastic limit is not exceeded.

For a fluid, the rate of strain is proportional to the applied stress and the strain in a fluid is independent of time over which the force is applied and, if the elastic limit is not exceeded.

A fluid continues to flow for as long as the force is applied and will not recover its original form when the force is removed.

(b) In solids, the molecules are very closely spaced.

In fluids, the spacing is relatively large. As such in a given volume, a solid contains a large number of molecules and a gas contains less number of molecules.

1.4 Difference between Liquids and Gases

(a) A liquid is difficult to compress and for many purposes, may be regarded as incompressible.

A gas is comparatively easy to compress.

(b) A given mass of liquid occupies a fixed volume irrespective of the size or shape of its container and a free surface is formed, if the volume of the container is greater than that of liquid.

Changes of volume with pressure is large, cannot normally be neglected and are related to changes of temperature. A given mass of a gas has no fixed volume and will expand continuously unless restrained by a containing vessel in which it is placed and therefore does not form a free surface.

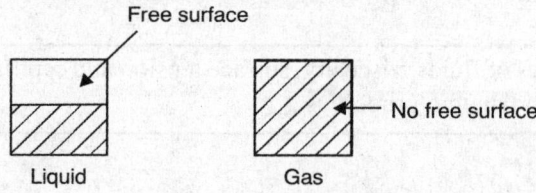

· **Fig. 1.1:** Gases and liquids

1.5 Properties of Fluids

1. **Pressure (P):** The normal stress on any plane through a fluid element at rest. The direction of pressure force will always be perpendicular to the surface of interest. It is force per unit area.

2. **Velocity:** It is the rate of change at a point in a flow field. It is used not only to specify flow field characteristics but also to specify flow rate, momentum and viscous effects for a fluid in motion unit m/s.

3. **Specific volume:** It is the volume of a fluid is the volume occupied by a unit mass.

4. **Density or mass density:** Density or mass density of a fluid is defined as the ratio of the mass of fluid to its volume. Mass per unit volume of a fluid is called density. (ρ) kg/m^3 (SI).

   Value of density of water is 1 gm/cm^3 or 1000 kg/m^3.

5. **Specific weight or weight density:** Ratio between the weight of a fluid to its volume.

$$W = \frac{\text{Weight of fluid}}{\text{Volume of fluid}} = \frac{\text{Mass of fluid} \times \text{Acceleration due to grains}}{\text{Volume of fluid}}$$

$$= \frac{\text{Mass of fluid} \times g}{\text{Volume of fluid}}$$

$$W = \rho \times g$$

for water, it is 9.81×1000 newton/m^3

6. **Specific Gravity:** (Relative density) is defined as the ratio of weight density of a fluid to the weight density of a standard fluid. It is a dimensionless property.

Example:

1. Calculate the specific weight, density and specific gravity of one litre of a liquid which weighs 7N.

$$\text{Volume} = 1 \text{ litre} = \frac{1}{1000}\,m^3 \left(1 \text{ litre} = \frac{1}{1000}\,m^3\right)$$

$$\text{Weight} = 7N.$$

(i) $\quad$ Specific weight $(W) = \dfrac{\text{Weight}}{\text{Volume}} = \dfrac{7N}{\left(\dfrac{1}{1000}\right)m^3} = 7000 \text{ N/m}^3$

(ii) $\qquad\qquad$ Density $(\rho) = \dfrac{W}{g} = \dfrac{7000}{9.81}\,kg/m^3 = 713.5\,kg/m^3$

(iii) $\qquad$ Specific gravity $= \dfrac{\text{Density of liquid}}{\text{Density of water}}$

$$= \frac{713.5}{1000}(\rho \text{ of water} = 1000\,kg/m) = 0.7135$$

7. **Viscosity:** Viscosity is defined as the property of a fluid which offers resistance to the movement of layer of fluid over another adjacent layer of the fluid.

   When two layers of a fluid, a distance '*dy*' apart, move one over the other at different velocities, (say *u* and *u* + *du*), the viscosity along with relative velocity causes a shear stress acting between the fluid layers.

   Top layer causes a shear stress on the adjacent lower layer while the lower layer causes a shear stress on the adjacent top layer.

   This shear stress is proportional to the rate of change of velocity with respect to *y*. It is denoted by symbol τ called Tau.

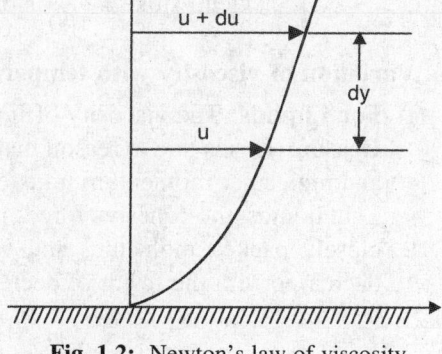

$$\tau \propto \frac{du}{dy}$$

$$\boxed{\tau = \mu\,\frac{d}{dy}}$$

Fig. 1.2: Newton's law of viscosity

This equation is known as Newton's law of viscosity

μ = constant of proportionality

$\quad$ = dynamic viscosity

$\dfrac{dy}{dx}$ = rate of shear strain or rate of shear deformation or velocity gradient.

$$\therefore \quad \mu = \frac{\tau}{\dfrac{du}{dy}}$$

∴ shear stress of fluid element is directly proportional to rate of shear strain

$$\mu = \frac{\text{Shear stress}}{\text{Change of velocity/Change of distance}}$$

$$= \frac{\text{Force} \times \text{time}}{(\text{Length})^2}$$

In S.I. units, μ is $\dfrac{\text{Newton-sec}}{\text{m}^3} = \dfrac{\text{Ns}}{\text{m}^2} = \text{PaS (Pascal)}.$

Kinematic Viscosity (ν)

It is defined as the ratio between dynamic viscosity and density of fluid, nu(ν)

$$\nu = \frac{\mu}{\rho}$$

$$\nu = \frac{(\text{length})^2}{\text{time}}$$

$$\text{One stoke} = \text{cm}^2/\text{s} = \left(\frac{1}{100}\right)^2 \text{m}^2/\text{s} = 10^{-4}\,\text{m}^2/\text{s}$$

$$\text{Centi stoke} = \frac{1}{100}\ \text{stoke}.$$

a. Variation of viscosity with temperature.

(i) **For Liquids:** The viscosity of liquids decreases with the increase of temperature.
Reason: This is due to reason that the viscous forces in a fluid are due to cohesive forces and molecular momentum transfer.

In liquids, the cohesive forces predominates the molecular momentum transfer, due to closely packed molecules and with the increase in temperature, the cohesive forces decreases with the result of decreasing viscosity.

$$\mu = \mu_0\left(\frac{1}{1 + \alpha t + \beta t^2}\right)$$

μ = viscosity of liquid at t°C in poise.

μ_0 = viscosity of liquid at °C, in poise

α, β = are constants for the liquids.

$\mu_0 = 1.79 \times 10^{-3}$ poise, $\alpha = 0.03368$ and $\beta = 0.000221$.

(ii) For a gas, $\mu = \mu_0 + \alpha t - \beta t^2$

$\mu_0 = 0.000017$, $\alpha = 0.0000000\,5C$, $\beta = 0.1189 \times 10^{-9}$

But in case of gases the cohesive force are small and molecular momentum transfer predominates with the increase in temperature molecular momentum transfer increases and hence viscosity increases.

b. Types of Fluids

1. Ideal fluid 2. Real fluid 3. Newtonian fluid 4. Non-newtonian fluid 5. Ideal plastic fluid.

1. **Ideal fluid:** A fluid which is incompressible and is having no viscosity, is known as ideal Fluid. Ideal Fluid is only an imaginary fluid.
2. **Real fluid:** A fluid, which possesses viscosity, is known as real fluid.
3. **Newtonian fluid:** A real fluid, in which the shear stress is directly, proportional to the rate of shear strain (or velocity gradient) is known as a newtonian fluid.
4. **Non-newtonian fluid:** A real fluid, in which the shear stress is not proportional to the rate of shear strain (or velocity gradient), known as a non-newtonian fluid.
5. **Ideal plastic fluid:** A fluid, in which the shear stress is more than the yield value and shear stress is proportional to rate of shear stress, is known as ideal plastic fluid.

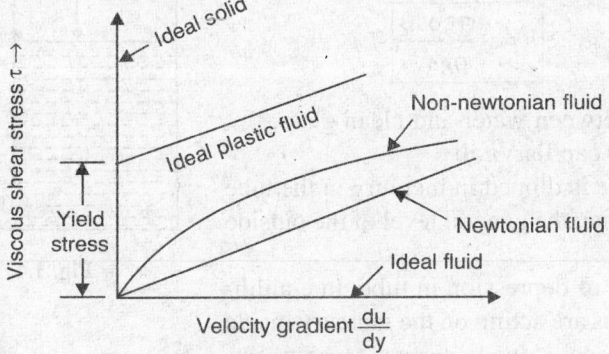

Fig. 1.3: Types of fluids

1.6 Capillarity

Capillarity is defined as phenomenon of rise or fall of a liquid surface in a small tube relative to the adjacent general level of liquid when the tube is held vertically in the liquid.

The rise of liquid surface is known as *capillarity rise* while the fall of liquid surface is known as *capillary depression*.

Capillarity is expressed in terms of cm or mm of liquid. It's value depends upon the specific weight of the liquid diameter of the tube and surface of the liquid.

1.6.1 Expression for capillary rise

(a) Consider a glass tube of small diameter d opened at both ends and is inserted in a liquid, say water. The liquid will rise in the tube above the level of the liquid.

Let h = height of the liquid in the tube under a state of equilibrium and the weight of liquid of height h is balanced by the force at the surface of the liquid in the tube.

But the force at the surface of the liquid in the tube is due to surface tension.

Let σ = surface tension of liquid.

θ = angle of contact between liquid and glass tube.

The wt. of liquid of height h in the tube = (area of tube × h) ρ × g

$$\frac{\pi d^2}{4} \times h \times \rho g \qquad \qquad \ldots(1)$$

Vertical component of the surface tensile force

$$= (\sigma \times \text{circumference}) \times \cos\theta$$
$$= \sigma \times \pi d \cos\theta \qquad \qquad \ldots(2)$$

For equilibrium, equation (1) and (2) are equal.

$$\frac{\pi}{4} d^2 \times h\rho g = \sigma \pi d \cos\theta$$

$$h = \frac{\sigma \pi d \cos\theta}{\frac{\pi}{4} d^2 \rho g} = \frac{\sigma \cos\theta}{\rho g d}$$

$$\boxed{h = \frac{4\sigma \cos\theta}{\rho g d}}$$

∴ Value of θ = 0 between water and clean glass,

(b) Expression for capillary fall

If the glass tube is dipped in mercury in the tube will be lower than the general level of the outside liquid.

Let h = Height of depression in tube. In equilibrium, two forces are acting on the mercury inside the tube. First one is due to surface tension, acting in downward direction and is equal to $\sigma \pi d \cos\theta$. Second force is due to hydrostatic force acting upward and is equal to intensity of pressure at a depth h × area

Force due to surface tension = $\sigma \pi d \cos\theta$ ...(1)

Force due to hydrostatic force = $\rho \times A$

$$\rho A = \rho g h \frac{\pi d^2}{4} \qquad \ldots(2)$$

Equating these two,

$$\sigma \times \pi d \times \cos\theta = \rho g h \frac{\pi d^2}{4}$$

$$\boxed{h = \frac{4\sigma \cos\theta}{\rho g d}}$$

Value of θ for mercury and glass tube = 128°

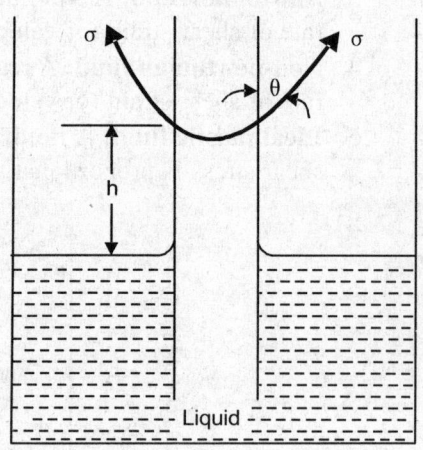

Fig. 1.4: Capillary rise

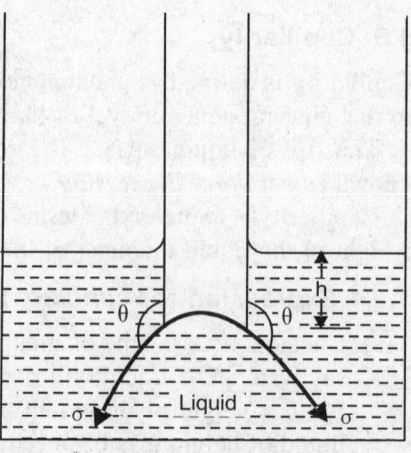

Fig. 1.5: Capillary fall

1.7 Surface Tension

Surface tension is defined as the tensile force acting on the surface of a liquid in contact with a gas or on the surface between two immiscible liquid.

Surface tension has the dimension of force per unit length, or energy per unit area. This two are equivalent—but when referring to energy per unit of area, the term surface energy is used which is more general term applicable to solids, not just liquids.

1.7.1 Cause

Surface tension is caused by the attraction between the liquid's molecules by various inter molecular forces.

It is denoted by the symbol σ (sigma).

1.7.2 Effects in everyday life

1. Formation of drop occurs when a mass of liquid is stretched.
2. Boarding of rain water on the surface of a waxed automobile.

1.7.3 Surface tension on liquid droplet

Considering a small spherical droplet of a liquid of radius r. On the entire surface of liquid, droplet, the tensile force due to surface tension, will be acting.

Let,
σ (Sigma) = Surface tension.

p = Pressure inside the droplet.

d = diameter of the droplet.

Let the droplet is imagined to cut into two halves.

Now, forces acting on one half.

1. Tensile force due to surface tension, acting around the circumference of the cut portion as shown:

$$\text{Tensile force} = \sigma \times \text{circumference}$$
$$= \sigma \times \pi d \qquad \ldots(1)$$

Pressure force on area = pressure intensity × area

$$= P \times \frac{\pi}{4} d^2 \qquad \ldots(2)$$

Two forces of equation (1) and (2) will be equal and opposite under equivalent condition.

$$\sigma \times \pi d = P \times \frac{\pi}{4} d^2$$

$$\boxed{P = \frac{4\sigma}{d}}$$

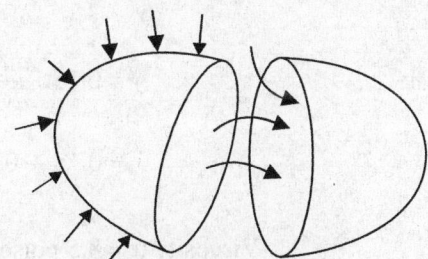

Fig. 1.6: Surface tension

On hollow bubble

Two surfaces in contact with air, so for one is acting towards inside, and other is acting towards outside.

$$\therefore \qquad P \times \frac{\pi}{4} d^2 = 2 \times (\sigma \times \pi d)$$

$$\boxed{P = \frac{8\sigma}{d}}$$

On a liquid jet

One liquid jet of diameter d and length L is considered, P = pressure intensity inside the liquid jet above the outside pressure.

Considering the equilibrium,

Force due to pressure = pressure intensity × area of semijet

$$= P \times L \times d$$

Force due to surface tension,

Equating the forces,

$$P \times L \times d = \sigma \times 2L$$

$$P = \frac{\sigma \times 2L}{L \times d}$$

$$\boxed{P = \frac{2\sigma}{d}}$$

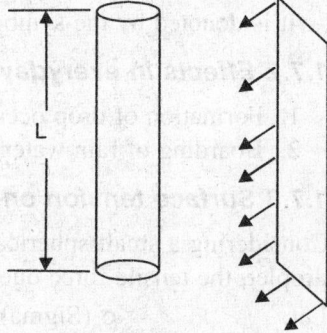

Fig. 1.7

Example:

1. Velocity distribution for flow over a flat plate is given by $u = \frac{3}{4} y - y^2$ in which u is the velocity in m per second at a distance y metre above the plate. Determine the shear stress at $y = 0.15$ m. Take dynamic viscosity of fluid as 8.6 poise.

$$u = \frac{3}{4} y - y^2$$

$$\frac{du}{dy} = \frac{3}{4} - 2y$$

at
$$y = 0.15, \quad \frac{du}{dy} = \frac{3}{4} - 2 \times 0.15$$

$$= 0.75 - 0.30 = 0.45$$

Viscosity $\mu = 8.5$ poise $= \frac{8.5}{10} \frac{\text{NS}}{\text{m}^2} = \left(10 \text{ poise} = 1 \frac{\text{NS}}{\text{m}^2} \right)$

$$\tau = \mu \frac{du}{dy} = \frac{8.5}{10} \times 0.45 \frac{\text{N}}{\text{m}^2} = 0.3825 \frac{\text{N}}{\text{m}^2}$$

2. Determine the viscosity of a liquid having kinematic viscosity 6 stokes and specific gravity 1.9.

$$\text{Kinematic viscosity} = 6 \text{ stokes} = 6 \text{ cm}^2/\text{s} = 6 \times 10^{-4} \text{ m}^2/\text{s}$$

Sp. gn of liquid = 1.9

Viscosity of liquid = μ

$$\text{Sp. gn of liquid} = \frac{\text{DS of liquid}}{\text{D of water}}$$

$$1.9 = \frac{D \text{ of } L}{1000} = 19000 \text{ kg}/\text{m}^3$$

$$\nu = \frac{\mu}{\rho}$$

$$6 \times 10^{-4} = \frac{\mu}{1900}$$

$$\mu = 114 \text{ NS/m}^2 = 1.14 \times 10 = 11.40 \text{ poise.}$$

3. An oil of viscosity 5 poise is used for lubrication between a shaft and sleeve. The diameter of shaft is 0.5 m and it rotates at 200 r.p.m. Calculate power cost in oil for a sleeve length of 100 mm. The thickness of oil film 1.0 mm.

Viscosity, $\mu = 5$ poise,

$$= \frac{5}{10} = 0.5 \text{ NS}/\text{m}^2$$

Dia. of shaft $D = 0.5$ m

Speed of shaft $N = 200$ N.p.m.

Sleeve length, $L = 100$ mm $= 100 \times 10^{-3}$ m $= 0.1$ m

Thickness of oil film,

$$t = 1.0 \text{ mm} = 1 \times 10^{-3} \text{m}$$

Tangential velocity of shaft $u = \dfrac{\pi D N}{60}$

$$= \frac{\pi \times 0.5 \times 200}{60}$$

$$= 5.235 \text{ m/s}$$

$$\tau = \mu \frac{du}{dy}$$

$du = 4 - 0 = 4$

$dy = t = 1 \times 10^{-3} \text{ m}$

$\tau = \dfrac{0.5 \times 5.235}{1 \times 10^7} = 2617.5 \text{SN}/\text{m}^2$

Power cost $= T \times W$

$$= T \times \frac{2\pi N}{60} W$$

$$= 2.15 \text{ kW}$$

$$\tau \times A$$

$$\text{Shear force} = \text{shear stress} \times \text{area}$$

$$= 2617.5 \times \pi D \times L$$

$$= 0.5 \times 0.1 = 410.95 \text{ N}$$

$$T = \text{Force} \times \frac{D}{2\pi} = 107.74 \text{ Nm}$$

4. Calculate the capillary rise in a glass tube of 2.5 mm dia when immersed vertically in (a) water (b) mercury.

   Take surface tension $\sigma = 0.0725$ N/m for water and $\sigma = 0.52$ N/m for mercury in contact with air. The specific gravity for mercury is given as 13.6, angle of contact = 130°

$$d = 2.5 \text{ mm} = 2.5 \times 10^{-3} \text{ m}$$

Surface tension, σ for water = 0.0725 N/m

$$\sigma \text{ for mercury} = 0.52 \text{ N/m}$$

sp gr for mercury = 13.6

$$\rho \text{ for mercury} = 13.6 \times 1000 \text{ kg/m}^3$$

(a)   Capillary rise for water $\theta = 0$

$$h = \frac{4\sigma}{\rho \times g \times d} = \frac{4 \times 0.0725}{1000 \times 9.81 \times 2.5 \times 10^{-3}}$$

$$= 0.0118 \text{ m} = 1.18 \text{ cm}$$

(b) For mercury,

$$\theta = 130°$$

$$h = \frac{4\sigma \cos\theta}{\rho \times g \times d} = \frac{4 \times 0.52 \times \cos 130°}{13.6 \times 1000 \times 981 \times 2.5 \times 10^{-3}}$$

$$= -0.004 \text{ m} = -0.4 \text{ cm}$$

$$-\text{ve indicates depression.}$$

5. Surface tension of water in contact with air at 20°C is 0.0725 N/m. The pressure inside a droplet of water is to be 0.02 N/cm^2 greater than the outside pressure. Calculate the diameter of droplet of water.

   Surface tension, $\sigma = 0.0725$ N/m.

   Pressure intensity, P in excess of outside pressure in

$$P = 0.02 \text{ N/cm}^2 = 0.02 \times 10^4 \frac{\text{N}}{\text{m}^2}$$

$$d = \text{dia of droplet,}$$

$$P = \frac{4\sigma}{d} \Big/ r, \, 0.02 \times 10^4 = \frac{4 \times 0.0725}{d}$$

$$d = \frac{4 \times 0.0725}{0.02 \times 10^4} = .00145 \text{ m} = 0.00145 \times 1000$$

$$= 1.45 \text{ mm.}$$

6. Find the surface tension in a soap bubble of 40 mm dia when the inside pressure is 2.5 N/m^2 above atmospheric pressure dia of bubble, $d = 40$ mm $= 40 \times 10^{-3}$ m.

Pressure in excess of outside, $P = 2.5$ N/m^2. For a soap bubble,

$$P = \frac{8\sigma}{d} \text{ or, } 2.5 = \frac{8 \times \sigma}{40 \times 10^{-3}}$$

$$\sigma = \frac{2.5 \times 40 \times 10^{-3}}{8} \text{ N/m}^2 = 0.0125 \text{ N/m}$$

1.8 Vapour Pressure (also known as equilibrium vapour pressure)

Definition: Vapour pressure is the pressure of a vapour in equilibrium with its non-vapour phases.

Importance: Every solid and liquid have a tendency to evaporate to a gaseous form. All gases also have a tendency to condense back into their original form (either liquid or solid).

At any given temperature, for a particular substance, there is a pressure at which the gas of that substance is in dynamic equilibrium with its liquid or solid forms. This is the vapour pressure of that substance at that temperature.

The equilibrium vapour pressure is an indication of a liquid's evaporation rate.

It relates to the tendency of molecules and atoms to escape from a liquid or a solid.

A substance with a high vapour pressure at normal temperatures is often referred to as volatile.

Relation between vapour pressure and normal boiling points of liquids

The higher the vapour pressure of a liquid at a given temperature, the lower the normal boiling point (i.e. the boiling point at atmospheric pressure) of the liquid.

For example, at any given temperature, propane has the higher vapour pressure of any of the liquids. It also has the lowest normal boiling point (−42.1°C).

1.9 Cavitation

According to the Bernoulli Equation, cavitation occurs when fluid accelerates in a control valve or around a pump impeller.

Cavities or bubbles are formed in the liquid when the pressure of fluid falls below vapour pressure. When the bubbles pass into higher region of pressure, they force liquid energy into very small volumes, thereby creating spots of high temperature and emitting shock waves.

1.9.1 Definition

The phenomenon of forming the cavity and bursting of it to emit shock waves which are a source of noise is called cavitation.

1.9.2 Reasons

The cavities form for five basic reasons

(a) Vapourization

(b) Internal recirculation

(c) Flow turbulence

(d) Air ingestion

(e) The vane passing syndrome.

This can be done by:

(a) Reengineering components initiating high speed velocities and low static pressures.

(b) Increasing the total or local static pressure in the system.

(c) Reducing the temperature of the fluid.

Cavitation Damage

(a) Cavitation may happen when the fluid accelerates in a control valve or around a pump impeller at the low pressure or suction side of the pump, causing several things to happen all at once.

(b) Cavitation, causes undesirable occurrence, a damage to many of components, loss of capacity, less lead, drop of efficiency, in turbines.

(c) The noise created by cavitation is a particular problem for military submarines, as it increases the chances of being detected by passive sonar.

Avoiding Cavitation

Cavitation can in general be avoided by increasing the distance between the actual static pressure in the fluid and the vapour pressure of the fluid at the actual temperature.

Application

Although the collapse of a cavity is a relatively low energy causes a great deal of noise, some cavitation may be used as follows:

1. **Chemical engineering application:** In industry, cavitation is often used to homogenize, or mix and break down, suspended particles in a colloidal liquid compound such as point mixtures' or milk.

2. **Biomedical application:** Cavitation plays an important role for destruction of kidney stones in shock wave lithotripsy.

3. **Cleaning application:** In industrial cleaning application, cavitation has sufficient power to overcome the particle to substate adhesion forces, loosening contaminants.

1.10 Thermodynamic Properties

Liquids and gases behaves differently with the change in temperature. As gases are compressible fluids, thermodynamic properties will change with the change of pressure and temperature. With the change of pressure and temperature, the gases undergo large variation in density.

Relationship between absolute pressure, specific volume and absolute temperature of a gas is given by

$$PV = RT$$

or

$$\frac{P}{\rho} = RT$$

where,
$$P = \text{Absolute pressure of a gas in N/m}^2$$

$$v = \text{Specific volume} = \frac{1}{\rho}$$

$$R = \text{Gas constant}$$

$$T = \text{Absolute temperature is OK.}$$

$$\rho = \text{Density of gas.}$$

1.10.1 Isothermal process

If the changes in density occurs at constant temperature, then the process is called isothermal and relationship between pressure p and density ρ is given by,

$$\frac{P}{\rho} = \text{constant.}$$

1.10.2 Adiabatic process

If the change in density occurs with no heat exchange to and from the gas, the process is called adiabatic.

If no heat is generated within the gas due to friction, the relationship pressure and density, is given by

$$\frac{P}{\rho^k} = \text{constant.}$$

where, K = Ratio of specific heat of a gas at constant pressure and constant = 1.4

SOLVED PROBLEMS

---

1. A lubricating oil of viscosity undergoes steady shear between a fixed lower plate and an upper plate moving at speed v. The clearance between the plates is h. Show that a linear velocity profile results if the fluid does not slip at either plate.

   The shear stress τ is constant throughout the fluid for the given geometry and motion and from Newtons law of viscosity.

   $$\frac{du}{dy} = \frac{\tau}{\mu} = \text{constant}$$

   $$u = a + by \qquad \qquad \ldots(1)$$

   The constants a and b are calculated from no slip conditions at the upper and lower plates.

   $$u = 0 \text{ at } y = 0, a = 0$$
   $$u = v \text{ at } y = h, v = a + bh$$

Hence $a = 0$, $b = \dfrac{V}{h}$, the velocity profile between the plates is then given by $u = \dfrac{Vy}{h}$, is linear as indicated

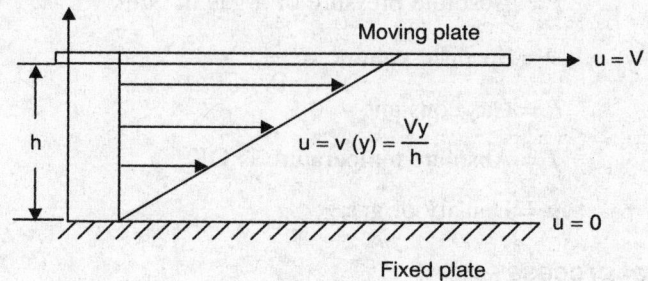

Fig. 1.8

Hence $a = 0$ and $b = \dfrac{V}{h}$, velocity profile between the plates is then given by

$$b = \frac{Vy}{h},$$

Viscous shear stress is given by Newton's law of viscosity

$$\tau = \mu \frac{du}{dy}$$

$$u = 14 \text{ poise} = 1.4 \text{ NS/m}^2$$

$$du = 2.5 \text{ m/s}, \quad dy = 1.25 \times 10^{-2} \text{ m}.$$

$$\tau = 1.4 \times \frac{2.5}{1.24 \times 10^{-2}}$$

$$= 280 \text{ N/m}^2 = 280 \text{ pa.}$$

2. A fluid of absolute viscosity 8 poise flows past a flat plate and has a velocity 100 cm/s at the vertex which is at 20 cm from the plate surface. Find velocity gradient and shear stress at 3 different points 5, 10 and 15 cm from the boundary.

Ans. $u = 8 \text{ poise} = 0.8 \text{ NS/m}^2$

For a straight line velocity distribution, the velocity gradient at the boundary, *i.e.* at $y = 0$ is

$$\frac{du}{dy} = \frac{100 - \sigma}{20 - 0} = 5 \text{ per sec.}$$

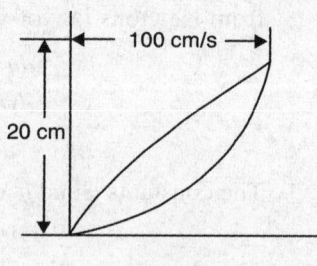

Shear stress $\tau = \mu \dfrac{du}{dy} = 0.8 \times 5 = 4 \text{ N/m}^2.$

For $y = 5$, 10 and 15 cm, the velocity gradient $\dfrac{du}{dy}$ and

Fig. 1.9

the shear stress τ would be also 5 per sec and 4 N/m^2.

3. Air is introduced through a nozzle into a tank of water to form a stream of bubbles. If the bubbles are intended to have a diameter of 2 mm calculate the pressure of the air at the

nozzle. Assuming the surface tension of water is 0.073 N/m, what would be absolute pressure inside the bubble if the surrounding water is at 100 Kps?

Ans. Excess pressure intensity of air over that surrounding water is,

$$P_i - P_0 = \frac{4\sigma}{d} = \frac{4 \times 0.023}{2 \times 10^{-3}} = 146 \, \text{N/m}^2.$$

Pressure outside the droplet P_0 = 100 Kps = 10,000 N/m².

Pressure inside the droplet = 146 + 100000

$$= 100146 \, \text{N/m}^2 = 100146 \, \text{KPA}.$$

4. Two square flat plates with side 50 cm are spaced 15 mm apart. If the lower plate is stationary and the upper plate requires a force of 100 N to keep it moving with a velocity of 2.5 m/s. The oil film between the plates has the same velocity as that of plates at the surface of constant.

Assuming a linear velocity distribution, determine kinematic and dynamic viscosity of the oil in stokes in poise.

$$\text{Shearing stress } \tau = \mu \frac{du}{dy}$$

$$= \mu \frac{u}{t}$$

$$= \mu \times \frac{2.5}{0.025}$$

$$= 100 \, \mu$$

Shearing area = $0.5 \times 0.5 = 0.25 \, \text{m}^2$

Shearing force = $\tau \times A = 100 \, u \times 0.25 = 25 \, u$

But, given force = 100 N

$$25 \, u = 100$$

$$u = 4 \, \text{Ns/m}^2 = 40 \, \text{poise}$$

Dynamic viscosity = 40 poise

Kinematic viscosity of the oil

$$v = \frac{\mu}{\rho} = \frac{4}{0.95 \times 10000} = 4.2 \times 10^{-3} \, \text{m}^2/\text{s}$$

$$= 42 \times 10^{-4} \, \text{m}^2/\text{s}$$

$$= 42 \, \text{cm}^2/\text{s}$$

$$= 42 \, \text{stokes}.$$

5. Calculate the capillary effects in millimeter in a glass tube of 4 mm diameter, when immersed in (i) water (ii) mercury. The temperature of the liquid in 20° C and the values of surface tension of water and mercury at 20° C in contact with air are 0.0735 N/m and 0.48 N/m. The contact angle of water $\theta = 0°$ and mercury $\theta = 130°$.

Ans. The rise or depression h of a liquid in a capillary tube

$$h = \frac{4\sigma \cos \theta}{wd}$$

(i) Capillary effect in water,

$\sigma = 0.0735$ N/m, angle of contact $\theta = 0$,

$w = 9800$ N/m^2 at 20° C.

$$h = \frac{4 \times 0.0735 \times 0.50^2}{9800 \times 0.0004}$$

$$= 7.50 \times 10^{-3} \text{ m}$$

(ii) Capillary effect in mercury,

$\sigma = 0.48$ N/m, angle of contact $\theta = 130°$

$w = (9800 \times 13.6)$ N/m^2.

$$h = \frac{4 \times 0.48 \times \cos 136°}{9800 \times 13.6 \times 0.004}$$

$$= -2.31 \times 10^{-3} \text{ m} = 2.31 \text{ mm}.$$

EXERCISE

1. A cylinder roller gate 3 m in diameter is placed on the dam in such a way that water is just going to spill. If the length of gate is 6 m, calculate the magnitude and direction.
   [Ans.: 336.73 KN, $\theta = 38.13°$]

2. A sliding gate 2 m wide × 1 m height and weighing 20 KN has been located in a vertical plane with its upper edge at a depth of 5 m from the surface of water. Calculate the vertical force required to raise the gate. Coefficient of friction of 0.12. **[Ans.: 32.95 KN]**

3. A rectangular box with base 2.5 m × 4 m is filled with kerosene oil of specific gravity 0.8 to a depth of 6 m. Determine the resultant pressure and its point of application on the base and on each vertical face of the box. **[Ans.: 470880 N, 353160 N, 565056 N]**

4. A square plate 4 m × 4 m hangs in water from one of its corners and its centroid lies at a depth of 8 m from the free water surface. Workout the total pressure on the plate and locate the position of centre of pressure with respect to the plate centroid.
   [Ans.: 1255680 N, 8.167 m]

Unit

2

Fluid Statics

Introduction, fluid pressure at a point, Pascal's law, pressure variation in a static fluid, absolute, gauge, atmospheric and vacuum pressures, simple manometers, differential manometers, total pressure and center of pressure, vertical plane surface submerged in liquid, horizontal plane surface submerged in liquid, inclined plane surface submerged in liquid, curved surface submerged in liquid, buoyancy, center of buoyancy, metacenter and metacentric height, conditions of equilibrium of floating and submerged bodies.

2. INTRODUCTION

2.1 Fluid Pressure at a Point

The general rules of statics apply to fluids at rest.

- A static fluid can have no shearing force acting on it.
- Any force between the fluid and the boundary must be acting at right angles to the boundary.

2.2 Fluid Pressure at a Point (Pascal's law)

Considering a small area dA in large mass of fluid.

Force exerted by the surrounding fluid on the area dF, will always be perpendicular to the surface dA.

$$\frac{dF}{dA} = \text{intensity at pressure.}$$

$$P = \frac{dF}{dA}$$

$$\boxed{P = \frac{F}{A} = \frac{\text{Force}}{\text{Area}}}$$

Newton/m^2 or N/m^2

N/mm^2 → SI

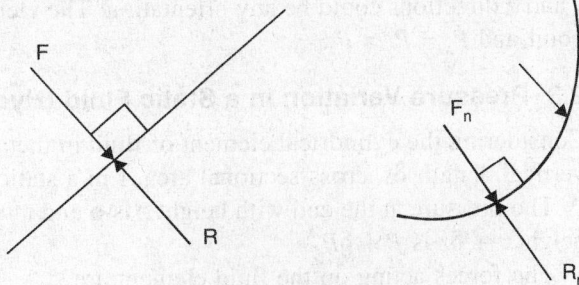

Fig. 2.1: Fluid pressure

Kgf/m² and kgf/cm² → MKS
KPa = Kilo Pascal = 1000 N/m²
1 bar = 100 kpa = 10^5 N/m⁻².
Dimensions $ML^{-1}T^{-2}$

Pascal's Law

Statement: It states that pressure of or intensity of pressure at a point in a static fluid is equal in all directions.

Considering an arbitrary fluid element of wedge shape.

P_x, P_y, P_z are the acting on force *AB*, *AC* and *BC* respectively, $\angle ABC = \theta$.

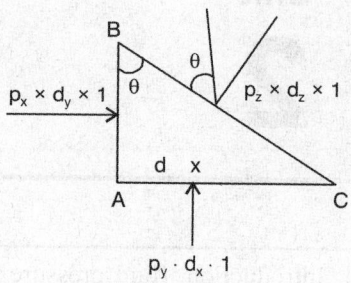

2.2.1 Forces acting on the element

1. Pressure forces normal to surface.
2. Weight of element in vertical direction.

Force on face $AB = P_x \times$ Area of face *AB*

$= P_x \times d_y \times 1$

Force on face $AC = P_y \times d_x \times 1$

Force on face $BC = P_z \times d_z \times 1$

Weight of element = (Mass of element) × g

= (Volume × ρ) × g

Resolving forces in *x* direction,

$$P_x \times d_y \times 1 - p_z\, d_z \sin(90° - \theta) = 0$$
$$d_z \cos\theta = AB = d_y$$
$$P_x = P_z$$

As,
$$d_s \sin\theta = d_x.$$
$$P_y\, d_x - P_z \times d_x = 0$$
$$P_y = P_z$$
$$P_x = P_y = P_z$$

Fig. 2.1: Pascal's law

This proves that pressure at any point is the same in all directions, This is known as Pascal's law.

Considering the prismatic element again, P_s is the pressure on a plane at any angle θ, the *x*, *y* and *z* directions could be any orientation. The element is so small that it can be considered a point and $P_x = P_y = P_z$.

2.3 Pressure Variation in a Static Fluid (Hydrostatic Law)

Considering the cylindrical element of fluid in the above figure, inclined at any angle θ to the vertical, length δs, cross-sectional area *A* in a static fluid of mass density ρ.

The pressure at the end with height *z* is ρ and at the end with height *z* is *P* and at the end of height *z* + δz is *P* + δ*P*.

The forces acting on the fluid element are

(a) Pressure force *PA* acting at right angles to the end of the face at *z*.

(b) Pressure force (*P* + δ*P*) *A* acting at right angles to the end of the face at *z* + *Sz*.

(c) Weight of the fluid element = Density × g × volume = ρ*gA*δ*S* cosθ.

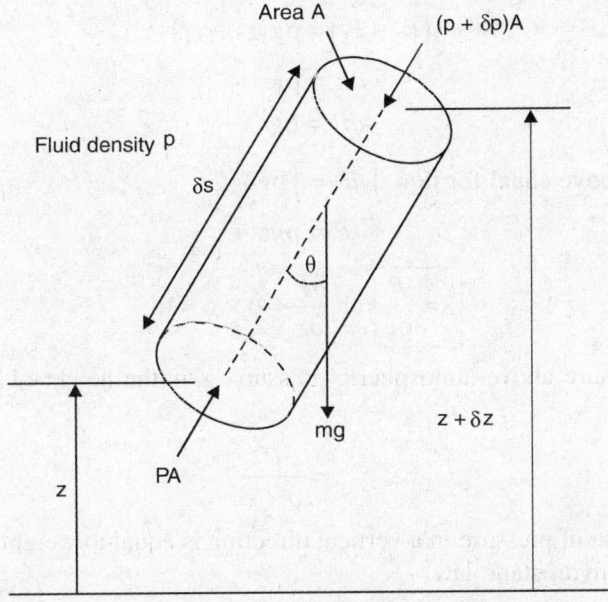

Fig. 2.2: A cylindrical element of fluid at an arbitrary orientation

For equilibrium of the element, the resultant of forces in any direction is zero.
Resolving the forces in the direction along the central axis,

$$PA - (P + \delta P)A - \rho g A \delta S \cos\theta = 0$$

$$\delta P = -\rho g \delta S \cos\theta$$

$$\frac{\delta P}{\delta S} = -\rho g \cos\theta$$

or, in the differential form,

$$\frac{dP}{dS} = -\rho g \cos\theta$$

if $\theta = 90°$, then s in the x or y directions (i.e. horizontal) so

$$\left(\frac{dP}{ds}\right)_{\theta=90°} = \frac{dP}{dx} = \frac{dP}{dy} = 0$$

confirming that pressure on any horizontal plane is zero.
if $\theta = 0°$, then s is in the z direction (vertical),

$$\left(\frac{dP}{ds}\right)_{\theta=0} = \frac{dP}{dz} = -\rho g$$

confirming the result,

$$\frac{P_2 - P_1}{z_2 - z_1} = \rho g$$

$\Rightarrow$ $$(P_2 - P_1) = \rho g (z_2 - z_1)$$

$\Rightarrow$ $$dP = \rho gz$$

$\Rightarrow$ $$dP = \rho g dz$$

by integrating, the above equal for liqn. $\int dP = \int \rho g dz$

$$P = \rho gz$$

$$\boxed{z = \frac{P}{\rho g}} \text{ or } \frac{\partial P}{\partial z} = \rho \times g = W$$

where, P is the pressure above atmospheric pressure, Z is the height of the point from free surface,

$$z = \frac{P}{\rho \times g}$$

The rate of increase of pressure in a vertical direction is equal to weight density of the fluid at that point. This is hydrostatic law.

2.4 Absolute, Gauge, Atmospheric and Vacuum Pressure

In a static fluid of constant density,

$$\frac{dP}{dz} = -\rho g,$$

This can be integrated to give,

$$p = -\rho gz + \text{constant},$$

In a liquid with a free surface, the pressure at any depth z, measured from the free surface, so that $z = -h$,

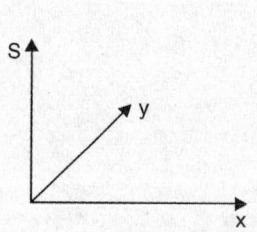
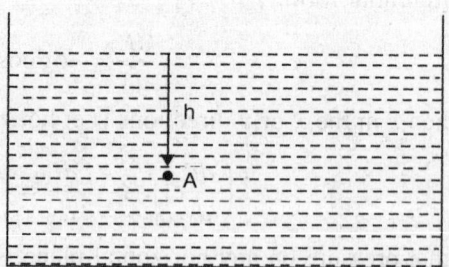

Fig. 2.3

(i) Atmospheric pressure:

In case of fluid head measurment in a tank, at the surface of fluids, the pressure is atmospheric pressure, $P_{atmosphere}$

So,

$$P = \rho gh + P_{atmos}.$$

Atmospheric pressure ($P_{atm.}$) varies with altitude, because the air nearer the earth's surface is compressed by air above. At sea level, value of atmospheric pressure is close to 1.01325 bar or 760 mm of Hg column or 10.55 m of water column.

(ii) Absolute Pressure

Pressure is force per unit area due to interaction of fluid particles amongst them selves. A zero pressure intensity will occur when molecular momentum is zero.

Pressure intensity measured from this state of vacuum or zero pressure is called absolute pressure.

(iii) Gauge pressure (Pg.) and Vacuum pressure.

When the unknown pressure is more than atmospheric pressure, the pressure recorded by the instrument is called gauge pressure. A pressure reading below the atmospheric pressure is known as vacuum, rarefaction or negative pressure.

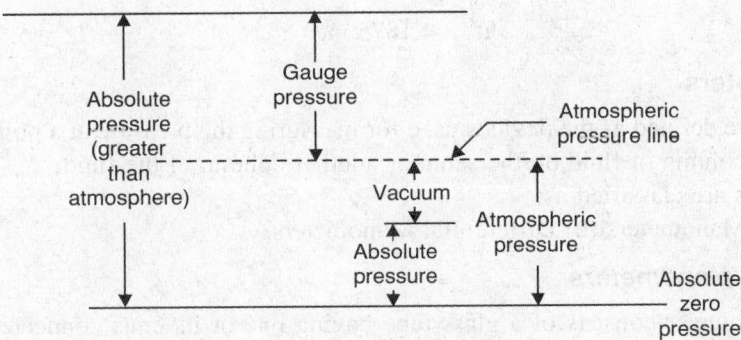

Fig. 2.4: Different types of pressure

Problem 1. A rectangular plane surface is 2 m wide and 3 m deep. It lies in vertical plane in water. Determine the total pr and position of centre of pressure on the plane surface when its upper edge is horizontal (a) coincides with water surface (b) 2 m below the free water surface.

Width of plane surface $b = 2$ m

Depth of plane surface $d = 3$ m

(a) Upper edge coincide with water surface.

$$F = \rho g A \bar{h}$$

$$\rho = 1000 \text{ kg/m}^3, \ g = 9.81 \text{ m/s}^2$$

$$A = 3 \times 2 = 6 \text{ m}, \ \pi = \frac{1}{2} \times 3 = 1.5 \text{ m}$$

$$F = 1000 \times 9.81 \times 6 \times 1.5$$

$$= 88290 \text{ N}.$$

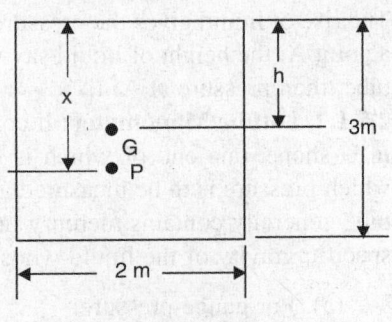

Depth of centre of pressure, $h^* = \dfrac{I_G}{A\bar{h}} + \bar{h}$

Fig. 2.5

$$I_G = \text{M.O.I.}$$

$$= \frac{bd^3}{12} = \frac{2 \times 3^3}{12} = 4.5 \text{ m}^4$$

$$h = \frac{4.5}{6 \times 1.5} + 1.5 = 0.5 + 1.5 = 2.0 \text{ m}$$

(b) Upper edge is 2.5 m blow with surface.

$$F = \rho g A \bar{h}$$

The distance of centre of pressure from free surface

$$= 2.5 + \frac{3}{2} = 4.0$$

$$F = 1000 \times 981 \times 6 \times 4.0 = 235440 W$$

$$I_N = 4.5, \ A = 6, \ \bar{h} = 4$$

$$h^A = 4.1875 \text{ m.}$$

2.5 Manometers

Manometers are defined as the devices used for measuring the pressure at a point in a fluid by balancing the column of fluid by the same or another column of the fluid.

Manometers are classified as:

(a) Simple Manometers (b) Differential Manometers.

2.5.1 Simple Manometers

A simple manometer consists of a glass tube having one of its ends connected point where pressure is to be measured and other end remains open to atmosphere.

Common types of simple manometers are:

1. Piezometer
2. U-tube Manometer and
3. Single Column Manometer

2.5.1.1 Piezometer: It is the simplest form of manometer used for measuring gauge pressures. One end of this mano-meter is connected to the point where pressure is to be measured and other end is open to the atmosphere as shown. The rise of liquid gives the pressure head at that point. If at a point A: the height of liquid say water is h in piezometer tube, then pressure at A: $(\rho \times g \times h)$ M/m^2

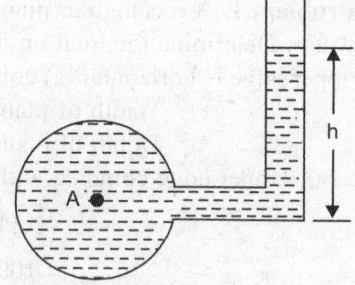

Fig. 2.6: Piezometer used for measuring gauge pressures

2.5.1.2 U-tube Manometer: It consists of glass tube bent in U-shape, one end of which is connected to a point at which pressure is to be measured and other end remains open to be atmosphere as shown. The tube generally contains mercury or any other liquid whose specific gravity is greater than the specific gravity of the liquid whose pressure is to be measured.

(a) For gauge pressure:

$$P = \left(\rho_2 g h_2 - \rho_1 g h_1 \right)$$

(b) For Vacuum pressure:

$$P = -\left(\rho_2 g h_2 + \rho_1 g h_1 \right)$$

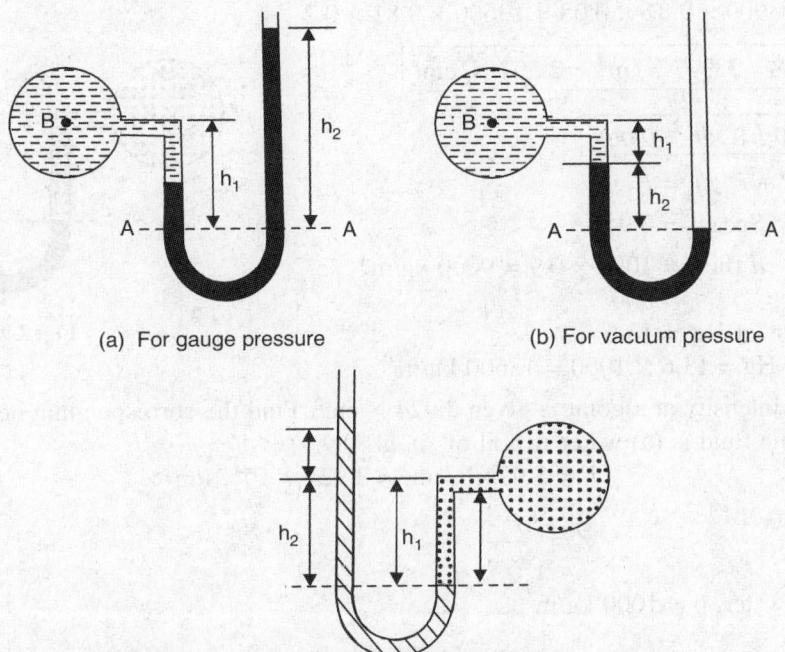

Fig. 2.7: U-tube manometer

1. The right limb of a simple U-tube manometer containing mercury is open to the atmosphere while the left limb is connected to a pipe in which a fluid of sp. gr. 0.9 is flowing. The centre of the pipe is 12 cm below the level of mercury in the height limb. Find the pressure of fluid in the pipe if the diff. of mercury level the two limbs is 20 cm.

Sp. gr. of fluid = 0.9.

Density of fluid, $\rho_1 e$ = density of water × sp. gr. of fluid

$$= 100\emptyset \times \frac{0.9}{1\emptyset}$$

$$\boxed{\therefore \rho_2 = 900 \ \text{kg/m}^3}$$

Sp. gravity of Hg = 13.6

∴ Density of mercury P_2 = 1000 × 13.6
= 13600 kg/mb³

Diff. of mercury level = h_2 = 20 cm = 0.2 m

Height of fluid from $A - A$;

h_1 = 20 − 12 = 8 cm = 0.08 m

Let P = pressure of fluid in the pipe.

Equating the pressure above $A - A$, we get:

$$P + P_1 g h_1 = P_2 g h_2$$

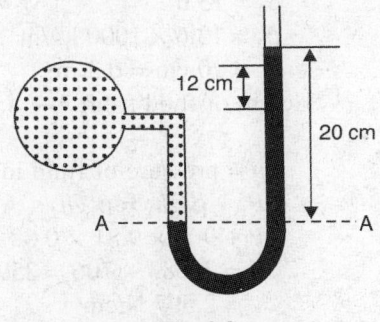

Fig. 2.8

$$\Rightarrow \quad P + 900 \times 9.81 \times 0.08 = 13600 \times 9.81 \times 0.2$$

$$\Rightarrow \quad \boxed{\therefore P = 25977 \ N/m^2 = 2.597 \ N/cm^2}$$

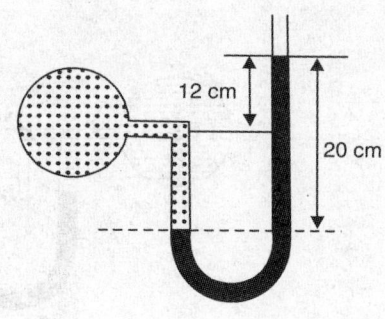

Fig. 2.9

$$\boxed{\rho + \rho_1 gh_1 = \rho_2 gh_2}$$
$$P + P_1 gh_1 = P_2 gh_2$$
$$Sp. \ gr_1 = 0.9$$
$$d \ \text{fluid} = 1000 \times 0.9 = 9000 \ kg/m^3$$
$$= 900$$
$$\text{sp. gr. of Hg} = 13.6$$
$$\therefore \quad d \ Hg = 13.6 \times 1000 = 13600 \ kg/m^3$$

1. The pr intensity at a point is given 3.924 N/cm². Find the corresponding height of fluid when the fluid is (a) water (b) oil of sp. gr. 0.9.

$$P = 3.924 \ N/cm^2 = 3.924 \times 10^4 \ N/m^2$$

$$Z = \frac{P}{\rho \times g}$$

(a) For water, $\rho = 1000 \ kg/m^3$

$$Z = \frac{P}{\rho \times g} = \frac{3.924 \times 10^4}{1000 \times 981} = 4 \ \text{m of water}$$

(b) For oil sp. gr. = 0.9
$\rho_0 = 0.9 \times 1000 = 900 \ kg/m^3$

$$Z = \frac{P}{\rho_0 \times g} = \frac{3.924 \times 10^4}{900 \times 9.81} = 4.44 \ \text{m at or 1.}$$

2. The right limb of a simple U tube manometer containing mercury is open to the atmosphere while the left limb is connected to a pipe in which a fluid of sp. gr. 0.9 is flowing. The centre of pipe is 12 cm below the level of mercury in the right limb. Find the pr of fluid in the pipe if the difference of mercury level in the two limbs is 20 cm.

$S_1 = 0.9$
$\rho_1 = S_1 \times 1000 = 0.7 \times 1000 = 900 \ kg/m^3$
$S_2 = 13.6$
$\rho_2 = 13.6 \times 1000 \ kg/m^3$
$h_2 = 20 \ cm = 0.2 \ m$

Height of fluid from $A - A$, $h_1 = 20 - 12 = 8 \ cm$
$$= 0.08 \ m$$

p = pressure of fluid in pipe
$P_1 + \rho_1 gh_1 = \rho_2 gh_2$
$P + 900 \times 9.81 \times 0.08 = 13.6 \times 1000 \times 9.81 \times 0.2$
$P = 26683 - 706 = 25977 \ N/m^2$
$$= 2.597 \ N/cm^2$$

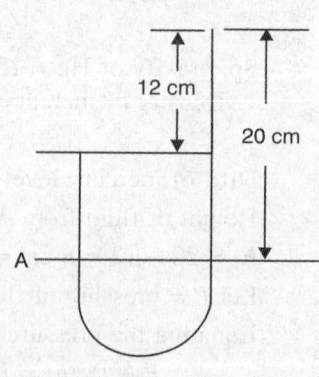

Fig. 2.10

Example: We can quote a pressure of 500 KNm^{-2} in terms of the height of a column of water of density, $p = 1000$ kgm^{-3}. Using $p = pgh$,

$$h = \frac{p}{\rho g} = \frac{500 \times 10^3}{1000 \times 9.81} = 50.95 \text{ m of water}$$

And in terms of Mercury with density, $\rho = 13.6 \times 10^3$ kgm^{-3}.

$$h = \frac{500 \times 10^3}{13.6 \times 10^3 \times 9.81} = 3.75 \text{ m of Mercury.}$$

2.5.1.3 Single Column Manometer: Single column manometer is a modified form of a U-tube manometer in which a reservoir, having a large cross-sectional area (about 100 times) as compared to the area of the tube is connected to one of the limbs (say left limb) of the manometer as shown. Due to large cross sectional area of the reservoir, for any variation in pressure, the change in the liquid level in the reservoir will be very small which may be neglected and hence the pressure is given by the height of liquid in the other limb. The other limb may be vertical or inclined.

Thus; there are two types of single column manometer as:
1. Vertical Single Column Manometer
2. Inclined Single Column Manometer

(i) *Vertical single column manometer*

Let $\Delta h \rightarrow$ fall of heavy liq. in reservoir

$\quad h \cdot 2 \rightarrow$ Rise of heavy liq. in right limb

$\quad P_A \rightarrow$ Pressure at A which is to be measured.

$\quad A \rightarrow$ Cross sectional area of the reservoir

$\quad S_1 \rightarrow$ sp. gr. of liq. in pipe

$\quad S_2 \rightarrow$ sp. gr. of heavy liq. in reservoir and right limb

$\quad \rho_1 \rightarrow$ Density of liquid in pipe

$\quad \rho_2 \rightarrow$ Density of liquid in reservoir

So; $\boxed{\Delta h = \dfrac{a \times h_2}{A}}$ $\quad \Leftarrow A \times \Delta h = a \times h_2$]

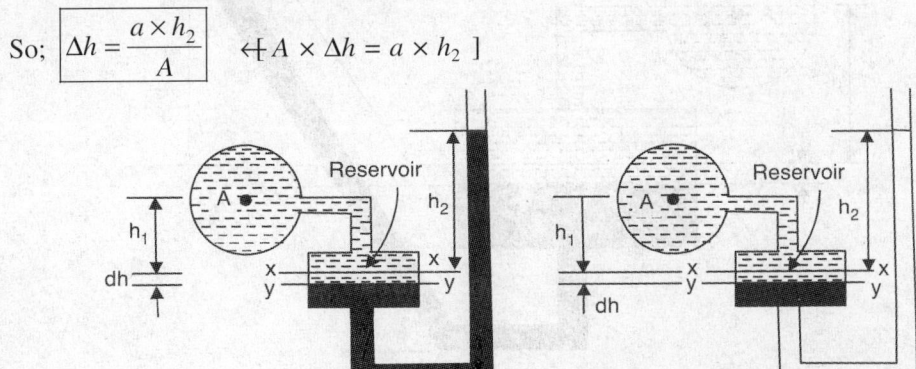

Fig. 2.11: Vertical single column manometer

and press in right limb above $y - y := P_2 \times g \times (\Delta h + h_2)$
and press in the left limb above $y - y = \rho_1 g(\Delta h + h_1) + P_A$

Equating pressures, we have:

$$\rho_2 \times g \times (\Delta h + h_2) = \rho_1 \times g \times (\Delta h + h_1) + p_A$$

But;
$$\Delta h = \frac{a \times h_2}{A}$$

$\therefore$
$$P_A = \frac{a \times h_2}{A} [\rho_2 g - \rho_1 g] + h_2 \rho_2 g - h_1 \rho_1 g$$

if A is very large as compared to $Q \dfrac{Q}{A} \approx 0$

$\therefore$
$$\boxed{\therefore P_A = h_2 \rho_2 g - h_1 \rho_1 g}$$

(ii) Inclined single column manometer

Figure shows the inclined single column manometer. This manometer is more sensitive. Due to inclination of the distance moved by the heavy liquid in the right limb will be more.

Let L = Length of heavy liquid moved in right limb from $X - X$

q = Inclination of right limb with horizontal

h_2 = Vertical rise of heavy liquid in right limb from $X - X = LX \sin\theta$

But; The pressure at A is:

$$P_A = h_2 \rho_2 g - h_1 \rho_1 g$$

Substituting the value of h_2; we get;

$$\boxed{P_A = \sin\theta \times \rho_2 g - h_1 \rho_1 g}$$

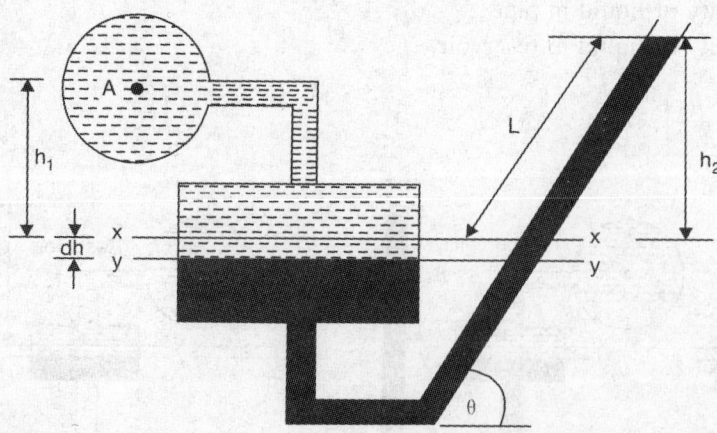

Fig. 2.12

2.5.2 Differential Manometer

Differential manometers are the devices used for measuring the difference of pressures between two points in a pipe or in two different pipes.

A differential manometer consists of a U-tube, containing a heavy liquid, whose two ends are connected to the points, whose difference of pressure is to be measured. Most commonly types of differential manometers are:

1. U-tube differential manometer.
2. Inverted U-tube differential manometer.

2.5.2.I U-Tube differential Manometer

Let (a) Two points are at different level.

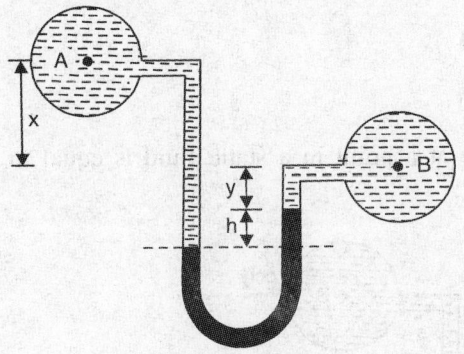 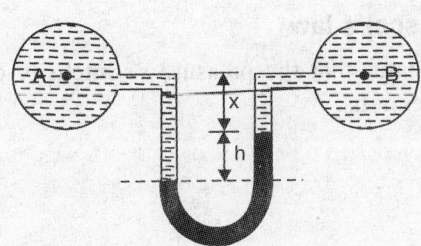

(a) Two types at different levels
U-tube differential manometers

(b) A → B are at the same level

Fig. 2.13

$h \to$ Different of Hg level in U-tube

$y \to$ Distance of centre of B, from the Hg level in right limb

$x \to$ Distance of Centre of A from the Hg level in right limb.

$\rho_1 \to$ Density of liq. at A

$\rho_2 \to$ Density of liq. at B

$\rho_g \to$ Density of heavy liq. or Hg.

(a) Two types at different levels
     U-tube different manometers

$$\rho_A - \rho_B = h \times g(\rho_g - \rho_1) + \rho_2 gy - \rho_1 gx$$

(b) A and B are at same level:

$$\rho_A - \rho_B = g \times h(\rho_g - \rho_1)$$

Taking datum line at X - X

Press above XX in left limb $= P_1 g \,(h + x) + P_A$ and press above XX in right limb $= P_y \times g \times h + \rho_2 \times g \times y + P_B$

Differential manometers are devices used for measuring the difference of pressures between two points in a pipe or in two different pipes.

It consists of a U-tube containing *a heavy liquid*, whose two ends are connected to the points, whose different of pressure is to be mean.

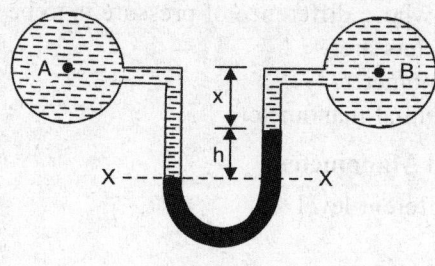

Fig. 2.14

Pascal's law

It states that the pressure or intensity of pressure at a point in a static fluid is equal in all directions.

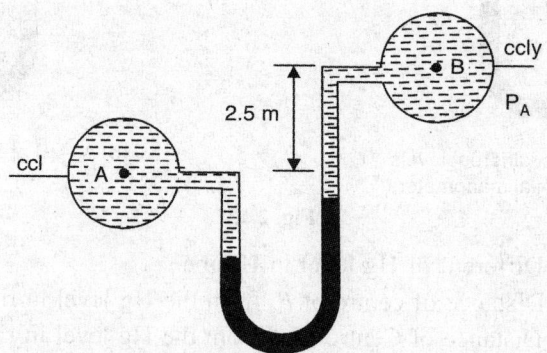

Fig. 2.15

Proof. Consider a fluid element of small dimensions d_x, d_y and d_s.

Consider an arbitrary fluid element of wedge shape in a fluid mass at rest. Let the width of the element is unity and *px*, *py* and *pz* are the pressure acting on faces *AB*, *AC*, *BC* respectively.

Let $\angle ABC = \theta$

Then; the forces acting on the element are:

1. Press forces normal to the surfaces
2. Weight of the element in the vertical direction.

The forces on the faces are:

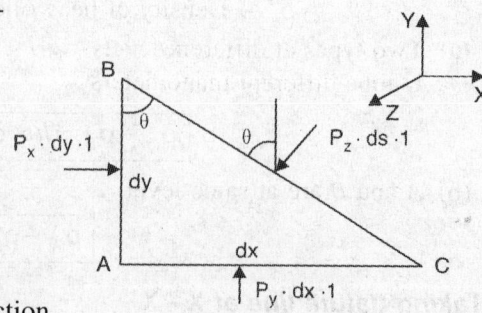

Fig. 2.16

Forces on face $AB = p_x \cdot dy \cdot 1$

Forces on face $AC = p_y \cdot dx \cdot 1$

Forces on face $BC = p_z \cdot ds \cdot 1$

and weight of element = (Mass of the element) $\times g$

$$= \text{vol.} \times \rho \times g = \left(\frac{AB \times AC}{2} \cdot 1 \right) \rho g$$

$$= \left(\frac{dx \cdot dy}{2} \cdot 1 \right) \rho g$$

Resolving the forces in x-direction:

$$P_x \cdot dy \cdot 1 - P_z ds \cdot 1 \cdot \sin(90 - \theta°) = 0$$

$$P_x dy - P_z ds \cos\theta = 0$$

$$P_x dy - P_z \cancel{ds} \cdot \frac{dy}{\cancel{ds}} = 0 \qquad \text{(Putting the value of cos } \theta)$$

$$P_x \cancel{dy} = P_z \cancel{dy}$$

$$\boxed{P_x = P_z}$$

$$\cos\theta = \frac{AB}{BC} = \frac{dy}{ds}$$

Now, resolving forces in y-direction

$$P_y \cdot dx \cdot 1 - P_2 ds \sin\theta - \frac{dx \cdot dy}{2} \cdot 1 \rho g = 0$$

if, $\sin\theta = \dfrac{AC}{BC} = \dfrac{dx}{ds}$

$$\boxed{\therefore P_y = P_z}$$

Advantages

Some advantage of manometers

- They are very simple.
- No calibration.

Disadvantages

- Slow Response—Only really useful for very slowly varying pressures—no use at all for fluctuating pressures.
- It is often difficult to measure small variations in pressure.
- For very accurate work, the temperature and relationship between temperature and density must be known.

2.6 Total Pressure and Centre of Total Pressure Force

It is defined as the force exerted by a static fluid on a surface either plane or curved when the fluid comes in contact with the surface. This force always acts normal to the surface.

2.6.1 Centre of Pressure

It is described as the point of application of the total pressure on the surface.

Four cases of submerged surfaces are here on which the total pressure force and centre of pressure is to be determined.

(a) Vertical plane surface

(b) Horizontal plane surface

(c) Inclined plane surface

(d) Curved surface.

(a) Vertical plane surface submerged in liquid

Consider a plane vertical surface of arbitrary shape immersed in a liquid.

Let A = Total area of the surface

$\bar{h}$ = Distance of C.G. of the area from free surface of liquid.

G = Centre of gravity of plane surface.

P = Centre of pressure

h^* = Distance of Centre of pressure from free surface of liquid.

(i) Total Pressure (F):

The total pressure on the surface may be determined by dividing the surface into a number of small parallel strips. The force on small strip is then calculated and the total pressure force on the whole area is calculated by integrating the force on small strip.

Consider a strip of thickness dh and width b at a depth of h from free surface of liquid as shown:

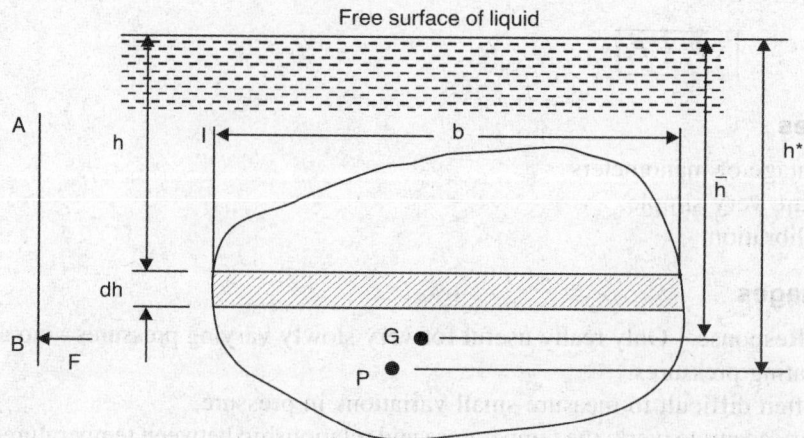

Fig. 2.17: Vertical plane surface

(b) Horizontal plane surface

A plane surface submerged and held in a horizontal position at depth y below the free surface of the liquid. Since every point on the surface is at the same depth, the pressure intensity is constant over the entire plane surface.

From hydrostatic equation $p = wh$, and A is the total area of the surface, then total pressure force on the horizontal surface is,

$$F = pA = wya = A\left(wy\right) = A\left(wy_c\right)$$

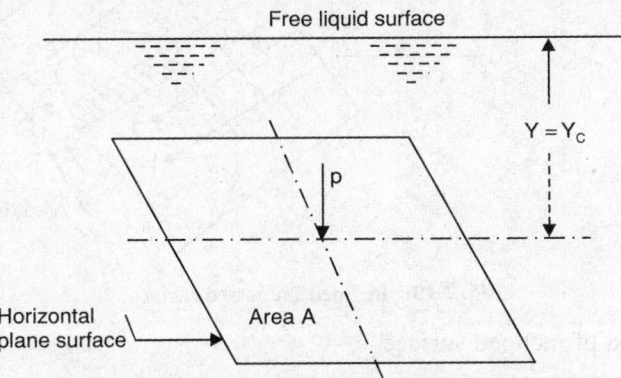

Fig. 2.18: Horizontal plane surface

Pressure intensity on the strip; $P = \rho gh$.

Area of strip, $dA = b \times dh$

Total press force on strip, $dF = P \times \text{Area} = Pgh \times b \times dh$

∴ Total pressure force on the whole surface:

$$F = \int dF = \int \rho gh \times b \times dh = \rho g \int b \times h \times dh$$

But;

$$\int b \times h \times dh = \int h \times dA$$

= Moment of surface area about the free surface of liquid.

= Area of surface × Distance of C.G. from free surface.

$$= A \times \overline{h}$$

$$\boxed{\therefore F = \rho g A \overline{h}} \quad \text{For water: } \rho = 1000 \text{ kg/m}^3, g = 9.8 \text{ m/sec}^2$$

(ii) Centre of Pressure (h*):

$$\boxed{h^* = \frac{I_G}{A\overline{h}} + \overline{h}}$$

$I_G \rightarrow$ moment of inertia of area about an axis passing through the C.G. of the area and parallel to the free surface of the liquid.

(c) Inclined plane surface submerged in liquid

Consider a plane surface of arbitrary shape immersed in a liquid in such a way that the plane of the surface makes an angle 'θ' with the free surface of the liquid as shown:

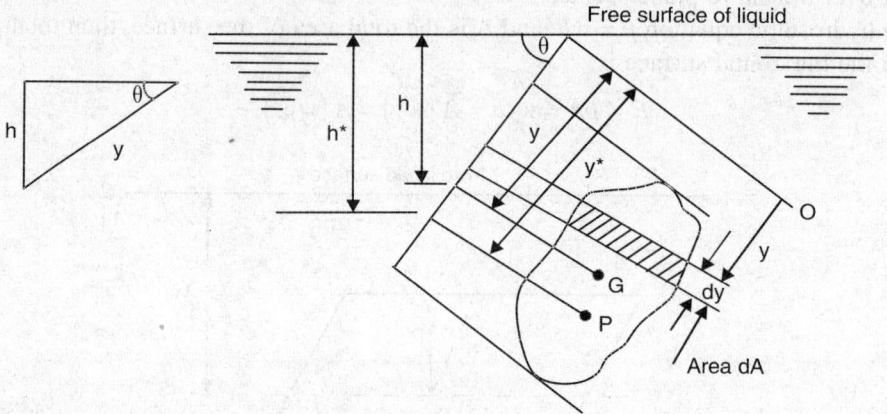

Fig. 2.19: Inclined immersed surface

Let A = Total area of inclined surface

$\bar{h}$ = Depth of C.G. of inclined area from free surface.

h^* = Distance of Centre of Pressure from free Surface of Liquid.

θ = Angle made by the plane of the surface with free liquid surface.

Let the plane of the surface, if produced meet the free liquid surface at 0. Then, $0 - 0$ is the axis ⊥, to the plane of the surface.

Let $\bar{y}$ = distance of the C.G. of the inclined surface from $0 - 0$.

y^* = distance of the centre of pressure from $0 - 0$.

Considering a plane surface submerged and held in a horizontal position. Since, every point of the surface is at the same depth from the free surface of the liquid, the pressure intensity will be equal on the entire surface and equal to $p = \rho g h$, h is the depth of surface,

Let, A = total area of surface

Total force

$$F = \text{Pressure intensity} \times \text{area}$$

$$= \rho g \times h \times A$$

$$= \rho g A \bar{h}$$

Here,

$\bar{h}$ = depth of centre of gravity from free surface of liquid = h

h^* = Depth of centre of pressure from free surface = h,

Now, consider a small strip of area dA at depth 'h' from free surface and a distance y from the axis $0 - 0$, as shown.

Pressure Intensity on the strip $P = \rho gh$.

∴ Press force, dF on the strip $= p \times$ Area of strip $= \rho gh \times dA$

Total pressure force on the whole area, $F = \int dF = \int \rho g h \, dA$

But; from fig: $\dfrac{h}{y} = \dfrac{\overline{h}}{\overline{y}} = \dfrac{h^*}{y^*} = \sin\theta$

$$\boxed{\therefore h = y\sin\theta}$$

∴
$$F = \int \rho g \times y \times \sin\theta \times dA = \rho g \sin\theta \int y \, dA$$

But; $\int y \, dA = A\overline{y}$

where; $\overline{y}$ = Distance of C.G. from axis $0 - 0$.

$$F = \rho g \cdot \sin\theta \, \overline{y} \times A$$

$$\boxed{\therefore F = \rho g A\overline{h}} \text{ where; } \boxed{\overline{h} = \overline{y}\sin\theta}$$

Centre of pressure (h*)

$$\boxed{h^* = \frac{I_G \cdot \sin^2\theta}{A\overline{h}} + \overline{h}}$$

2.7 Buoyancy

When a body is immersed in a fluid an upward force is exerted by the fluid on the body. Buoyancy force is defined as that upward force which is equal to the weight of the fluid displaced by the body.

Buoyancy force is a vertical force which is equal to the weight of the fluid displaced by the body.

So the centre of Buoyancy is defined as the point, through which the buoyancy force is supposed to act. It is the centre of gravity of fluid displaced.

Depending on the ratio of the weight W of a body/object and the buoyant force Fb, three cases are possible.

(i) $W > Fb$; the body tends to move downwards and sinks

(ii) $W = Fb$; the body/object floats and partially submerged

(iii) $W < Fb$; the body is lifted upwards and rises to the surface.

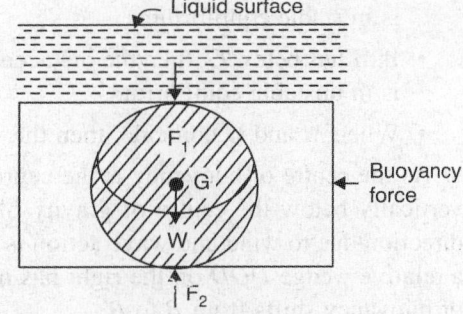

Fig. 2.20

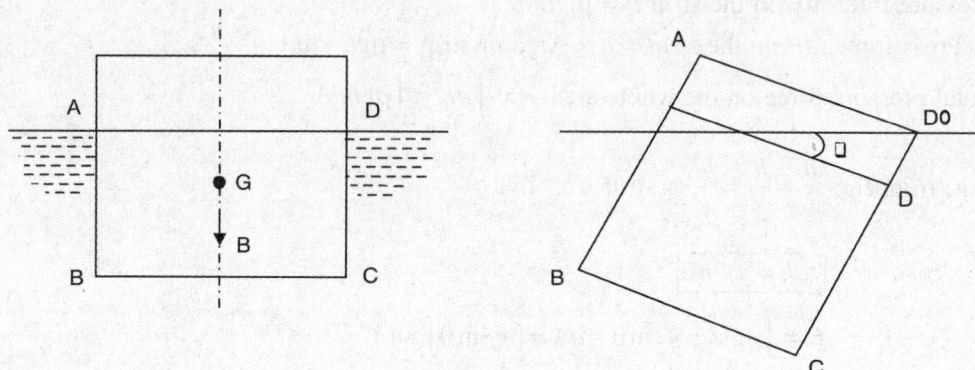

Fig. 2.21

The weight of the object/body/ship W acts vertically downwards through centre of gravity G and Buoyancy force F_B, acts through centre of Buoyancy B.

Conditions of equilibrium of floating and submerged bodies:

Condition	Submerged Body	Floating Body
Stable	Centre of gravity lies below centre of buoyancy.	Centre of gravity lies below metacentre.
Unstable	Centre of gravity, lies above centre of buoyancy.	Centre of gravity lies above metacentre.
Neutral	Centre of gravity coincides with centre of buoyancy.	Centre of gravity lies at metacentre.

2.8 Metacenter and Metacentric height

2.8.1 Stability of floating bodies

The stability of a floating is governed by the position of meta centre M relative to centre of gravity G (Fig. 2.22).

- If M lies above G, then the metacentric height GM is regarded as positive and the system is in stable equilibrium.

- If M lies below G, then the metacentric height GM is regarded as negative and the system is in unstable equilibrium.

- When M and G coincide, then the object/body in state.

As the centre of buoyancy is the centre of gravity of the immersed portion ($ABCD$), it lies vertically below the centre of gravity of the ship. If a tilting of small angle θ in clockwise direction due to wind and wave action is applied to $ABCD$ and submerged portion is $A'B'C'D'$ a relative wedge DOD on the right has moved into the field. Due to redistribution, the centre of buoyancy shifts from B to B'.

Stable, Unstable and Neutral Equilibrium

2.8.2 Definition

The point of intersection of the line of action of the buoyant force before and after the heel is defined as metacentre (M) and the distance between centre of gravity and meta centre is called as metacentric height (GM).

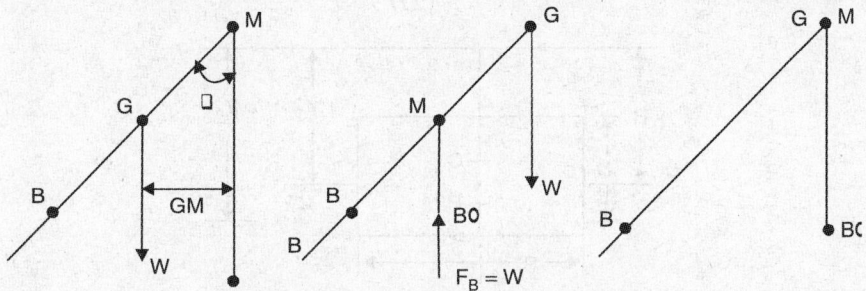

Fig. 2.22: Stable, unstable and neutral equilibrium

Position of meta centre for stability of floating bodies

The stabilizing/righting moment is $W \times GM$

where M = Displacement of centre of buoyancy.

$\qquad \theta$ = heel

$\therefore \quad WGM \sin \theta = WGM\, \theta = WGM \tan \theta$

2.8.3 Stability of submerged bodies

Stability of a floating or submerged body can have three possible conditions of equilibrium.

Difference and co-relation between these three conditions of equilibrium.

Property	Stable Equilibrium	Unstable Equilibrium	Neutral Equilibrium
1. Definition	A small displacement from the equilibrium position produces a righting moment tending to restore the body to original equilibrium position.	A small displacement produce a over turning moment tending to restore the body further to a condition different from the initial equilibrium position.	The body remains remains at rest in any position to which it may be displaced.
2. Position of MG	G is located below B.	G is located above B.	G is located above G.
3. Configuration			

SOLVED PROBLEMS

1. A rectangular sluice gate of Breadth B and depth D has been provided in the vertical is H units below the free water surface, show that the depth of pressure is given by

$$Y_P = \left(H + \frac{D^2}{12H} \right)$$

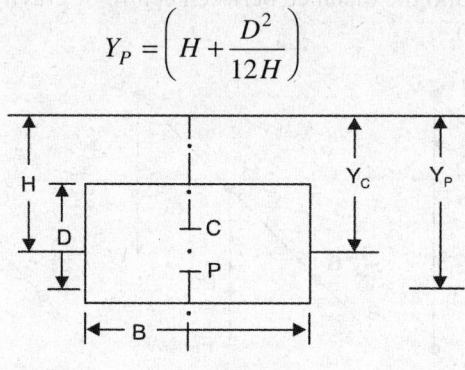

Fig. 2.23

Area of sluice gate $= B \times D$

Depth of centroid from the free surface $Y_C = H$.

Moment of inertia of the area about centre of gravity

$$I_C = \frac{BD^3}{12}$$

position of centre of pressure is given by,

$$Y_P = Y_C + \frac{I_C}{AY_C}$$

$$= H + \frac{BD^3/12}{BD \times H}$$

$$= \left(H + \frac{D^2}{12H} \right)$$

2. A rectangular plate of 2 m length and 1 m height lies immersed vertically in a liquid of relative density 0.75 such that 2 m side is parallel to and at a depth of 0.7 m from the free liquid surface of the plate has a circular hole of 0.5 m diameter drilled at its centre, represent the total pressure in metre.

Ans. Specific weight of liquid

$$= 0.75 \times 9810 = 7357.5 \text{ N/m}^3$$

$$\text{Area of plate, } A = 2 \times 1 - \frac{\pi}{4} (0.5)^2$$

$$= 1.804 \text{ m}^2$$

Depth of centroid of plate from the free surface,

$$Y_C = 0.7 + \frac{1}{2} = 1.2 \text{ m}.$$

∴   Total pressure force = 2g AY_C

$$= 7357.5 \times 1.804 \times 1.2$$
$$= 15927 \text{ N}.$$

Moment of inertia about the centroid axis parallel to free surface,

$$I_C = \frac{2 \times 1^3}{12} - \frac{\pi}{64}(0.5)^2$$
$$= 0.1636 \text{ m}^4$$

∴ Depth of centre of pressure,

Total pressure force = $F = w\, AY$

$$= (1000 \times 9.81) \times 3 \times 3.268$$
$$= 96177 \text{ N}.$$

Depth of centre of pressure from the water surface is,

$$Y_y = Y_C + \frac{I_C \sin^2 \theta}{AY_C}$$

where, $I_C = \dfrac{bh^3}{36} = \dfrac{2 \times 3^3}{36} = 1.5 \text{ m}^4$

$$\sin \theta = \sin 60° = 0.866$$

$$Y_p = 3.268 + \frac{1.5 \times (0.866)^2}{3 \times 3.268}$$
$$= 3.382 \text{ m}$$

3. A triangular plate of base width 2 m and height 3 m is immersed in water with its plan making an angle of 60° with the free surface of water. Determine the hydrostatic pressure force and the position of centre of pressure when the apex of the triangle lies 5 m below the free water surface.

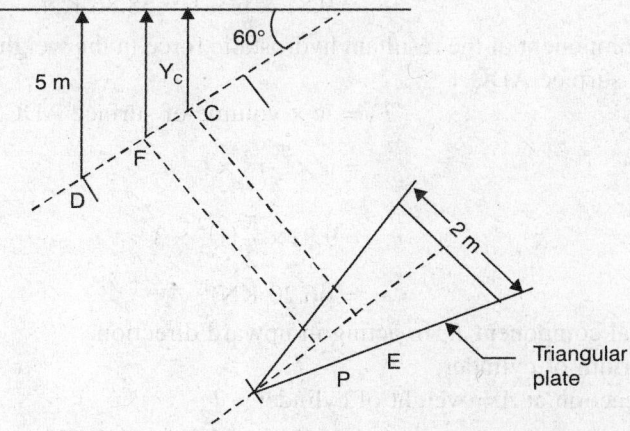

Fig. 2.24

Area of the plate,

$$A = \frac{1}{2} \times \text{base} \times \text{height}$$

$$= \frac{1}{2} \times 2 \times 3 = 3 \text{ m}^2$$

The diameter of centre of gravity of a triangular element from the apex is:

$$OC = \frac{2}{3} \times 3 = 2 \text{ m.}$$

∴ Depth of centroid from free water surface is, $Y_C = S - oc \times \sin 60°$
$$= S - 2 \times 0.866$$
$$= 3.268 \text{ m}$$

4. A cylinder 2 m in diameter × 3 m in length and supported retains water on one side. If the cylinder weighs 150 KN make calculations for the vertical and horizontal reaction ignoring the functional effects.

The horizontal component of the resultant hydrostatic force acting on the horizontal force on the curved surface on a vertical plane.

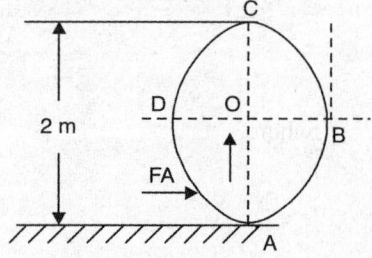

Fig. 2.25

F_b = Hydrostatic pressure force on the curved area ADC projected area of the curved surface on a vertical plane

$$= wAY$$

The projected area of the curved surface ADC on the vertical plane AOC is

$$A = \text{distance } AOC \times \text{length of cylinder}$$

$$= 2 \times 3 = 6 \text{ m}^2$$

Depth of centroid of AOC, $Y_C = \frac{1}{2} \times 2 = 1$ m.

∴ $$F_b = 9.81 \times 6 \times 1 = 58.86 \text{ KN.}$$

The vertical component of the resultant hydrostatic force in the weight of water supported by the curved surface ADC.

$$F_Y = w \times \text{volume of surface ADC}$$

$$= w \times \frac{\pi}{2} r^2 \times 1$$

$$= 9.81 \times \frac{\pi}{2} (1)^2 \times 3$$

$$= 46.20 \text{ KN.}$$

The vertical component F_v in acting in upward direction.

At equilibrium of cylinder,

Vertical reaction at A = weight of cylinder $- F_v$
$$= 150 - 46.20 = 103.8 \text{ KN.}$$

Horizontal reaction at B = 58.86 KN.

EXERCISE

1. A pressure gauge at deviation 10 m on the side of a tank reads 52.5 KN/m² 68 KN/m². Make calculations for specific weight, main density and specific growth of the fluid.

   [**Ans.:** 6200 N/m², 632 kg/m³, 0.632]

2. A rectangular tank 6 m long, 2 m wide and 2 m deep contains water to a depth of 1 m. If it is accelerated horizontally at 2.5 m/s² in the direction of its length, determine slope of free surface. [**Ans.:** 14°-18′ to the horizontal]

3. A triangular plate of base 3 m and height 3 m is immersed in water in such a way that plan of the plane makes an angle of 60° with the free surface. The base of the plane is parallel in water surface and at a depth of 2 m from water surface. Make calculations for the total pressure and depth of centre of pressure on the plate. [**Ans.:** 126.5 KN, 3 m]

4. A rectangular tank 10 m × 5 m and 3 meters deep is divided by a partition wall parallel to the shorter wall of the tank. One of the compact contains water to a deep of 3 meters and the other oil of specific gravity 0.75 to a depth of 2 meters. Find the resultant pressure on the partition. [**Ans.:** 147.1 KN]

5. A cylindrical buoy 2 m in diameter, 2.5 m high, weighs 18 KN. Examine whether the buoy will or will not float with its axis vertical in sea water of specific gravity 1.025.

   [**Ans.:** metacentric height = − 0.527 m, unstable]

6. Pipe 10 cm in diameter and 1000 m long is used to pump oil of viscosity 3.5 poise and specific gravity 0.92 at the rate of 1200 wt/min. The first 300 m of the pipe is laid along the sloping upwards at 15° to the horizontal. Find out the nature of the flow. Determine the pressure required for the pump and power of the efficiencies of the pump as 60%.

   [**Ans.:** laminar, p_π = 100125 m of oil, power = 3012 KW]

7. Crude oil of dynamic viscosity 1.5 poise and specific gravity 0.9 flows through a 20 mm diameter vertical pipe. Two presume gauges have been fixed at 20 m apart. The pressure gauge fixed at 20 m apart. The pressure gauges gives the reading of 10. Find out the pressure required.

Unit
3

Fluid of Kinematics

Introduction, types of fluid flow, continuity equation, continuity equation in three dimensions (Cartesian co-ordinate system only), velocity and acceleration, velocity potential function and stream function.

3.1 Introduction

Eulerian and Lagrangian approaches, classification of fluid flow as steady and unsteady flow; uniform and non uniform flow, laminar and turbulent flow, path line, steam line, streak line and stream tube, one, two, and three dimensional flow, velocity and accelerations in steady and unsteady flow.

Basic Hydrodynamics: Ideal fluids, equations of continuity in the differential form rotational and irrotational flow, circulation and vorticity, stream function, velocity potential, one dimensional flow along a stream line, Bernoulli's equation and its limitations, measurement of velocity, pitot tube and Pitot-Static tube, venturi meter, orifice meter, flow nozzles, notches and weirs.

3.2 Kinematics of Fluid Flow

This deals with the geometry of motions of fluid particles. This also gives an idea about the velocity and acceleration of fluid particles in motion. The motion of a fluid can be analysed on the same principles as those applied in the motion of a solid. There, however, exists a basic difference between the motion of a solid and the motion of a fluid. A solid body is compact and moves as one mass. There is no relative motion between the particle of a solid. But in the case of a fluid body, the fluid particles are all separately mobile and have motions independently. A fluid particle may have a motion different from those surrounding it. However it may be possible to obtain a relationship between the motions of neighbouring fluid particles.

3.2.1 Methods of describing Fluid Motion

Two methods are adapted to describe fluid motion namely Lagrangian method and Eulerian method.

The Lagrangian method describes the motion of a single fluid particle. The Eulerian method describes the velocity, pressure and all such characteristics at a given point in space.

The Eulerian method is commonly adopted.

3.2.2 Lagrangian method

In this method study of one fluid particle is made as it passes in space. Description can be given for paths, velocities, acceleration and other properties. In the System, any variable can be expressed in terms of four independent variable, x, y, z and t say in Cartesian coordinates. For an individual particle n, y, z are to be kept constant, say at time, $t = 0$ the coordinates of particle be n_0, y_0, z_0 any particle $P = P (x_0, y_0, z_0, t)$. As a fluid particle moves, the property may vary with time, but its identity x_0, y_0, z_0 remains invariant. By varying x_0, y_0, z_0 any particle of a continuum can be described. Study of properties by tracing the particles is difficult and hence this approach is not used in practice in fluid mechanics.

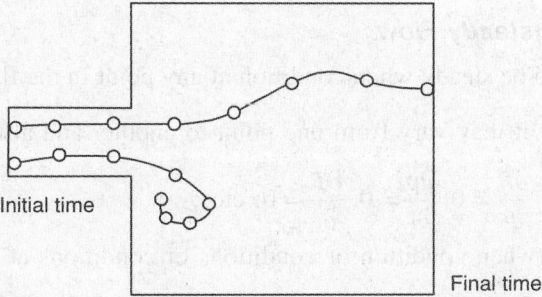

Initial time

Final time

Fig. 3.1: Lagrangian approach—study of each particle with time

3.2.3 Eulerian method

In this method one point in space is fined and study of various particles passing through that point is made, i.e., it refers to a description of the velocity, pressure and other characteristics at certain points or section in the fluid at each instant of time. In Cartesian coordinates any physical property of a continuum is represented by four independent variables x, y, z and t or say $P = P (x, y, z, t)$.

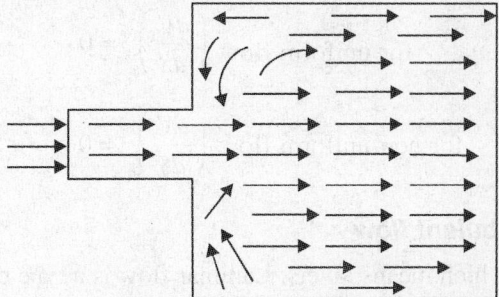

Fig. 3.2: Eulerian approach—study at fined station in space

For example the velocity of a particle is the function of both space (x, y, z) in rectangular coordinates and time (t) i.e.,

$$\left.\begin{array}{l} u = u\left(x,\, y,\, z,\, t\right) \\ v = v\left(x,\, y,\, z,\, t\right) \\ w = w\left(x,\, y,\, z,\, t\right) \end{array}\right\}$$

In vector form the velocity

$$V = ui + vj + wk$$
$$= \sqrt{u^2 + v^2 + w^2}$$

Similar expression can be written for acceleration pressure and other characteristics in this method of approach. The fluid motion at all point is determined by applying the laws of mechanics at all fined station. Eulerian method of analysis is easier to Lagrangian method and so it is commonly used.

3.3 Classification of Fluid Flow

3.3.1 Steady and Unsteady Flow

A flow is considered to be steady when condition at any point in the fluid do not change with time, i.e., $\dfrac{\partial V}{\partial t} = 0$, but it may vary from one point to another and also the properties do not change with time, i.e., $\dfrac{\partial P}{\partial t} = 0,\ \dfrac{\partial \rho}{\partial t} = 0,\ \dfrac{\partial T}{\partial t} = 0$, etc.

A flow is unsteady when condition or conditions or conditions at any point change with time, i.e.,

$$\frac{\partial}{\partial t}(\ ) \neq 0$$

3.3.2 Uniform and non-uniform Flow

If the flow characteristic like velocity, density and pressure at any given time remain the same at all points the flow is uniform.

If the flow characteristic have different values at different points, at any particular given instant, the flow is non-uniform fluid

$$\text{for uniform flow} \quad \left(\frac{dV}{dS}\right)_{t_1} = 0$$

$$\text{for non-uniform flow} \quad \left(\frac{dV}{dS}\right)_{t_1} \neq 0$$

3.3.3 Laminar and turbulent flow

Lamina is a greek work which means layers. Laminar flow is a type of flow in which the fluid particles moves in layers. Even if, there is no transportation of fluid particles from one layer to another, the fluid particles in any layer move along well-defined paths or stream lines. Hence this flow is called stream lines flow. This type of flow is also called viscous flow and can exist only at low velocities.

Turbulent flow is the most common type of flow that occurs in nature. The flow shows eddy current and the velocity of flow changes in direction and magnitude from point to point. There is a general mixing up of the fluid particles in motion.

The velocity at which laminar flow changes to turbulent flow in a pipe is called the *critical velocity*.

The type of flow, whether it is laminar or turbulent, is classified by a non-dimensional number called the Reynolds number (Re). It is the ratio of the inertial force to the viscous force acting on the particle.

$$Re = \frac{\rho vl}{\mu}$$

ρ = density of fluid kg/m

v = average velocity m/s

l = length characteristic: (for pipes diameter d and for plates length l)

μ = dynamic viscosity of fluid kg/ms.

For flow in pipes if the Reynold's number is less than 2000, the flow is called laminar and if it is more than 4000, the flow is called turbulent, if the Reynolds number lies between 2000 and 4000, the flow may be laminar or turbulent (also known as transition period). In general the flow is assumed to be turbulent if Reynold number exceeds 2300.

For flow on plates, if the Reynolds number is less than 3×10^5 the flow is called laminar and if it is more than 5×10^5 the flow is called turbulent if the Reynolds number lies between 3×10^5 to 5×10^5 the flow may be laminar or turbulent.

3.3.4 Compressible and incompressible flows

Compressible flow is that type of flow in which the density of the fluid changes from point to point or in other words the density (ρ) is not constant for the fluid. Thus: for compressible flow:

$$\rho \neq constant.$$

For compressible flow: In compressible flow is that type of flow in which the density is constant for the fluid flow. Liquids are generally incompressible, while gases are compressible.

For incompressible flow: ρ = constant

3.3.5 Rotational and irrotational flows

Rotational flow is that type of flow in which the fluid particles while flowing along stream-lines, also rotate about their own axis, and if the fluid particles while flowing along stream lines, do not rotate about their axis, that type of flow is called Irrotational flow.

3.3.6 One, two and three dimensional flow

A fluid flow is said to be one, two or three dimensional depending upon the number of independent space coordinates, i.e. one, two or three respectively required to describe the nature of flow.

One-dimensional flow: It neglects variations or changes in velocity, pressure, etc. transverse to the main flow direction. Conditions at a cross-section are expressed in terms of average values of velocity, density and other properties. In rectangular coordinates for one-dimensional flow, say $u = f(x)$, $v = w = 0$ or flow through a pipe, for example may be characterized one-dimensional flow.

Flow parameter such as velocity is a function of time and one space coordinate such as x, $u = f(x)$, $v = 0$, $w = 0$.

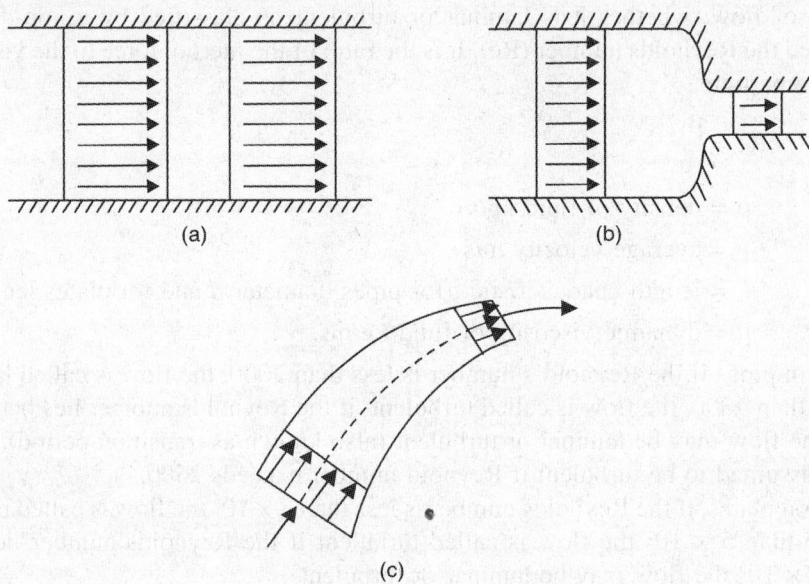

Fig. 3.3: One-dimensional flow in pipe

Two-dimensional flow: The variable like velocity, pressure, etc. vary with only two space coordinates, and do not vary along the third coordinate. All the fluid particles are assumed to flow in parallel planes along identical paths in each of these planes and hence there are no changes in flow normal to these planes. The flow around a circular, cylinder of infinite length, the flow part an aerofoil of uniform cross-section and infinite span the flow over a weir of uniform cross-section and infinite width, are the examples of two-dimensional flow.

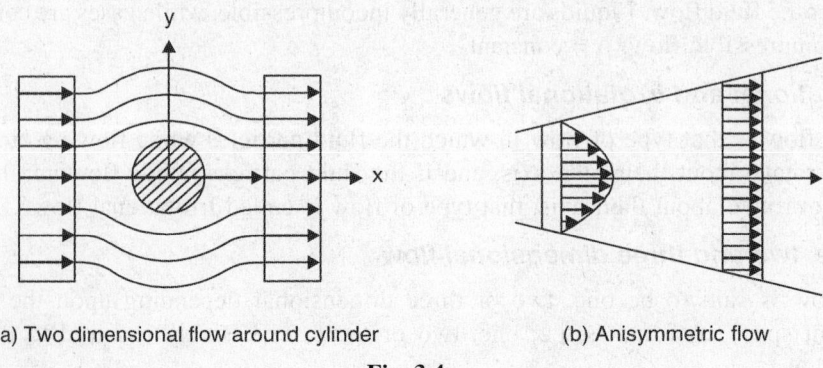

(a) Two dimensional flow around cylinder   (b) Anisymmetric flow

Fig. 3.4

In case of two-dimensional flows, if the velocity profile is symmetrical about the axis of symmetry the flow is called *Anisymmetrical flow.*

Three-dimensional flow: It is the most general flow. In this flow the velocity components u, v, w are in mutually perpendicular directions and are functions of space coordinates x, y, z and for unsteady flow conditions are also functions of time t.

One two and three dimensional steady and unsteady flow

Dimensional	Steady	Unsteady
1. one-dimensional	$u = f(x)$	$u = f(x, t)$
2. two-dimensional	$u = f(x, y)$	$u = f(x, y, t)$
3. three-dimensional	$u = f(x, y, z)$	$u = f(x, y, z, t)$

3.4 Rate of Flow or Discharge (Q)

It is defined as the quantity of a fluid flowing per second through a section of a pipe or a channel: For an incompressible fluid (or liquid) the rate of flow or discharge is expressed as the volume of fluid flowing across the section/second. For compressible fluids, the rate of flow is usually expressed as the weight of fluid flowing across the section. Thus:

(i) For liquids; the Units of Q are m³/s or lit/sec

(ii) For gases; the Units of Q is kgf/sec or N/sec.

Consider, a liquid flowing through a pipe in which:

$$A \rightarrow \text{Cross-sectional area of pipe}$$

$$V \rightarrow \text{Average velocity of fluid across the section}$$

Then: discharge, $Q = A \times V$

3.4.1 Path line

The path followed by a single fluid particle in motion is called a path line. Thus the path line shows the direction of a particle, for a certain period of time or between two given sections.

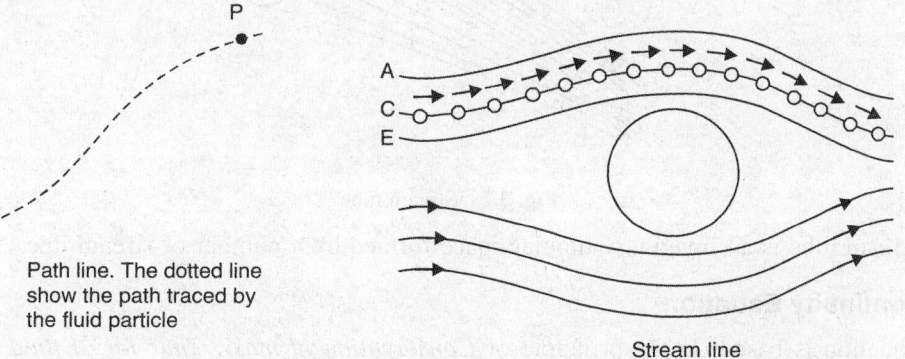

Path line. The dotted line show the path traced by the fluid particle

Stream line

Fig. 3.5: Path line

3.4.2 Stream line

A stream line is a line which shows the direction of the velocity of the fluid at each point along the line. The tangent to the stream line at any point is the direction of the velocity of the fluid at that point.

Note: The path line differs from a stream line. A path line refers to a single particle, but the stream lines shows the direction of motion of a number of particles at the same time.

3.4.3 Streak line

The streakline is the locus of the position of fluid particles which have passed through a given point in space in succession. For instance, consider a space point $A(x_1\ y_1\ z_1)$. The various fluid particles which have passed through A in succession lie in some position at a certain instant. The locus of these points constitutes the streak line.

In the case of a steady flow a streak line coincides with a stream line.

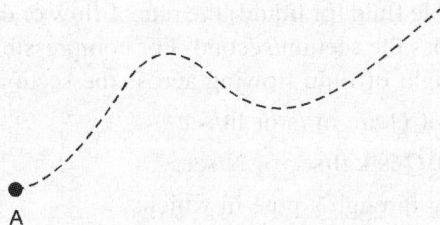

A

Fig. 3.6: Streak line

Stream tube

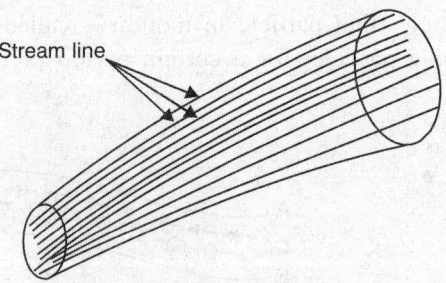

Stream line

Fig. 3.7: Stream tube

A stream tube is an imaginary tubular space formed by a number of stream line.

3.5 Continuity Equation

This equation is based on the principle of *Conservation of mass. Thus for, a fluid flowing through the pipe at all the cross-section, the quantity of fluid per second is constant.*

Let $\qquad V_1 \rightarrow$ Average velocity at cross-section 1-1.

$\rho_1 \rightarrow$ Density at section 1-1.

$A_1 \rightarrow$ Area of pipe at section 1-1.

V_2, ρ_2 and A_2 are corresponding values at section: 2-2:

Then; rate of flow at section 1-1 $= \rho_1 A_1 V_1$

Rate of flow at section 2-2 $= \rho_2 A_2 V_2$

But according to law of conservation of mass:

Rate of flow at section 1-1 = Rate of flow at section 2-2

$\Rightarrow$
$$\rho_1 A_1 V_1 = \rho_2 A_2 V_2$$

This equation is applicable to the compressible as well as incompressible fluids and is called *continuity equation.*

If the fluid is in compressible; then $\rho_1 = \rho_2$ and continuity equation reduces to:

$$A_1 V_1 = A_2 V_2$$

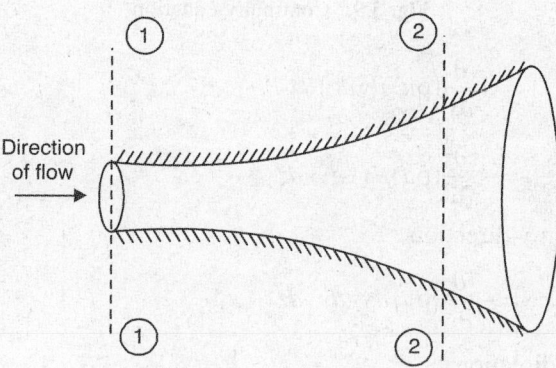

Fig. 3.8: Fluid flowing through a pipe

3.5.1 Continuity equation in three-dimensions

Consider a fluid element of lengths dx, dy and dz in the direction and Z. Let u, v and w are the inlet velocity components in x, y and z respectively. *Mass of fluid entering the face ABCD per second:*

$= \rho \times$ velocity in x-direction $\times$ Area of *ABCD.*

$= \rho \times u \times (dy \times dz)$

Then, *mass of fluid, leaving the face EFGH per second* $= \rho u \cdot dy \cdot dz + \dfrac{\partial}{\partial x}\left(\rho u\, dy\, dz\right) dx$

$\therefore$ Gain of mass in x-directions:

$=$ mass of fluid entering the face ABCD + change of mass of fluid

$=$ mass through *ABCD* $-$ Mass through *EFGH* per second

$= \rho u \cdot dy \cdot dz - \rho u dy\, dz - \dfrac{\partial}{\partial x}\left(\rho u\, dy\, dz\right) dx$

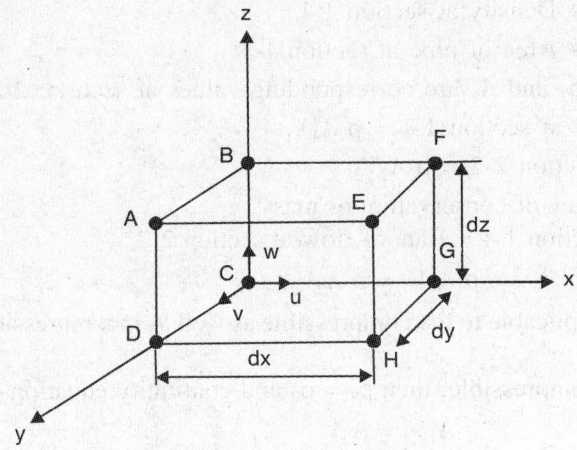

Fig. 3.9: Continuity equation

$$= -\frac{\partial}{\partial x}(\rho u \cdot dy\,dz)\,dx$$

$$= -\frac{\partial}{\partial x}(\rho u)\,dx \cdot dy \cdot dz \qquad\qquad [\because dy \cdot dz \text{ is const.}]$$

The net gain of mass in *y*-direction

$$= -\frac{\partial}{\partial y}(\rho v)\,dx \cdot dy \cdot dz$$

Net gain of mass in *y*-direction

$$= -\frac{\partial}{\partial y}(\rho v) \cdot dx \cdot dy \cdot dz$$

In *Z*-direction

$$= -\frac{\partial}{\partial z}(\rho\omega) \cdot dx \cdot dy \cdot dz$$

$$\therefore \quad \text{Net gain of masses} = \left[\frac{\partial}{\partial x}(\rho u) + \frac{\partial}{\partial y}(\rho v) + \frac{\partial}{\partial z}(\rho w)\right] \cdot dx \cdot dy \cdot dz$$

Since, the mass is neither created nor destroyed in the fluid element, the net increase of mass per unit time in the fluid element must be equal to the rate of increase of mass of fluid in the element. But, mass of fluid in the element is: $\rho \cdot dx \cdot dy \cdot dz$ and its rate of increase with time

is $\frac{\partial}{\partial t}(\rho \cdot dx \cdot dy \cdot dz)$ or $\frac{\partial p}{\partial t} \cdot dx \cdot dy \cdot dz$.

Equating the two expressions

or, $\qquad -\left[\frac{\partial}{\partial x}(\rho u) + \frac{\partial}{\partial y}(\rho v) + \frac{\partial}{\partial z}(\rho w)\right]dx \cdot dy \cdot dz = \frac{\partial \rho}{dt} \cdot dx \cdot dy \cdot dz$

or,
$$\boxed{\frac{\partial p}{\partial t} + \frac{\partial}{\partial x}(\rho u) + \frac{\partial}{\partial y}(\rho v) + \frac{\partial}{\partial z}(\rho w) = 0}$$ (Cancelling $dx \cdot dy \cdot dz$ to both sides)

This equation is the continuity equation in Cartesian co-ordinates in its most general form. This equation is applicable to:

$$\frac{\partial p}{\partial t} + \frac{\partial}{\partial x}(\rho u) + \frac{\partial}{\partial y}(\rho v) + \frac{\partial}{\partial z}(\rho w) = 0$$

(i)   Steady and unsteady flow
(ii) Uniform and non-uniform flow and
(iii) Compressible and in compressible fluids

For steady flow: $\frac{\partial \rho}{\partial t} = 0$ and hence equation becomes as:

$$\frac{\partial}{\partial x}(\rho u) + \frac{\partial}{\partial y}(\rho v) + \frac{\partial}{\partial z}(\rho \omega) = 0$$

If the fluid is in compressible, then ρ is constant and the above equation becomes as:

$$\boxed{\frac{\partial u}{\partial x} + \frac{\partial v}{\partial y} + \frac{\partial w}{\partial z} = 0} \qquad \qquad \ldots(1)$$

Equation 1 is the continuity equation in three-dimensions. For a two-dimensional flow, the component $w = 0$ and hence continuity equation becomes as:

$$\boxed{\frac{\partial u}{\partial x} + \frac{\partial v}{\partial y} = 0}$$

3.6 Velocity and Acceleration

Let V is the resultant velocity at any point in a fluid flow. Let u, v and w are its component in x, y and z directions. The velocity components are functions of space co-ordinates and time. Mathematically, the velocity components are given as:

$$u = f_1(x,\ y,\ z,\ t)$$
$$v = f_2(x,\ y,\ z,\ t)$$
$$w = f_3(x,\ y,\ z,\ t)$$

and Resultant Velocity: $V = ui + vj + wk = \sqrt{u^2 + v^2 + w^2}$

Now:

Let a_x, a_y and a_z are the *total acceleration* in x, y and z directions respectively. Then, by the chain rule of differentiation, we have:

as,
$$a_x = \frac{du}{dt} = \frac{\partial u}{\partial x} \cdot \frac{dx}{dt} + \frac{\partial u}{\partial y} \cdot \frac{\partial y}{\partial t} + \frac{\partial u}{\partial z} \cdot \frac{\partial z}{\partial t} + \frac{\partial u}{\partial t}$$

$\Rightarrow$ $\qquad a_x = \dfrac{du}{dt} = u\dfrac{\partial u}{\partial x} + v\dfrac{\partial u}{\partial y} + w\dfrac{\partial y}{\partial z}$ $\qquad\qquad$ [Putting the value of u]

where, $\qquad \dfrac{dx}{dt} = u,\ \dfrac{dy}{dt} = v$ and $\dfrac{dz}{dt} = w$

Again, $\qquad\begin{aligned}a_x &= \dfrac{du}{dt} = u\cdot\dfrac{\partial u}{\partial x} + v\cdot\dfrac{\partial u}{\partial y} + w\cdot\dfrac{\partial u}{\partial z} + \dfrac{\partial u}{\partial t};\\[2mm] a_y &= \dfrac{dv}{dt} = u\cdot\dfrac{\partial u}{\partial x} + v\cdot\dfrac{\partial v}{\partial y} + w\cdot\dfrac{\partial v}{\partial z} + \dfrac{\partial v}{\partial t}\\[2mm] \text{and }\ a_z &= \dfrac{dw}{dt} = u\cdot\dfrac{\partial w}{\partial x} + v\cdot\dfrac{\partial w}{\partial y} + w\cdot\dfrac{\partial w}{\partial z} + \dfrac{\partial w}{\partial t}\end{aligned}\Biggr]$ $\qquad$ ...(2)

But, for steady flow, $\dfrac{\partial v}{\partial t} = 0$, where V is resultant velocity.

or, $\qquad \dfrac{\partial u}{\partial t} = 0,\ \dfrac{\partial v}{\partial t} = 0$ and $\dfrac{\partial w}{\partial t} = 0$

Hence, acceleration in x, y and z directions becomes:

$\qquad\begin{aligned}a_x &= \dfrac{du}{dt} = u\cdot\dfrac{\partial u}{\partial x} + v\cdot\dfrac{\partial u}{\partial y} + w\cdot\dfrac{\partial u}{\partial z}\\[2mm] a_y &= \dfrac{dv}{dt} = u\cdot\dfrac{\partial u}{\partial x} + v\cdot\dfrac{\partial v}{\partial y} + w\cdot\dfrac{\partial v}{\partial x}\\[2mm] a_z &= \dfrac{dw}{dt} = u\cdot\dfrac{\partial w}{\partial x} + v\cdot\dfrac{\partial w}{\partial y} + w\cdot\dfrac{\partial w}{\partial z}\end{aligned}\Biggr]$ $\qquad$ ...(3)

$\therefore$ Acceleration vector: $A = a_x i + a_y j + a_z k$

$\qquad\qquad\qquad = \sqrt{a_x^2 + a_y^2 + a_z^2}$

3.7 Local Acceleration and Convective Acceleration

Local Acceleration: It is defined as the rate of increase of velocity with respect to time at a given point in a flow field. In the equation 3 (3.6). The expression $\dfrac{\partial u}{\partial t},\ \dfrac{\partial v}{\partial t}$ or $\dfrac{\partial w}{\partial t}$ is known as local acceleration.

Convective Acceleration: It is defined as the rate of change of velocity due to the change of position of fluid particles in a fluid flow. The expressions $\dfrac{\partial u}{\partial t},\ \dfrac{\partial v}{\partial t}$ and $\dfrac{\partial w}{\partial t}$ in the equation are known as *Convective Acceleration*.

Example 1: $\qquad\qquad V = 4x^3i - 10x^2yj + 2tk$

Find the velocity and acceleration of a fluid particle at $(2, 1, 3)$ at time $t = 1$.

Solution: The velocity components u, v and w are:

$$u = 4\,x^3,\ v = -10\,x^2y,\ w = 2t$$

For the point $(2, 1, 3)$; we have $x = 2$, $y = 1$ and $z = 3$ at $t = 1$.

Hence, velocity components at $(2, 1, 3)$ at $t = 1$ as

$$u = 4x^3 = 4(2)^3 = 32 \text{ units}$$
$$v = -10x^2y = -10(2)^2 \times 1 = -40 \text{ units}$$
$$w = 2t = 2 \times 1 = 2 \text{ units}$$

$\therefore$ Velocity vector V at $(2, 1, 3) = 32i - 40j + 2k$

$\therefore$ Resultant velocity $= \sqrt{(32)^2 + (40)^2 + (2)^2} = 51.26$ units.

Now, acceleration is given by:

$$a_x = u\frac{\partial u}{\partial x} + v\frac{\partial u}{\partial y} + w\frac{\partial u}{\partial z} + \frac{\partial u}{\partial t}$$

$$a_y = u\frac{\partial v}{\partial x} + v\frac{\partial v}{\partial y} + w\frac{\partial v}{\partial z} + \frac{\partial v}{\partial t}$$

$$a_z = u\frac{\partial w}{\partial x} + v\frac{\partial w}{\partial y} + w\frac{\partial w}{\partial z} + \frac{\partial w}{\partial t}$$

But, $\qquad\qquad u = 4x^3,\ v = -10x^2y,\quad w = 2t$

So,

$$\frac{\partial u}{\partial x} = 12x^2,\quad \frac{\partial u}{\partial y} = 0,\quad \frac{\partial u}{\partial z} = 0,\quad \frac{\partial u}{\partial t} = 0$$

$$\frac{\partial v}{\partial x} = -20xy,\quad \frac{\partial v}{\partial y} = -10x^2,\quad \frac{\partial v}{\partial z} = 0,\quad \frac{\partial v}{\partial t} = 0$$

$$\frac{\partial w}{\partial x} = 0,\quad \frac{\partial w}{\partial y} = 0,\quad \frac{\partial w}{\partial z} = 0,\quad \frac{\partial w}{\partial t} = 2.$$

Putting these values, the acceleration components are $(2, 1, 3)$ at time $t = 1$ are

$$a_x = 4x^3(12x^2) + 0 + 0 + 0$$
$$a_n = 48x^5 = 40\,(2)^5 = 1536 \text{ units}$$
$$a_y = 4x^3(-20\,xy) - 10x^2y\,(-10x^2) + 2t(0) + 0$$
$$= -80x^4y + 100x^4y$$
$$= -80(2)^4(1) + 100(2)^4\,1 = 320 \text{ units}$$
$$a_2 = 4x^3(0) + 0 + 0 + 2$$

$\therefore\qquad\qquad a_2 = 2$ units

$\therefore\qquad\qquad A = a_xi + a_yj + a_zk = 1536i + 320j + 2k$

$$\text{Resultant } A = \sqrt{(1536)^2 + (320)^2 + (2)^2}$$

$\therefore$ $\qquad A = 1568.9$ units

2. Find the convective acceleration at the middle of a pipe which converges uniformly from 0.4 m diameter to 0.2 m diameter over 2 m length. The rate of flow is 20 lit/sec. If the rate of flow changes uniformly from 20 l/s to 40 l/s in 30 sec, find total acceleration at the middle of the pipe at 15th seconds.

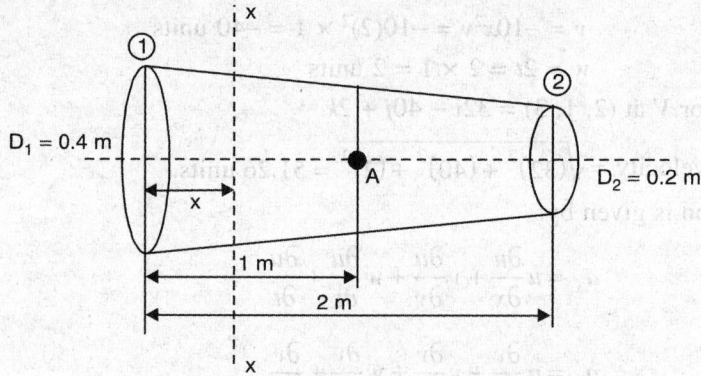

Fig. 3.10

Solution: $\qquad D_1 = 0.4$m

$\qquad\qquad D_2 = 0.2$m

$\qquad\qquad l = 2$m

$$Q = 20 \; l/\text{sec} = \frac{20}{1000} = 0.02 \; \text{m}^3/\text{sec}$$

$$1l = 1000 \; \text{cm}^3$$

$$1l = \frac{1}{1000} = \text{m}^3$$

(i) Convective acceleration at middle, i.e. at A when $Q = 20 \; l$/s.

In this case,

rate of flow = const. = 0.02 m³/sec

The velocity of flow is in x-direction only.

Hence, this is one-dimensional flow and velocity components in y and z directions are zero, i.e. $v = 0$, $w = 0$

$\therefore$ Convective acceleration $= u\dfrac{\partial u}{\partial x}$ only

Now, we find the value of u and $\dfrac{\partial u}{\partial x}$ at a distance x from inlet.

The diameter (D_x) at a distance x from inlet or at section $x - x$ is given by:

$$D_x = 0.4 - \frac{(0.4 - 0.2)}{2} x = (0.4 - 0.1x)\,\text{m}$$

$$a_x = \frac{du}{dt} = u\frac{\partial u}{\partial x} + v\frac{\partial u}{\partial y} + \frac{w\partial u}{\partial z} + \left(\frac{\partial u}{\partial t}\right)$$

$$a_y = \frac{dv}{dt} = u\frac{dv}{dx} + v\frac{\partial v}{\partial y} + w\frac{\partial v}{\partial z} + \left(\frac{\partial v}{\partial t}\right)$$

Then, the area of cross section (A_x) at section $x - x$ is given by:

$$A_x = \frac{\pi}{4}D_n^2 = \frac{\pi}{4}(0.4 - 0.1x)^2$$

$$a_2 = \frac{\partial w}{\partial t} = u\frac{\partial w}{\partial x} + v\frac{\partial w}{\partial y} + w\frac{\partial w}{\partial z} + \frac{\partial w}{\partial t}$$

and velocity (u) at the section $x - x$ in terms of Q (i.e. in terms of rate of flow)

$$u = \frac{Q}{\text{Area}} = \frac{Q}{A_x} = \frac{Q}{\pi/4 D_n^2} = \frac{4Q}{\pi(0.4 - 0.1x)^2}$$

$$u = \frac{1.273Q}{(0.4 - 0.1x)^2} = 1.273Q(0.4 - 0.1x)^{-2}\,\text{m/s}$$

$$D_x = 0.4 - \frac{(0.4 - 0.2)}{2}x$$

$$D_x = (0.4 - 0.1x)$$

$$\therefore \quad A_x = \frac{\pi}{4}D_x^2$$

$$= \frac{\pi}{4}(0.4 - 0.1x)^2$$

Now: To find: $u\dfrac{\partial u}{\partial x}$; we differentiate equation (ii) with respect to x;

$$\frac{\partial u}{\partial x} = 1.273Q(-2)(0.4 - 0.1x)^{-3}(-0.1)$$

$$= 0.2546Q(0.4 - 0.11)^{-3}$$

$\therefore$ Connective acceleration $u\dfrac{\partial u}{\partial x}$

$$= 1.273\ Q\ (0.04 - 0.1x)^{-2} \times 0.254\ Q\ (0.4 - 0.1\ x)^{-3}$$

But at middle $\qquad = 1.273 \times 0.02 \times 0.\ 254 \times 0.02\ (0.4 - 0.1\ x)^{-5}$

$n = 1$ so: $\qquad = 1.273 \times 0.254 \times 0.02 \times 0.02\ (0.4 - 0.1 \times 1)^{-5}$

$$= 0.0048\ \text{m/s}^2$$

3. Check the continuity for the following equations:

(i) $u = x^2 + y^2 + z^2$, $v = xy^2 - y_z^2 + xy$

(ii) $v = 2y^2$, $w = 2xyz$.

Solution: The continuity equation for incompressible fluid is given by equation:

$$\boxed{\frac{\partial u}{\partial x} + \frac{\partial v}{\partial y} + \frac{\partial w}{\partial z} = 0}$$

Case I: $u = x^2 + y^2 + z^2$

$$\frac{\partial u}{\partial x} = 2x$$

$$v = xy^2 - yz^2 + xy$$

$$\frac{\partial v}{\partial y} = 2xy - z^2 + x$$

Now: $2x + 2xy - z^2 + x + \dfrac{\partial w}{\partial z} = 0$

$$3x + 2xy - z^2 + \frac{\partial w}{\partial z} = 0$$

$$\boxed{\therefore \ \partial w = (z^2 - 3x - 2xy)\partial z}$$

Integrating both sides:

$$\int dw = \int \left(z^2 - 3x - 2xy\right) dz$$

$$\Rightarrow \qquad w = \int z^2 dz - 3x \int dz - 2xy \int dz + c$$

$$\Rightarrow \qquad w = \frac{z^3}{3} - 3xz - 2xyz + c$$

$$\boxed{\because \ w = -3xz - 2xyz + \frac{z^3}{3} + f(x, y)}$$

The rate of flow at $t = 15$ sec is given

$$Q = Q_1 + \frac{Q_2 - Q_1}{30} \times 15$$

$$Q_2 = 0.04 \ \text{m}^3/\text{sec}$$

$$Q_1 = 0.02 \ \text{m}^3/\text{sec}$$

$$\Rightarrow \qquad Q = 0.02 + \left(\frac{0.04 - 0.02}{\underset{2}{\cancel{30}}} \times \cancel{15}\right)$$

$\therefore \qquad\qquad Q = 0.03\,\text{m}^3/\text{sec.}$

$$Q = Q_1 + \left(\frac{Q_2 - Q_1}{30}\right)15$$

$$Q_2 = 0.04$$

$$Q_1 = 0.02$$

Rate of flow at $t = 15$ sec is given,

$$Q = Q_1 + \left(\frac{Q_2 - Q_1}{30}\right) \times 15$$

4. $V = x^2yi + y^2zj - (2xyz + yz^2)k$

At point (2, 1, 3)

$$u = x^2y, \ v = y^2z, \ w = -(2xyz + yz^2)$$

$$\frac{\partial u}{\partial x} = 2xy; \ \frac{\partial v}{\partial y} = 2yz, \ \frac{\partial w}{\partial z} = -2xy - 2yz$$

$$\frac{\partial u}{\partial x} + \frac{\partial v}{\partial y} + \frac{\partial w}{\partial z} = 2\,xy + 2\,yz - 2\,xy - 2\,yz = 0$$

Hence it is proved that it is a case of steady continuity.

Now: $\dfrac{\partial u}{\partial x}$ $\qquad u = x^2y = (2)^2 1 = 4$

$$a_n = u\frac{\partial y}{\partial x} + v\frac{\partial u}{\partial y} + w\frac{\partial u}{\partial z} + \frac{\partial u}{\partial t} \ \text{for a steady flow}, \ \frac{\partial u}{\partial t} = 0$$

$$v = y^2z = (1)^2 \times 3 = 3$$

$$w = -(2xyz + yz^2) = -(2 \times 2 \times 1 \times 3 + 1 \times 3^2)$$

$$w = -(12 + 9) = -21$$

$\therefore \qquad\qquad V = \sqrt{4^2 + 3^2 + (-x)^2} = 21.587 \ \text{unit}$

3.8 Velocity, Potential Function and Stream Function

3.8.1 Velocity Potential Function

It is defined as a scalar function of space and time such that its negative derivative with respect to any direction gives the fluid velocity in that direction. It is defined by ϕ(phi). Mathematically, the velocity, potential is defined as: $\phi = f(x, y, z)$ for steady flow, such that:

$$\left.\begin{aligned} u &= -\frac{\partial \phi}{\partial x} \\ v &= -\frac{\partial \phi}{\partial y} \\ w &= -\frac{\partial \phi}{\partial z} \end{aligned}\right\} \quad \text{where: } u, v \text{ and } w \text{ are this components of velocity in } x, y \text{ and } z\text{-directions respectively.} \qquad \dots(1)$$

The velocity components in cylindrical polar co-ordinates in terms of velocity potential function are given by:

$$u_r = \frac{\partial \phi}{\partial r}$$
$$u_\theta = \frac{1}{r} \cdot \frac{\partial \phi}{\partial \theta}$$

where, $u_r \rightarrow$ velocity component in radial direction [i.e., in r direction]

$u_\theta \rightarrow$ [velocity component in tangential direction]

The continuity equation for an incompressible steady flow is:

$$\frac{\partial u}{\partial x} + \frac{\partial v}{\partial y} + \frac{\partial w}{\partial z} = 0$$

Substituting the values of u, v and w from (1): we get:

$$\frac{\partial}{\partial x}\left(-\frac{\partial \phi}{\partial x}\right) + \frac{\partial}{\partial y}\left(-\frac{\partial \phi}{\partial y}\right) + \frac{\partial}{\partial z}\left(-\frac{\partial \phi}{\partial z}\right) = 0$$

$\Rightarrow$

$$\frac{\partial^2 \phi}{\partial x^2} + \frac{\partial^2 \phi}{\partial y^2} + \frac{\partial^2 \phi}{\partial z^2} = 0 \qquad \qquad \dots(2)$$

This equation is known as Laplace equation.

For two-dimensional case, the equation 2 reduces to: $\dfrac{\partial^2 \phi}{\partial x^2} + \dfrac{\partial^2 \phi}{\partial y^2} = 0$ $\qquad \dots(3)$

If any value of ϕ that satisfies the Laplace equation, it will correspond to some case of fluid flow.

Properties of the Potential Function

The rotational components are given by:

$$w_z = \frac{1}{2}\left(\frac{\partial v}{\partial x} - \frac{\partial u}{\partial y}\right)$$
$$w_y = \frac{1}{2}\left(\frac{\partial u}{\partial z} - \frac{\partial w}{\partial v}\right)$$
and $$w_x = \frac{1}{2}\left(\frac{\partial w}{\partial y} - \frac{\partial v}{\partial z}\right)$$

Substituting the values of u, v and w from (1) in above rotational components:

$$w_z = \frac{1}{2}\left[\frac{\partial}{\partial x}\left(-\frac{\partial \phi}{\partial y}\right) - \frac{\partial}{\partial y}\left(-\frac{\partial \phi}{\partial x}\right)\right] = \frac{1}{2}\left[\frac{-\partial^2 \phi}{\partial x \cdot \partial y} + \frac{\partial^2 \phi}{\partial y \cdot \partial x}\right] = 0$$

$$w_y = \frac{1}{2}\left[\frac{\partial}{dz}\left(-\frac{\partial \phi}{\partial x}\right) - \frac{\partial}{\partial x}\left(-\frac{\partial \phi}{\partial z}\right)\right] = \frac{1}{2}\left[-\frac{\partial^2 \phi}{\partial z \cdot \partial x} + \frac{\partial^2 \phi}{\partial x \cdot \partial z}\right] = 0$$

$$w_x = \frac{1}{2}\left[\frac{\partial}{\partial y}\left(-\frac{\partial \phi}{\partial z}\right) - \frac{\partial}{\partial z}\left(-\frac{\partial \phi}{\partial y}\right)\right] = \frac{1}{2}\left[-\frac{\partial^2 \phi}{\partial y \cdot \partial z} + \frac{\partial^2 \phi}{\partial z \cdot \partial y}\right] = 0$$

If 'ϕ' is a continuous function, then: $\dfrac{\partial^2 \phi}{dx \cdot dy} = \dfrac{\partial^2 \phi}{\partial y \partial x}$

$$\frac{\partial^2 \phi}{\partial z \cdot \partial x} = \frac{\partial^2 \phi}{dx \cdot dz} \text{ etc.}$$

$\therefore$ $$\boxed{\therefore \ w_z = w_y = w_x = 0}$$

when rotational components are zero, the flow is called irrotational. Hence; the properties of the potential function are:

1. If velocity potential (ϕ) exists, the flow should be irrotational.
2. If velocity potential (ϕ) satisfies the Laplace equation, it represents the possible steady incompressible irrotational flow.

$$u_z - \frac{\partial \phi}{\partial x} \quad\bigg|\quad v = \frac{d\psi}{dx}$$

$$v_z - \frac{\partial \phi}{\partial y} \quad\bigg|\quad -u = \frac{\partial \psi}{\partial y}$$

3.8.2 Stream Function

It is defined as the scalar function of space and time, such that its partial derivative with respect to any direction gives the velocity component at right angles to that direction. It is denoted by ψ (Psi) and defined only for two dimensional flow.

Mathematically, for steady flow it is defined as:

$$\psi = f(x, y); \text{ such that:}$$

$$\left.\begin{array}{l} \dfrac{\partial \psi}{\partial x} = v \\[2mm] \text{and } \quad \dfrac{\partial \psi}{\partial y} = -u \end{array}\right\} \qquad \text{...(4)}$$

$$\frac{\partial \psi}{\partial x} = v \text{ and } \frac{\partial \psi}{\partial y} = -u$$

The velocity components in cylindrical polar-coordinates in terms of stream f_x are given as:

$$\boxed{u_r = \frac{1}{r} \cdot \frac{\partial \psi}{\partial \theta}} \text{ and } \boxed{u_\theta = \frac{-\partial \psi}{\partial r}}$$

where; u_r = radial velocity and u_θ = tangential velocity.

The continuity equation for two-dimensional flow is: $\dfrac{\partial u}{\partial x} + \dfrac{\partial v}{\partial y} = 0$.

Now; substituting the values of u and v from equation (4);

$$\frac{\partial}{\partial x}\left(-\frac{\partial \psi}{\partial y}\right) + \frac{\partial}{\partial y}\left(\frac{\partial \psi}{\partial x}\right) = 0$$

or,
$$\frac{-\partial^2 \psi}{\partial x \cdot \partial y} + \frac{\partial^2 \psi}{\partial x \cdot \partial y} = 0.$$

Hence; existence of ψ means a possible case of fluid flow. The flow may be *rotational or irrotational*.

The rotational component w is given by: $w_z = \dfrac{1}{2}\left[\dfrac{\partial v}{\partial x} - \dfrac{\partial u}{\partial y}\right]$. Again; Substituting the values of u and v from equation (4), in above rotational component we have:

$$w_z = \frac{1}{2}\left[\frac{\partial}{\partial x}\left(\frac{\partial \psi}{\partial x}\right) - \frac{\partial}{\partial y}\left(-\frac{\partial \psi}{\partial y}\right)\right] = \frac{1}{2}\left[\frac{\partial^2 \psi}{\partial x^2} + \frac{\partial^2 \psi}{\partial y^2}\right]$$

For irrotational flow: $\boxed{w_2 = 0}$

Hence, above equation becomes as $\boxed{\dfrac{\partial^2 \psi}{\partial x^2} + \dfrac{\partial^2 \psi}{\partial y^2} = 0}$ which is Laplace equation for ψ.

The properties of stream function (ψ) are:

1. The stream function (ψ) exists, it is a possible case of fluid flow which may be *rotational or irrotational*.
2. If stream function (ψ) satisfies the Laplace equation, it is possible case of an irrotational flow.

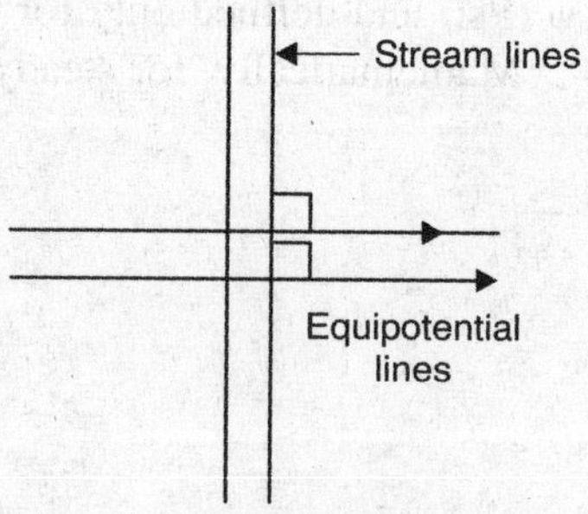

3.8.3 Equipotential Line

A line along which the velocity potential 'ϕ' is constant, is called the equipotential line.

For equipotential line; ϕ = constant

$$\Rightarrow d\phi = 0$$
$$\Rightarrow \phi = f(x, y) \text{ for steady flow.}$$

Fig. 3.11

Then;
$$d\phi = \frac{\partial \phi}{\partial x} \cdot dx + \frac{\partial \phi}{\partial y} \cdot dy$$

$$= -u\,dx - v\,dy$$

$$\left[\because \ \frac{\partial \phi}{\partial x} = -u, \frac{\partial \phi}{\partial y} = -v\right]$$

$$= -(u\,dx + v\,dy)$$

$$d\phi = \frac{\partial \phi}{\partial x} \cdot dx + \frac{\partial \phi}{\partial y} \cdot dy$$

$$0 = -udx - vdy$$

$$udx = -vdy$$

$$\frac{dy}{dx} = -\frac{u}{v} = m.$$

$$\therefore \qquad \boxed{\frac{dy}{dx} = \text{slope of equipotential line} = -\frac{u}{v}}$$

Line of Constant Stream Function

$\psi = $ constant

Then; $d\psi = 0$

When $\qquad\qquad \psi = $ const.

$$d\psi = 0$$

But; $d\psi = \dfrac{\partial \psi}{\partial x} \cdot dx + \dfrac{\partial \psi}{\partial y} \cdot dx = vdx - udy$ $\qquad\qquad \left[\because \dfrac{\partial \psi}{\partial x} = v, \dfrac{\partial \psi}{\partial y} = -u \right]$

$$d\psi = \frac{\partial \psi}{\partial x} \cdot dx + \frac{\partial \psi}{\partial y} dy$$

$$0 = vdx - udy$$

$$vdx = udy$$

$$\therefore \qquad \frac{dy}{dx} = \frac{v}{u} = m. \qquad\qquad\qquad\qquad\qquad\qquad\qquad\qquad \ldots(5)$$

For a Line of Constant Stream Function

i.e., $\qquad d\psi = 0$ or; $vdx - udy = 0$

or; $\qquad d\psi = \dfrac{\partial \psi}{\partial x} dx + \dfrac{\partial \psi}{\partial y} dy = 0$

But; $\dfrac{dy}{dx}$ is slope of stream Line.

$$d\phi = \frac{\partial \phi}{\partial x} \cdot dx + \frac{\partial \phi}{\partial y} \cdot dy = 0$$

$$= -\left(udx + vdy\right) = 0$$

$$= \left(udx + vdy\right) = 0$$

$$\frac{dy}{dx} = -\frac{u}{v} = m_2 \qquad\qquad\qquad\qquad\qquad\qquad\qquad\qquad\qquad \ldots(6)$$

$$\therefore \quad \boxed{m_1 \times m_2 = -1}$$

From equation (5) and (6); it is clear that the product of the slope of the equipotential line and the slope of the stream line at the point of intersection is equal to -1. Thus; the equipotential lines are orthogonal to the stream lines at all points of intersection.

Flow Net: A grid obtained by drawing a series of equipotential lines and stream lines is called a flow net. The flow net is an important tool in analysing two dimensional irrotational flow problems.

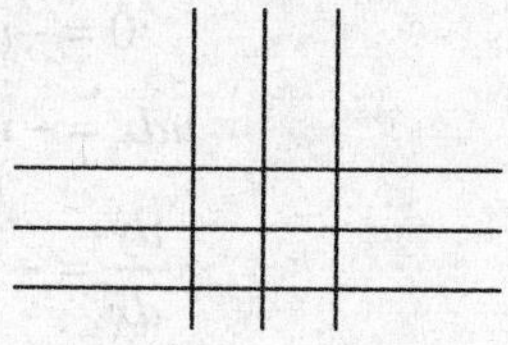

Fig. 3.12: Flow net

Relation between Stream Function and Velocity Potential Function

We have:
$$u = \frac{-\partial \phi}{\partial x} \text{ and } v = \frac{-\partial \phi}{\partial y}$$

As,
$$u = -\frac{\partial \phi}{\partial x}$$

$$v = -\frac{\partial \phi}{\partial y}$$

$$w = -\frac{\partial \phi}{\partial x}$$

Also; we have:
$$u = \frac{-\partial \psi}{\partial y} \text{ and } v = \frac{\partial \psi}{\partial x}$$

Thus; we have:
$$u = \frac{-\partial \phi}{\partial x} = -\frac{\partial \psi}{\partial y}$$

and
$$v = -\frac{\partial \phi}{\partial y} = \frac{\partial \psi}{\partial x}$$

Hence;
$$\boxed{\frac{\partial \phi}{\partial x} = \frac{\partial \psi}{\partial y}} \quad \text{and;} \quad \boxed{\frac{\partial \phi}{\partial y} = -\frac{\partial \psi}{\partial x}}.$$

Example 1. The velocity potential function (ϕ) is given by an expression:

$$\phi = -\frac{xy^3}{3} - x^2 + \frac{x^3 y}{3} + y^2$$

(i) Find the velocity components in x and y direction.

(ii) Show that 'ϕ' represents a possible case of flow: $\left[\dfrac{\partial^2 \phi}{dx^2} + \dfrac{\partial^2 \phi}{dy^2} + \dfrac{\partial^2 \phi}{dz^2} = 0 \right]$

Solution: (i) $\phi = -\dfrac{xy^3}{3} - x^2 + \dfrac{x^3 y}{z} + y^2$

We know; $\boxed{u = -\dfrac{\partial\phi}{dx}}$ Here; $\dfrac{\partial\phi}{\partial x} = -\dfrac{y^3}{3} - 2x + \dfrac{3x^2 y}{3} = -\dfrac{y^3}{3} - 2x + x^2 y$

$\therefore$ $-\dfrac{\partial\phi}{\partial x} = \dfrac{1}{3} y^3 + 2x - x^2 y$

$\therefore$ Velocity component in x-direction, $w = \dfrac{y^3}{3} + 2x - x^2 y$

$$v = -\dfrac{\partial\phi}{\partial y} = -\left[-\dfrac{3}{3} xy^2 + \dfrac{x^3}{3} + 2y \right] = xy^2 - \dfrac{x^3}{3} - 2y$$

$\therefore$ Velocity component in y-direction, $v = xy^2 - \dfrac{x^3}{3} - 2y$

(ii) The given value of ϕ, will represent a possible case of flow if it satisfies the laplace equation.

i.e. $\dfrac{\partial^2\phi}{\partial x^2} + \dfrac{\partial^2\phi}{\partial y^2} = 0$

But; $\dfrac{\partial\phi}{\partial x} = -\dfrac{y^3}{3} - 2x + x^2 y$

Then; $\dfrac{\partial^2\phi}{\partial x^2} = -2 + 2xy$

and $\dfrac{d\phi}{\partial y} = -xy^2 + \dfrac{x^3}{3} + 2y$

Then; $\dfrac{\partial^2\phi}{\partial y^2} = -2xy + 2$

Now; $-\cancel{2} + \cancel{2xy} - \cancel{2xy} + \cancel{2}$

$= 0.$

Hence; proved.

$\phi = x(2y - 1)$ ϕ at point p

at point p (4, 5)

For Velocity Potential

$$\phi = x(2y - 1) = 2xy - x$$

$$u = -\dfrac{\partial\phi}{\partial x} = -(2y - 1) = 1 - 2y = 1 - 2(5) = -9 \text{ units.}$$

$$v = -\frac{\partial \phi}{\partial y} = -(2x) = -2x = -2(4) = -8 \text{ units.}$$

∴

$$v = \sqrt{u^2 + v^2} = \sqrt{(-9)^2 + (-8)^2} = \sqrt{81 + 64} = \sqrt{145} = 12.04 \text{ m/sec}$$

(ii) Value of stream function at *P*.

We know; $\quad v = \dfrac{d\tau}{dx}$

and $\qquad -u = \dfrac{d\psi}{dy}$

But; $\qquad u = 1 - 2y$

∴ $\qquad -4 = -(1 - 2y) = 2y - 1$

Integrating with respect to *y*: $\displaystyle\int d\psi = \int (2y - 1)\, dy$

⇒ $\qquad \boxed{\psi = 2y^2 - y + c}$.

Now, differentiating the above equation with respect to *x*, we get:

$$\frac{d\psi}{dx} = \frac{dk}{dx}$$

But; $\qquad \dfrac{d\psi}{dx} = v = -2x$

∴ $\qquad \dfrac{dk}{dx} = -2x$

$$\int dk = -2 \int x\, dx$$

$$k = -\frac{\cancel{2}\, x^2}{\cancel{2}}$$

∴ $\qquad c = -x^2 = -(4)^2 = -16$

∴ Stream fn. ψ at $p(4, 5) = 5^2 - 5 - 4^2 = 25 - 5 - 16 = 4$ units.

3.9 Types of Motion

A fluid particle while moving may undergo anyone or combination of following four types of displacement:

(i) Linear translation or pure translation
(ii) Linear deformation
(iii) Angular deformation
(iv) Rotation

1. **Linear Translation:** It is defined as the movement of a fluid element in such a way that it moves bodily from one position to another position and the two axes *ab* and *cd* represented in new positions by *a′b′* and *c′d′* are parallel.

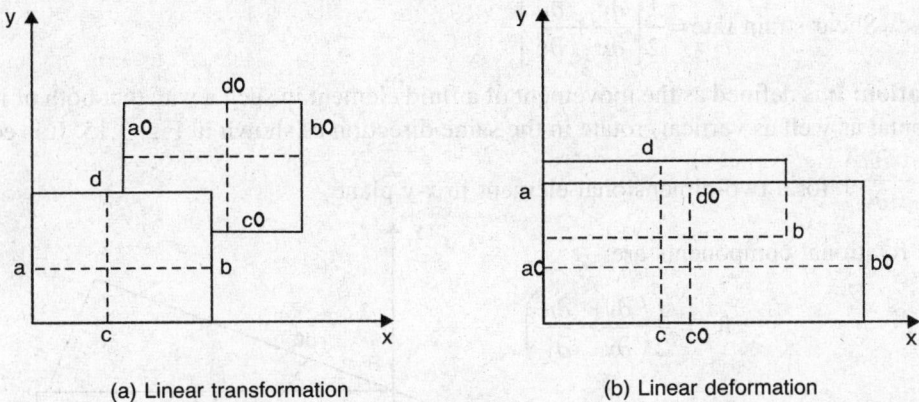

(a) Linear transformation   (b) Linear deformation

Fig. 3.13

2. **Linear Deformation:** It is defined as the deformation of a fluid element in linear direction when the element moves. The axis of the element in the deformed position and undeformed position are parallel, but their lengths change as shown in Fig. 3.13b.

3. **Angular Deformation or Shear Deformation:** It is defined as the average change in the angle contained by two adjacent sides. Let ΔQ_1 and ΔQ_2 is the change in angle between two adjacent side of a fluid element as shown; then angular deformation or shear strain rate (Fig. 3.14).

$$= \frac{1}{2}\left[\Delta Q_1 + \Delta Q_2\right]$$

Now; $\Delta Q_1 = \dfrac{\partial v}{\partial x} \times \dfrac{\Delta x}{\Delta x} = \dfrac{\partial v}{\partial x}$ and $\Delta Q_2 = \dfrac{\partial u}{\partial y} \cdot \dfrac{\Delta y}{\Delta y} = \dfrac{\partial u}{\partial y}$

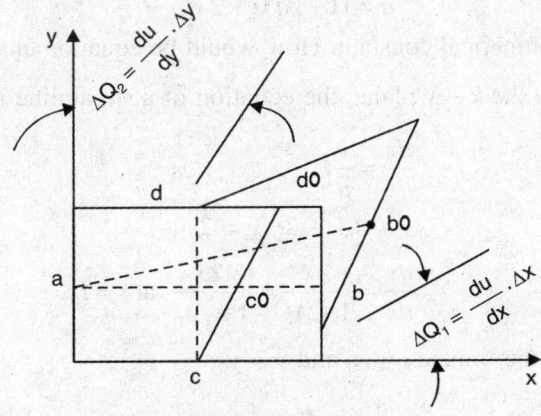

Angular deformation

Fig. 3.14

$\therefore$ Angular deformation $= \dfrac{1}{2}\left[\Delta Q_1 + \Delta Q_2\right]$

or;   Shear strain rate $= \dfrac{1}{2}\left[\dfrac{\partial v}{\partial x} + \dfrac{\partial u}{\partial y}\right]$

4. Rotation: It is defined as the movement of a fluid element in such a way that both of its axes (horizontal as well as vertical) rotate in the same direction as shown in Fig. 3.15. It is equal to $\dfrac{1}{2}\left(\dfrac{\partial v}{\partial x} - \dfrac{\partial u}{\partial y}\right)$ for a two-dimensional element in *x-y* plane.

The rotational components are:

$$w_z = \frac{1}{2}\left(\frac{\partial v}{\partial x} - \frac{\partial u}{\partial y}\right)$$

$$w_x = \frac{1}{2}\left(\frac{\partial w}{\partial y} - \frac{\partial v}{\partial z}\right)$$

and $w_y = \dfrac{1}{2}\left(\dfrac{\partial u}{\partial z} - \dfrac{\partial w}{\partial x}\right)$

Vorticity: It is defined as the value twice of the rotation and hence it is given as $= 2w$.

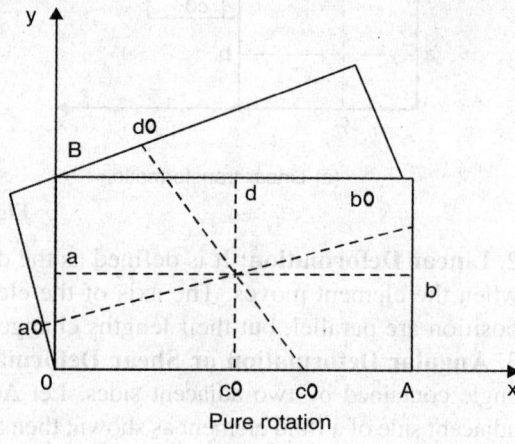

Pure rotation

Fig. 3.15

SOLVED PROBLEMS

1. Determine the equation of a stream line at $t = t_0$, passing through the (x_0, y_0) for the velocity field given by,

$$\vec{q} = (1 + At)\,i + 2xj$$

where A is some numerical constant, How would be equation change for a steady flow?

Ans. For a 2-D, flow in the $x - y$ plane, the equation of a streamline is given by:

$$\frac{dy}{dx} = \frac{v}{u}.$$

$$u = 1 + 2t, \; v = 2x$$

$\therefore$ $\dfrac{dy}{dx} = \dfrac{2x}{1 + At} = \dfrac{2x}{1 + At_0}$ at $t = t_0$,

Integrating within the limits, $x = x_0$ and $y = y_0$

$$\int_{y_0}^{y} dy = \frac{1}{1 + At_0}\int_{x_0}^{x} 2x\,dx$$

$$y - y_0 = \frac{1}{1 + At_0}\left(x^2 - x_0^2\right)$$

For a steady flow $A = 0$, consequently the streamline equation would take the form

$$y - y_0 = \left(x^2 - x_0^2\right)$$

2. An idealized flow is given by,

$u = a + by - cz$, $v = d - bx - ez$, $w = f + cx - ey$.

where a, b, c, d, e and f are arbitrary constants.

(a) Show that these are the velocity components of a fluid motion.

(b) Does these velocity components represent irrational flows? If not determine the velocity and rotation.

Ans. From the given velocity components

$$\frac{\partial u}{\partial x} = 0, \frac{\partial v}{\partial y} = 0, \frac{\partial w}{\partial z} = 0$$

$$\frac{\partial u}{\partial x} + \frac{\partial v}{\partial y} + \frac{\partial w}{\partial z} = 0,$$

The velocity components satisfy the flow continuity equation and therefore, the given field is a possible case of fluid flow.

(b) The curl of velocity vector is

$$\text{curl } V = (\nabla \times V) = \begin{vmatrix} i & j & k \\ \dfrac{\partial}{\partial x} & \dfrac{\partial}{\partial y} & \dfrac{\partial}{\partial z} \\ a + by - cz & d - bx - ez & f + cx - cy \end{vmatrix}$$

$$= 9\left[\frac{\partial}{\partial y}(f + cx - cy) - \frac{\partial}{\partial x}(d - bx - x)\right]$$

$$- j\left[\frac{\partial}{\partial x}\left(f + (x - cy) + \frac{\partial}{\partial z}(a + by - cz)\right)\right]$$

$$+ k\left[\frac{\partial}{\partial x}(d - bx - cz)\right]\left[\frac{\partial}{\partial y}(a + by - cz)\right]c/2$$

$$= i(-e + e) - j(e + c) + k(-b - b)$$

$$= -2(cj + bk)$$

Since $(\nabla \times v) = 0$, the flow is irrotational

vorticity $\xi = (\nabla \times V)$

$$= -2(cj + bk) = 2\sqrt{c^2 + b^2}$$

vorticity equals twice the value of rotation

$\therefore$ Rotation $w = \sqrt{c^2 + b^2}$

2. Water is flowing through a pipe of 0.5 m diameter with an average velocity of 1 m/s. What is the rate of discharge of water? The same flow then passes through another section where the diameter is 1 m. Water is the average flow velocity at this section?

$$\text{Discharge } Q = \text{area} \times \text{velocity}$$

$$= \frac{\pi}{4}(0.5)^2 \times 1 = 0.196 \text{ m}^3/\text{s}$$

Let v_2 be the velocity at the section where diameter is 1 m. From continuity equation.

$$Q = Av_1 = A_2 v_2$$

$\therefore$
$$v_2 = \frac{A_1}{A_2} V_1 = \frac{\dfrac{\pi}{4}(0.5)^2}{\dfrac{\pi}{4}(1)^2} \times 1 = 0.25 \text{ m/s}$$

3. Prove that stream function $\psi = 2xy$ is irrotational? If so determine the velocity potential. The velocity components for the given flow field are:

$$u = \frac{\partial \psi}{\partial y} = 2x, \; n = -\frac{\partial \psi}{\partial x} = -2y$$

Velocity vector $v = 2x_i - 2y_i$

$$(\nabla \times v) = \begin{vmatrix} i & j & k \\ \dfrac{\partial}{\partial x} & \dfrac{\partial}{\partial y} & \dfrac{\partial}{\partial z} \\ 2x & -2y & 0 \end{vmatrix}$$

$$= i\left[\frac{\partial}{\partial y}(0) - \frac{\partial}{\partial z}(-2y)\right] - j\left[\frac{\partial}{\partial x}(0) - \frac{\partial}{\partial z}(2x)\right] + k\left[\frac{\partial}{\partial x}(-2y) - \frac{\partial}{\partial y}(2x)\right] = 0$$

Thus the curl v is zero, the flow is irrotational. Hence the velocity potential does not exists.

Recalling the differential $d\phi$ for velocity potential function is:

$$d\phi = \frac{\partial \phi}{\partial x} \cdot dx + \frac{\partial \phi}{\partial y} \cdot dy$$

$$= u\,dx + v\,dy = 2x\,dx$$

4. For the velocity field given by, $\qquad \vec{q}(x, y, z, t) = 10xy_i + 5x^3 j + (t^2 x + z)k$

Find the velocity and acceleration of a fluid particle at position

$$\vec{r}(x, y, z) = (i + 2j + 3k) \text{ when time } t = 1.$$

Ans. The velocity components u, v and w are

$$u = 10xy, \quad v = 5x^2, \quad w = r^2x + z$$

From the position vector, the co-ordinates of the fluid particles are $x = 1$, $y = 2$ and $z = 3$. Hence, the velocity components at $(1, 2, 3)$ and at time $t = 1$ are:

$$u = 10 \times 1 \times 2 = 20 \text{ units}$$

$$v = 5 \times (1)^2 = 5 \text{ units}$$

$$w = (1)^2 \times 1 + 3 = 4 \text{ units}$$

$\therefore$ Velocity vector at $(1, 2, 3)$ and at $t = 1$ is: $20\,i + 15\,j + 4k$.

$\therefore$ Resultant velocity $= \sqrt{u^2 + v^2 + w^2} = \sqrt{(20)^2 + (5)^2 + (4)^2}$

$$= 21 \text{ units.}$$

The acceleration in x axis

$$a_x = \frac{\partial u}{\partial t} + u\frac{\partial u}{\partial x} + v\frac{\partial y}{\partial y} + w\frac{\partial y}{\partial z}$$

The x component of velocity gives,

$$\frac{\partial u}{\partial t} = 0; \quad \frac{\partial u}{\partial x} = 10y; \quad \frac{\partial u}{\partial y} = 10x; \quad \frac{\partial u}{\partial z} = 0.$$

Substituting these values in expression (1)

$$a_x = 0 + 10xy\,(10y) + 5x^2\,(10x) + (t^2x + z) \times 0$$

$$= 100\,xy^2 + 50\,x^2$$

at point $(1, 2, 3)$ the acceleration component a,

$$a_x = 100 \times 1 \times (2)^2 + 50 \times (1)^2 = 450 \text{ units}$$

The acceleration in the y-direction is

$$a_y = \frac{\partial v}{\partial t} + u\frac{\partial y}{\partial x} + v\frac{\partial v}{\partial y} + w\frac{\partial v}{\partial z}$$

The y component of velocity gives,

$$\frac{\partial v}{\partial t} = 0, \quad \frac{\partial v}{\partial x} = 10x, \quad \frac{\partial v}{\partial y} = 0, \quad \frac{\partial u}{\partial z} = 0.$$

$$a_y = 0 + 10xy(10x) + 5x^2 \times (0) + (t^2x + z) \times 0$$

at point $(1, 2, 3)$, $a_y = 100 \times (1)^2 \times 2 = 200 \text{ units}$,

The acceleration in z direction is

$$a_z = \frac{\partial w}{\partial t} + u\frac{\partial w}{\partial x} + v\frac{\partial w}{\partial y} + w\frac{\partial w}{\partial z}$$

The z component of velocity give,

$$\frac{\partial w}{\partial t} = 2tx; \quad \frac{\partial w}{\partial x} = t^2 ; \quad \frac{\partial w}{\partial y} = 0, \quad \frac{\partial w}{\partial z} = 1,$$

$$a_z = 2\lambda x + 10xy \times \left(r^2\right) + 5x^2 \times (0) + \left(t^2 x + z\right) \times 1$$

$$= 2tx + 10xyt^n + (rx + z)$$

At point (1,2,3) when $t = 1$

$$a_t = (2 \times 1 \times 1) + 10 \times 1 \times 2 \times (1)^2 \times \left(1^{-2} \times 1 + 3\right)$$

$$= 26 \text{ units}$$

3. velocity component of a particular 2D, steady, incompressible flow is given by,

$$u = e^{-x} \cosh y + 1,$$

Find the y component of velocity v presuming that $v = 0$ at $y = 0$ also work out the stream function and sketch the stream when $\psi = 0$ and $\psi = 1$.

4. A pipe branches into two pipes. The pipe has diameter of 45 cm at A, 30 cm at B, 20 cm at C and 15 cm at D, determine the discharge at start point if the velocity at that point is 2 m/s. Determine the velocities at any parts away from start point. [**Ans.:** $v_B = 4.5$ m/s]

5. Which of the following function represent a possible match for

   (i) $\psi = A\left(x^2 - y^2\right)$ (ii) $\phi = (r - 2/r)\sin\theta$

6. A 2 metre length of 10 cm diameter circular pipe in porous and the velocity at the inlet and exit of porous section have been measured to be 2.5 m/s and 1.6 m/s respectively. Make calculation for the discharge emitted through the poisonous wall and average velocity of emitted discharge.   [**Ans.:** 0.045 m/s]

The substantial acceleration at the given point and at the time of interval is

$$\bar{a} = 480u + 200j + 26k$$

$\therefore$ Resultant acceleration $= \sqrt{a_x^2 + ay^2 + az^2}$

$$= \sqrt{450^2 + 200^2 + 26^2}$$

$$= 493.13 \text{ units}$$

EXERCISE

1. The velocity components in flow field are known to be,

$$u = 6xt + y^2 z + 1s,$$

$$v = 3xy^2 + t^2 + y, \quad w = 2 + 3ty.$$

classify the velocity field as steady or unsteady, uniform / non uniform / 1 D or 2 D, (b) What is the velocity vector at point $p (x, y, z) = (3, 2, 4)$ at time $t = 35$? Also calculate the magnitude of this vector at this point and time.

                (Unsteady, non-uniform and 3D, $8si + 47j + 20k$ m/s)

                                    = 99.2 m/s

Dimensional Analysis

Introduction, derived quantities, dimensions of physical quantities, dimensional homogeneity, Buckingham's pi theorem, Rayleigh's method, dimensionless numbers, similitude, types of similitudes.

4.1 Introduction

The topic of dimensional analysis deals with the process in which all the important factors involved in a physical phenomenon, are symmetrical systematised into dimensionless numbers.

4.2 Difference between Dimension and Unit

(a) Dimensions refer to the qualitative characteristics for physical quantities.
Units are standards of comparison for the quantitative measure of dimensions.
(b) If distance between two points, the term length gives the qualitative concept of the physical quantity.

The term unit, however indicate the magnitude of distance. The distance may be quantitatively expressed or a metre as mile. The metre and mile will then be identical dimensions.

4.3 Primary and Secondary/Derived Quantity/Dimension

4.3.1 Primary dimension

The two most common systems of dimensioning a physical quantity are Face-length-time and Mass-length time systems referred to as *FLT* and *MLT* systems of units respectively.

The independent quantities which have no direct relations between them are called primary dimensions, e.g. Length L, mass M and Time T. Along with for compressible fluids, one more dimension namely the temperature θ.

4.3.2 Secondary dimension

The quantities which are expressed in terms of fundamental quantities or possess more than one fundamental dimensions are called as derived as secondary quantities.

For example,

(i) Density is defined by mass per unit volume, denoted by M/L^3 is a derived quantity.

(ii) Velocity and discharge are defined as mass per unit volume (Dimension = M/L^3), and area × velocity (Dimension = L^3T^{-1}) and called as derived quantity.

4.4 Dimension of Physical Quantity

Physical Quantities	Symbol	Dimensions	
		M-L-T System	F-L-T System
Fundamental Quantities			
Mass	M	M	$FL^{-1}T^2$
Length	L	L	L
Time	T	T	T
Force	F	MLT^{-2}	F
Geometric Quantities			
Area	A	L^2	L^2
Volume	$\forall$	L^3	L^3
Kinematic Quantities			
Linear velocity	u, V, U	LT^{-1}	LT^{-1}
Angular velocity	ω	T^{-1}	T^{-1}
Acceleraion	a	LT^{-2}	LT^{-2}
Discharge	q, Q	$L^3 T^{-1}$	$L^3 T^{-1}$
Gravity	g	LT^{-2}	LT^{-2}
Kinematic viscosity	v	$L^2 T^{-1}$	$L^2 T^{-1}$
Dynamic Quantities			
Density	ρ	ML^{-3}	$FL^{-4} T^2$
Specific weight	w	$ML^{-2} T^{-2}$	FL^{-3}
Surface tension	σ	MT^{-2}	FL^{-1}
Pressure intensity	p	$ML^{-1} T^{-2}$	FL^{-2}
Modulus of elasticity	E, K	$ML^{-1} T^{-2}$	FL^{-2}
Dynamic viscosity	μ	$ML^{-1} T^{-1}$	$FL^{-2} T$
Resisting force	F, R	MLT^{-2}	F
Thrust	T	MLT^{-2}	F
Torque	T	MLT^{-2}	FL
Work	W	$ML^2 T^{-2}$	FL
Energy	E	$ML^2 T^{-2}$	FL
Power	P	$ML^2 T^{-3}$	FLT

4.5 Dimensional Homogeneity

An equation is said to have dimensional homogeneity if the dimensions of various terms on the two sides of the equation are identical.

Example $Q = AV$ (discharge equation)

Properties

1. A dimensionally homogenous equation is independent of the fundamental units of measurement if the units therein are consistant.
2. A length dimension can be added to or subtracted from only a length dimension, etc.
4. If one term of an equation is measured in a particular unit, then all other terms in the equation must be measured in same unit.

Application

(i) The concept of dimensional homogeneity is a step towards dimensional analysis which is successfully employed to plan experiments and to present the results meaningfully.
(ii) Dimensional homogeneity helps to convert the unit from one system to another.

4.6 Buckingham's pi Theorem

Statement

If there are n variables in a dimensionally homogenous equation and if these variables contain m primary dimensions, then the variables can be grouped into (n-m) non-dimensional parameters. The non-dimensional groups arc called Pi (π terms).

Example: If a physical equation

$$f(x_1, x_2, x_4......x_n) = 0$$

where x's are dimensional physical quantities (such as velocity, density, viscosity, pressure and area etc.), then the $\phi[\pi, \pi_2, \pi_4... \pi_{n-m}] = 0$, where m represents the fundamental dimensions.

4.6.1 Choice of repeating variables

Generally the number of repeating variables are equal to the number of fundamental problem. The choice of repeating variables governed by the following considerations.

1. As far as possible, the dependent variable should not be selected as repeating variable.
2. The repeating variables should be chosen in such a way that one variable contains geometric property, other variable contains flow property and third variable contains fluid property.

Geometric Property	Flow Property	Fluid Property
Length l, diameter d, height H	Velocity (V), acceleration (a),	Density ρ, viscosity μ, ω

3. The repeating variables selected should not form a dimensionless group.
4. The repeating variables together must have the same number of fundamental dimensions.
5. No two repeating variables should have same dimension.

as example: In most of fluid mechanics problem, the choice of repeating variables may be
(i) d, v, ρ (ii) l, v, ρ (iii) l, v, μ (iv) d, v, μ

4.6.2 Procedure for solving

The resisting force R of a supersonic plane during flight can be considered as dependent upon the length of air craft l, velocity v, air viscosity μ, air density ρ and bulk modulus of air k. Express the functional relationship between these variables and resulting force.

Step 1. Resisting force R depends upon (i) l, (ii) v (iii) μ (iv) ρ and (iv) K, Hence R is a function of l, v, μ, ρ and K,

$$R = f(l, v, \mu, \rho, k),\qquad\qquad \text{...(i)}$$

or,
$$f_1(R, l, v, \rho, K) = 0 \qquad\qquad \text{...(ii)}$$

$\therefore$ Total No. of variables $n = 6$,

No. of fundamental dimension $m = 3$,

No, of dimensionless π term $= n - m = 6 - 3 = 3$,

Three π term are π_1, π_2 and π_3,

$$f_1(\pi_1, \pi_2, \pi_3) = 0 \qquad\qquad \text{...(1)}$$

Step 2. Each π term consists of $m + 1$ variables, where also called $m = 3$, repeating variables, out of 6 variables,

R, l, v, μ, ρ and K, 3 are to be selected as repeating variable,

R is a dependent variable $R.V.$ Out of after 5, one is Geometric, 2nd should be flow, 3rd should be fluid.

So, l, v, ρ are selected

1. They should not form dimensionless term.
2. They should have fundamental dimensions equal to m (3)

Dimension of $l = L$,

$$v = LT^{-1}, \rho = ML^{-3}$$

Hence, 3 fundamental dimensions exist as l, v and ρ and they themselves do not form dimensionless group.

Step 3. Each π term

$$\pi_1 = l^{a1} \cdot v^{b1} \cdot \rho^{c_1} \cdot R$$
$$\pi_2 = l^{a_2} \cdot v^{b2} \cdot \rho^{c_2} \cdot \mu$$
$$\pi_3 = l^{a_3} \cdot v^{b_3} \cdot \rho^{c_3} \cdot K$$

Step 4. Each π term is solved by the principle of dimensional homogeneity.

$$\pi_1 = M^0 L^0 T^0$$

$$M^0 L^0 T^0 = L^{a_1} \cdot (LT^{-1})^{b_1} (ML^{-3})^{c_1} \cdot MLT^{-2}$$

Equating the powers of M, L, T on both sides,

Power of M, $0 = c_1 + 1$, $c_1 = -1$

Power of L, $0 = a_1 + b_1 - 3c_1 + 1 = -2$

Power of T, $b_1 = -2$

$$\pi_1 = l^{-2}, v^{-2}, \rho^{-2}, R$$

$$\pi_1 = \frac{R}{l^2 v^2 \rho} = \frac{R}{\rho l^2 v^2}$$

Similarly, for 2nd π term,

$$\pi_2 = M^0 L^0 T^0 = L^{a_2} \cdot (LT^{-1})^{b_2} \cdot (ML^{-3})^{c_2} \cdot ML^{-1}T^{-1}$$
$$a_2 = -1, b_2 = -1, c_2 = -1$$

$$\pi_2 = l^{-1} \cdot v^{-1} \cdot \rho^{-1} \cdot \mu = \frac{\mu}{lv\rho}$$

3rd π term,

$$\pi_3 = l^{a_3} \cdot v^{b_3} \cdot \rho^{c_3} \cdot K$$

$$M L^0 T^0 = L^{a_3} \cdot \left(LT^{-3}\right)^{b_3} \cdot \left(ML^{-3}\right)^{a_3} \cdot ML^{-1}T^{-2}$$

$$a_3 = 0, \; b_3 = -2, \; c_3 = -1$$

$$\pi_3 = l^0 \cdot v^{-2} \cdot \rho^{-1} \cdot K = \frac{K}{v^2\rho}$$

Substituting the values of π_1, π_2 and π_3

$$f_1\left(\frac{R}{\rho l^2 v^2} \cdot \frac{\mu}{lv\rho} \cdot \frac{K}{v^2\rho}\right) = 0$$

or,

$$\frac{R}{\rho l^2 v^2} = \phi\left[\frac{\mu}{lv\rho}, \frac{K}{v^2\rho}\right]$$

$$R = \rho l^2 v^2 \phi\left[\frac{\mu}{lv\rho}, \frac{K}{v^2\rho}\right]$$

Example 1: The efficiencies as η of a fan depends on density ρ, dynamic viscosity μ of the fluid, angular velocity ω, diameter D of the rotor and discharge Q. Express η in term of dimensionless parameters.

Solution: η is a function of ρ, μ, ω, D and Q,

$$\eta = f(\rho, \mu, \omega, D, Q)$$

or, $f_1(\eta, \rho, \mu, \omega, D, Q) = 0$

Total No. of variables, $n = 6$,

Primary variable $m = 3$,

No. of π terms $= n - m = 6 - 3 = 3$

$$\therefore \; f_1(\pi_1, \pi_2, \pi_3) = 0,$$

Choosing, D, ω, ρ as repeating variable.

$$\pi_1 = D^{a_1} \cdot \omega^{b_1} \cdot \rho^{c_1} \cdot \eta$$

$$\pi_2 = D^{a_2} \cdot \omega^{b_2} \cdot \rho^{c_2} \cdot \mu$$

$$\pi_3 = D^{c_3} \cdot \omega^{b_3} \cdot \rho^{c_3} \cdot Q$$

$$a_1 = 0, \; b_1 = 0, \; c_1 = 0, \; \pi_1 = D^0 \omega^0 \rho^0, \; \eta = \eta$$

2nd term, $a_2 = -2, \; b_2 = -1, \; c_2 = -1, \; \pi_2 = D^{-2}, \; \omega^{-1}, \; \rho^{-1}, \; \mu$

3rd term,
$$= \frac{\mu}{D^r \omega \rho}$$

$$a_3 = -3, \ b_3 = -1, \ c_3 = 0$$

$$\pi_3 = D^{-3}, \ \omega^{-1} \cdot \rho^0 \cdot Q = \frac{Q}{D^r \omega}$$

$$f_1 \left(\eta, \frac{\mu}{D^r \omega \rho}, \frac{Q}{D^r \omega} \right) = 0$$

$$\eta = \phi \left[\frac{\mu}{D^r \omega \rho}, \frac{Q}{D^3 \omega} \right]$$

4.7 Rayleigh's Method

This method is used for determining the expression for a variable which depends upon maximum 3 or 4 variables only.

4.7.1 Disadvantage

If the no of independent variables becomes more than 4, it is very difficult to find the expression for dependent variable.

Let X is a variable, which depends on X_1, X_2 and X_3 variables. Then according to Rayleigh's method, X is a function of X_1, X_2 and X_3 and

$$X = f [X_1, X_2, X_3]$$

$$X = K X_1^a \cdot X_2^b \cdot X_3^c$$

K is a constant and a, b, c are arbitrarily powers.

SOLVED EXAMPLE

---

1. The time period (t) of a pendulum depends upon the length (L) of the pendulum and acceleration due to gravity (g). Derive an expression for the time period.

Solution. Time period t is a function of (i) L and (ii) g.

$t = K L^a \cdot g^b$ where K is a constant.

Substituting the dimensions on both sides,

$$T^1 = K L^a \cdot (L T^{-2})^b \qquad \qquad \text{...(1)}$$

Equating the powers of M, L, T on both sides,

$$T^1 = K L^a \cdot (L T^{-2})^b$$

Power of T, $1 = -2b$, $\therefore \ b = -\frac{1}{2}$

Power of L, $0 = a + b$, $\therefore \ a = -b = \left(-\dfrac{1}{2}\right) = \dfrac{1}{2}$

Substituting the values of a and b

$$t = KL^{\frac{1}{2}} \cdot g^{-\frac{1}{2}} = K\sqrt{\dfrac{L}{g}}$$

From experiments, value of K is determined as $K = 2\,T$

$$t = 2\pi\sqrt{\dfrac{L}{g}}$$

2. Find an expression for the drag force on smooth sphere of diameter D, moving with a uniform velocity V in a fluid of density ρ and dynamic viscosity μ Drag force F is a function of

$$F = f(D, v, \rho, \mu)$$
$$F = KD^a \cdot v^b \cdot \rho^c \cdot \mu^d$$

K is non dimensional field

$$MLT^{-2} = K \cdot L^a \cdot (LT^{-1})^b (ML^{-3})^c (ML^{-1}T^{-1})^d$$

Equating the powers of M, L and T on both sides,

Power of M, $1 = c + d$

Power of L, $1 = a + b - 3c - d$

Power of T, $-2 = -b - d$

$$c = 1 - d,\ b = 2 - d,\ a = 2 - d$$

$$F = KD^{2-d}v^{2-d}\rho^{1-d}\mu^{d}$$

$$= K\rho D^2 v^2 \left(\dfrac{\mu}{\rho vd}\right)d$$

$$= K\rho D^2 v^2 \phi\left(\dfrac{\mu}{\rho vD}\right)$$

4.8 Dimensionless Numbers and their Significance

By definition, dimensionless numbers are the numbers, which are generally ratio of two same dimension quantity (force).

Significance

1. Linear dimension l defines the geometrical boundary conditions, it might be the diameter of a pipe or the chord of an air foil.
2. Kinematic and dynamic characteristics of flow such as velocity v, pressure p, acceleration due to gravity g, explains relative importance of a particular dimension.

4.9 Main Dimensionless Groups are listed below with their application

S. No.	Dimensionless group	Definition and Symbol	Application
1.	Reynolds Number	$Re = \dfrac{\text{Inertia force}}{\text{Viscous force}}$ $Re = \dfrac{\rho v l}{\mu}$	(i) incompressible flow though small pipe (ii) low velocity motion around automobiles and aeroplanes (iii) motion of sub marine completely submerged under water.
2.	Froude Number	$Fr = \left(\dfrac{\text{Inertia force}}{\text{Gravity force}}\right)^{1/2}$ $= \dfrac{v}{\sqrt{Lg}}$	(i) Flow through open channels. (ii) Flow of liquid jets from orifice, notches, weir. (iii) Motion of shop in rough and turbulent.
3.	Mach Number	$M = \left(\dfrac{\text{Inertia force}}{\text{Elastic force}}\right)^{1/2}$ $= \dfrac{v}{C}$ $C = \sqrt{K/\rho} = \text{speed of sound}$	(i) Aerodynamic testing. (ii) Water hammer problem (iii) Flow of gases exceeding the velocity of sound.
4.	Weber number	$W = \dfrac{\text{Inertia force}}{\text{Surface tension force}}$ $= \dfrac{\rho l v^2}{\sigma}$	(i) Capillary tube flows. (ii) Capillary movement of water in soils. (iii) Flow of blood in veins and arteries, thin sheet flow. (iv) liquid atomisation.
5.	Euler Number	$E = \dfrac{\text{Pressure force}}{\text{Inertia force}}$ $= \dfrac{P}{\rho v^2}$	(i) Flow though pipes. (ii) Flow over submerged border. (iii) Discharge though orifices, month pieces, slices

4.9.1 Dimensionless Number in details

4.9.1.1 Reynolds Number (Re)

Reynolds number is defined as the ratio of inertia force F_i to the viscous friction force F_v.

Inertia force F_1 = mass × acceleration
$$= \rho l^2 v^2$$

$$l = \text{length, density}$$
$$\text{viscous force} = \rho vl,$$

$$\text{Re} = \frac{\text{Inertia force}}{\text{Viscous force}} = \frac{\rho vl}{\mu}$$

(i) Reynolds number is less than 2000, the flow is called laminar flow,

(ii) If Reynolds number is more than 4000, the flow is called turbulent flow.

(iii) If Reynolds number is 2000 and 4000, the flow is called transient flow.

4.9.1.2 Froude Number (Fr)

The Fraude's number is defined as the square root of the ratio of inertia force of a flowing fluid to the gravity force. Mathematically;

$$F_\pi = \sqrt{\frac{F_i}{F_g}}$$

where;

$$\boxed{F_i = \rho A V^2}$$

and

$$F_g = \text{mass} \times \text{accleration due to gravity}$$
$$= \rho \times \text{volume} \times g = \rho \cdot L^3 g$$
$$= \rho L^2 \cdot L_g$$

$$\boxed{\therefore\ F_g = \rho \cdot A \cdot L \cdot g}$$

$$\therefore \qquad F_\pi = \sqrt{\frac{F_i}{F_g}} = \sqrt{\frac{\rho A V^2}{\rho A Lg}}$$

$$\therefore \qquad \boxed{\therefore\ F_e = \frac{V}{\sqrt{Lg}}}$$

4.9.1.3 Euler Number (Eu)

It is defined as the square root of the ratio of the inertia force of a flowing fluid to the pressure force. Mathematically:

$$E_u = \sqrt{\frac{F_i}{F_p}}$$

But $\qquad F_i = \rho A V^2$

and $\qquad F_i = \text{Intensity of press} \times A = p \times A$

$$\therefore \qquad E_u = \sqrt{\frac{\rho A v^2}{p \times A}}$$

$$\therefore \qquad E_u = \sqrt{\frac{v^2}{p/\rho}} = \frac{V}{\sqrt{p/\rho}}$$

$$\therefore \qquad \boxed{\therefore E_u = \frac{v}{\sqrt{p/\rho}}}$$

4.9.1.4 Weber Number (We)

It is defined as the square root of the ratio of the inertia force of a flowing fluid to the surface tension force. Mathematically:

Weber's No.; $\qquad W_e = \sqrt{\dfrac{F_i}{F_s}}$

But; $\qquad F_s$ = Surface tension per unit length × length = $\sigma \times L$

$$\therefore \qquad F_i = \rho A V^2$$

$$\Rightarrow \qquad W_e = \sqrt{\frac{\rho A v^2}{\sigma L}} = \sqrt{\frac{\rho L^2 v^2}{\sigma L}}$$

$$\therefore \qquad W_e = \sqrt{\frac{v^2}{\sigma/\rho L}}$$

$$\boxed{\therefore W_e = \frac{v}{\sqrt{\sigma/\rho L}}}$$

4.9.1.5 Mach Number (M)

It is defined as the square root of the ratio of the inertia force of a flowing fluid to the elastic force. Mathematically, it is defined as:

$$M = \sqrt{\frac{\text{Inertia force}}{\text{Elastic force}}} = \sqrt{\frac{F_i}{F_e}}$$

where; $\qquad F_i = \rho A v^2$

and $\qquad F_e$ = Elastic force = Elastic stress × Area

$$\Rightarrow \qquad F_e = K \times A = K \times L^2 \ (K = \text{elastic stress})$$

$$\therefore \qquad M = \sqrt{\frac{\rho A V^2}{K L^2}} = \sqrt{\frac{\rho L^2 v^2}{K L^2}}$$

$$\therefore \qquad M = \frac{v}{\sqrt{K/\rho}}$$

But; $\qquad \sqrt{K/\rho} = c =$ Velocity of sound in the fluid.

$$\therefore M = \frac{V}{c}$$ M = Mach Number

$\qquad\qquad V$ = velocity of fluid, C = Velocity of sound

$\qquad\qquad M = 1$, The flow is called sonic flow.

$\qquad\qquad M < 1$, The flow is called sub sonic flow.

$\qquad\qquad M > 1$, The flow is called super sonic flow.

4.10 Model Analysis

For predicting the performance of the hydraulic structures (such as dams, spill ways etc.) or hydraulic machines (such as turbines, pumps etc.), before actually constructing or manufacturing, models of the structures or machines are made and tests are performed on them to obtain the desired information.

The Model is the small scale replica of the actual structure or machine. The actual structure or machine is called prototype. The study of models of actual machines is called Model Analysis. Model analysis is actually an experimental method of finding solutions of complex flow problems.

The following are the advantages of the dimensional and model analysis:

1. The performance of the hydraulic structure or hydraulic machine can be easily predicted, in advance, from its model.

2. With the help of dimensional analysis, a relationship between the variables influencing a flow problem in terms of dimensionless parameters is obtained. This relationship helps in conducting tests on the model.

3. The merits of alternative designs can be predicted with the help of model testing.

4.10.1 Similitude—Types of similarities

Similitude is defined as the similarity between the model and its prototype is every respect, which means that the model and prototype have similar properties or model and prototype are completely similar.

Three types of similarities most exist between the model and prototype. They are:

1. Geometric similarity
2. Kinematic similarity
3. Dynamic similarity

1. **Geometric Similarity:** The geometric similarity is said to exist between the model and prototype when the ratio of all corresponding linear dimension in the model and the prototype are equal:

Let L_m = Length of model, b_m = Breadth of model

$\qquad D_m$ = Diameter of model; A_m = Area of model

$\qquad V_m$ = Volume of model.

and L_p, b_p, D_p, A_p and V_p be the Corresponding values of the prototype.
Thus; For geometric Similarity between model and prototype; we must have:

$$\frac{L_p}{L_m} = \frac{b_p}{b_m} = \frac{D_p}{D_m} = L_r; \quad [\text{where } L_r \rightarrow \text{Scale Ratio}]$$

For geometric similarity between model and prototype, we must have the relation.
For area ratio and volume ratio the relation should be as given below:

$$\frac{A_p}{A_m} = \frac{L_p}{L_m} \times \frac{B_p}{B_m} = L_r \times L_r = L_r^2$$

So,
$$\frac{V_p}{V_m} = \left(\frac{L_p}{L_m}\right)^3 = \left(\frac{b_p}{b_m}\right)^3 = \left(\frac{D_p}{D_m}\right)^3 = L_r^3$$

2. **Kinematic Similarity:** Kinematic similarity means the similarity of motion between model and prototype. Thus, Kinematic similarity is said to exist between the model and the prototype; if the ratios of the velocity and acceleration at the corresponding points in the model and at the corresponding points in the prototype are the same.

The directions of velocity and accelerations at the corresponding points in the model and prototype also should be parallel.

Let V_{p1} = Velocity of fluid at point 1 in prototype.

V_{p2} = Velocity of fluid at point 2 in prototype.

a_{p1} = Acceleration of fluid at point 1 in prototype.

a_{p2} = Acceleration of fluid at point 2 in prototype.

V_{m1}, V_{m2}, a_{m1}, a_{m2} = Corresponding values at the corresponding points of fluid velocity and acceleration in the model.

For kinematic similarity, we must have:

$$\frac{V_{p_1}}{V_{m_1}} = \frac{V_{p_2}}{V_{m_2}} = V_r; \quad [\text{where } V_r \rightarrow \text{Velocity Ratio}]$$

For acceleration; we must have: $\boxed{\dfrac{a_{p_1}}{a_{m_1}} = \dfrac{a_{p_2}}{a_{m_2}} = a_r}$ $[a_r \rightarrow$ acceleration ratio]

Also, the directions of the velocities in the model and prototype should be same.

3. **Dynamic Similarity:** Dynamic similarity means the similarity of forces between the model and prototype. Thus, dynamic similarity is said to exist between the model and the prototype, if the ratios of the corresponding forces acting at the corresponding points are equal. Also the directions of the corresponding forces at the corresponding points should be same.

Let $(F_i)_p \rightarrow$ Inertia force at a point in prototype.

$(F_v)_p \rightarrow$ Viscous force at a point in prototype.

$(F_g)_p \rightarrow$ Gravity force at a point in prototype.

and $(F_i)_m$, $(F_v)_m$, $(F_g)_m \rightarrow$ Corresponding values of model.

Then, for dynamic similarity:

$$\frac{(F_i)_p}{(F_i)_m} = \frac{(F_v)_p}{(F_v)_m} = \frac{(F_g)_p}{(F_g)_m}$$

$$= F_r \rightarrow \text{force ratio}$$

4.10.2 Model's Laws or Similarity Laws

For the dynamic similarity between the model and the prototype the ratio of the corresponding forces at the corresponding points in the model and prototype should be equal. The ratio of the forces are dimensionless numbers. It means for dynamic similarity between the model and prototype, the dimensionless numbers should be same for model and the prototype. But, it is quite difficult to satisfy the condition that all the dimensionless numbers (i.e., R_e, F_e, W_e, E_u and M) are the same for the model and prototype. Hence, models are designed on the basis of ratio of the force, which is dominating in the phenomenon.

The Laws on which the models are designed for dynamic similarity are called model laws or Laws of similarity.

The following are the model laws:

1. Reynold's model law
2. Froude model law
3. Euler model law
4. Weber model law
5. Mach model law

4.10.2.1 Reynold's Model Law

1. Reynold's model law is the law in which models are based on Reynold's number. Models based on Reynold's number includes:

 (i) Pipe flow
 (ii) Resistance experienced by submarines, airplanes, fully immersed bodies etc.

Reynold's No. is the ratio of inertia force to the viscous force and hence fluid flow problems where viscous forces alone are predominant, the models are designed for a dynamic similarity on Reynold's law, which states that the Reynold number for the model must be equal to the Reynold number for the prototype.

Let $V_m \rightarrow$ Velocity of fluid in model

$\rho_m \rightarrow$ Density of fluid in model

$L_m \rightarrow$ Length or Linear dimension of the model

$l_{1m} \rightarrow$ Viscosity of fluid in model.

and; V_p, ρ_p, L_p, and $l_{1p} \rightarrow$ are the corresponding values of velocity, density, Linear dimension and viscosity of fluid in prototype. Then, Reynold's model law.

$$[Re]_m = [Re]_p \quad \text{or} \quad \frac{\rho_m V_m L_m}{l_{1m}} = \frac{\rho_p V_p L_p}{l_{1p}} \qquad \ldots\text{(i)}$$

$$\frac{\rho_p \cdot V_p \cdot L_p}{\rho_m \cdot V_m \cdot L_m} \times \frac{1}{\dfrac{l_{1p}}{l_{1m}}} = 1$$

or;

or;

$$\boxed{\frac{P_r \cdot V_r \cdot L_r}{l_1 r} = 1}$$

where;

$$\rho_r = \frac{\rho_p}{\rho_m}, \quad L_r = \frac{L_\rho}{L_m} = l_{1r} = \frac{l_{1p}}{l_{1m}}$$

$$V_r = \frac{V_\rho}{V_m}$$

Here; ρ_r, V_r, L_r and l_{1r} are called the scales ratios for density, velocity, linear dimension and viscosity.

The scale ratios for time, acceleration, force and discharge for Reynold's model law are obtained as:

$$t_r = \text{Time scale ratio} = \frac{L_r}{V_r}$$

$$\rho = \frac{m}{V}$$

$$a_r = \text{Acceleration scale ratio} = \frac{V_r}{t_r}$$

$$F_r = \text{Force scale ratio} = (\text{mass} \times \text{acceleration})$$
$$= m_r \times a_r = \rho_r \cdot A_r \cdot V_r \times a_r$$
$$= \rho_r \cdot L_r^2 \, V_r \times a_r$$

But;
$$Q_r = \text{Discharge scale ratio} = (\rho A V)_r$$
$$= \rho_r A_r V_r = \rho_r \cdot L_r^2 \cdot V_r$$

2. Diameter of prototype, $d_p = 1.5\text{m} \Rightarrow \Delta_\rho = \frac{\pi}{4}(1.5)^2$

Diameter of model, $d_m = 15$ cm $= 0.15$ m

Viscosity of fluid, $l_{1p} = 3 \times 10^{-2}$ poise $= 1l = 1000$ cm^3

Q for prototype $= 3000$ l/sec. $= 3$ m^3/sec $1l = \dfrac{1}{1000}$ cm^3

Sp. gravity of oil, $S_p = 0.9$ $w = \rho g$

∴ Density of oil $= P_p = S_p \times 1000$

$\Rightarrow P_p = 0.9 \times 1000 = 2900$ kg/m^3 Sp. gravity $= P \times 1000$

Now; diameter of model, $d_m = 0.15$ m

Viscosity of water at 20°C $= 0.1$ poise $= 1 \times 10^{-2}$ poise $\Rightarrow l_{1m} = 1 \times 10^{-2}$ poise

Density of water, $\rho_m = 1000$ kg/m^3

Now; For pipe flow, the dynamic similarity will be obtained if the Reynold's number in the model and prototype are equal.

So; $\dfrac{\rho_m V_m D_m}{l_{1m}} = \dfrac{\rho_p V_p D_p}{l_{1p}}$ [For pipe: Linear dimension is diameter]

$\Rightarrow \qquad \dfrac{V_m}{V_p} = \dfrac{\rho_p}{\rho_m} \cdot \dfrac{D_p}{D_m} \cdot \dfrac{l_{1m}}{l_{1p}}$

$\qquad\qquad = \dfrac{9\cancel{00}}{\cancel{1000}} \times \dfrac{\cancel{45}}{0.1\cancel{5} \times \cancel{10}} \times \dfrac{3 \times \cancel{10^{-2}}}{1 \times \cancel{10^{-2}}} \times \cancel{100}$

$\qquad\qquad = \dfrac{900}{100\cancel{0}} \times 1\cancel{0} \times 3 = 27$

But; $\qquad V_p = \dfrac{\text{Rate of flow in prototype}}{\text{Area of prototype}} = \dfrac{3}{\dfrac{\pi}{4}(1.5)^2} = 1.697\,\text{m/s}$

$\therefore \qquad V_m = 27 \times V_P = 27 \times 1.697$

$\therefore \qquad V_m = \dfrac{\pi}{4}(0.15)^2 \cdot \left(V_m\right)$

$\qquad Q_m = A_m \times V_m$

4.10.3 Froude Model Law

Froude model law is the law in which the models are based on Froude number which means for dynamic similarity between the model and prototype, the Froude number for both of them should be equal. Froude model law is applicable when the gravity force is only predominant force which controls the flow in addition to the force of inertia. Froude model law is applied in the following fluid flow problems:

1. Free surface flows such as flow over spillways, weirs, sluices, channels etc.
2. Flow of jet from an orifice or nozzle.
4. Where waves are likely to be formed on surface.
4. Where fluids of different densities flow over one another.

Let V_m = Velocity of fluid in model

$\qquad L_m$ = Linear dimension or Length of model

$\qquad g_m$ = Acceleration due to gravity at a place where model is tested.

and V_p, L_p and g_p are the corresponding values of the velocity, length and acceleration due to gravity for the prototype. Then, according to Froude model law:

$$(Fe)_{\text{model}} = (Fe)_{\text{prototype}} \Rightarrow \dfrac{V_m}{\sqrt{g_m \cdot L_m}} = \dfrac{V_p}{\sqrt{g_p \cdot L_p}} \qquad \ldots(1)$$

If the tests on the model are performed on the same place where prototype is to operate, then: $g_m = g_p$ and equation becomes:

$$\frac{V_m}{\sqrt{L_m}} = \frac{V_p}{L_p}$$

or:

$$\frac{V_m}{V_p} \times \frac{1}{\sqrt{\dfrac{L_m}{L_p}}} = 1$$

$$\therefore \quad \boxed{\frac{V_p}{V_m} = \sqrt{\frac{L_p}{L_m}} = \sqrt{L_r}}$$

$$\frac{V_p}{V_m} = \sqrt{L_r}$$

$$\frac{L_p}{L_m} = L_r$$

$$\boxed{Fe = \frac{V}{\sqrt{gh}}}$$

where; $L_r \rightarrow$ Scale Ratio for Length

$$\frac{V_p}{V_m} = V_r = \text{Scale ratio for velocity}$$

$$\therefore \qquad \frac{V_p}{V_m} = V_r = \sqrt{L_r}.$$

Scales ratios for various physical quantities based on Froude's model law are:

(a) Scale ratio for time

As time $= \dfrac{\text{Length}}{\text{Velocity}}$

Then; ratio of time for prototype and model is:

$$T_r = \frac{T_p}{T_m} = \frac{\left(\frac{L}{V}\right)_p}{\left(\frac{L}{V}\right)_m} = \frac{\frac{L_p}{V_p}}{\frac{L_m}{V_m}} = \left(\frac{L_p}{L_m}\right) \times \left(\frac{V_m}{V_p}\right) = L_r \times \frac{1}{\sqrt{L_r}} \quad \left| \begin{array}{l} u_r = \sqrt{L_r} \\ T_r = \sqrt{L_r} \\ \boxed{V_r = T_r} \end{array} \right| a_r = 1$$

$$\boxed{\therefore \ T_r = \sqrt{L_r}}$$

(b) Scale ratio for acceleration

$$\text{Acceleration} = \frac{V}{T}$$

$$a_r = \frac{a_p}{a_m} = \frac{\left(\frac{V}{T}\right)_P}{\left(\frac{V}{T}\right)_m} = \frac{V_p}{T_p} \times \frac{T_m}{V_m} = \left(\frac{V_p}{V_m}\right) \times \left(\frac{T_m}{T_p}\right) = \sqrt{L_r} \times \frac{1}{\sqrt{L_r}}$$

$$\boxed{\therefore \ a_r = 1}$$

(c) Scale ratio for discharge

$$Q = A \times V = L^2 \cdot \frac{L}{T} = \frac{L^3}{T}$$

Then;
$$Q_r = \frac{Q_p}{Q_m} = \frac{\left(\frac{L}{T}\right)^3_p}{\left(\frac{L^3}{T}\right)_m} = \left(\frac{L_p}{L_m}\right)^3 \times \left(\frac{T_m}{T_P}\right) = L_r^3 \cdot \frac{1}{\sqrt{L_r}}$$

$$\boxed{\therefore \ Q_r = L_r^{2.5}}$$

(d) Scale ratio for force

As force = Mass × Acceleration = $\rho L^3 \times \frac{V}{T} = \rho L^2 \cdot \frac{L}{T} \cdot V = \rho L^3 \frac{V}{T} = \rho L^2 \cdot \frac{L}{T} \cdot V$

$$\text{Force} = \rho L^2 \cdot V^2 \qquad \left[\frac{L}{T} = V\right]$$

Now, ratio for force; $F_r = \dfrac{F_P}{F_m} = \dfrac{\rho_p L_p^2 \cdot V_p^2}{\rho_m \cdot L_m^2 V_m^2} = \dfrac{\rho_p}{\rho_m} \times \left(\dfrac{L_p}{L_m}\right)^2 \times \left(\dfrac{V_p}{V_m}\right)^2$

If the fluid used in model and prototype is same, then:

$$\frac{\rho_p}{\rho_m} = 1 \quad \text{or} \quad \rho_p = \rho_m$$

Hence, $F_r = \left(\dfrac{L_p}{L_m}\right)^2 \times \left(\dfrac{V_p}{V_m}\right)^2 = L_r^2 \times \left(\sqrt{L_r}\right)^2 = L_r^3$

Given:

Scale ratio of length, $L_r = 40$
Velocity in model, $V_m = 2$ m/s
discharge in model, $Q_m \doteq 2.5$ m^3/sec
Then; $V_p = ?$ $Q_p = ?$

We know; $\dfrac{V_p}{V_m} = \sqrt{L_r} \Rightarrow V_p = V_m \sqrt{L_r}$

$\therefore \qquad V_p = 2 \times \sqrt{40} = 12.65$ m/sec

$\dfrac{Q_p}{Q_m} = Lr^{2.5} \left[\left(\dfrac{V_p}{V_m}\right) \dfrac{A_p}{A_m} \right] = \sqrt{L_r} \times \dfrac{L_p^2}{L_m^2} = L_r^2 \cdot L_r^{1/2}$

$\therefore \qquad Q_p = Q_m L_r^{2.5} = 2.5(40)^{2.5} = 25298.2$ m^3/sec

20. $\qquad L_p = 50$

$\qquad L_m = 1$

$\qquad V_m = 1$ m/s; $V_p = ?$

$\qquad F_m = 2N, F_p = ?$

Now; $\qquad L_r = 50$

We know; $\dfrac{V_p}{V_m} = V_r = \sqrt{L_r}$

$\therefore \qquad V_p = V_m \cdot \sqrt{L_r} = 1 \times \sqrt{50}$

$\therefore \qquad \boxed{\therefore V_p = 7.071 \text{ m/s}}$

Now; $\qquad F_r = \dfrac{F_p}{F_m} = L_r^3$

$\therefore$
$$F_p = F_m \cdot L_r^3$$
$$= 2 \times (50)^3$$
$$= 250000 \text{ N}$$

21. $\quad Q_m = 2\text{m}^3/\text{sec}, \ V_m = 1.5 \text{ m/sec}$

$\qquad L_p = 36x, \ L_m = x$

$\therefore \qquad L_r = \dfrac{L_p}{L_m} = \dfrac{36 \cdot \cancel{x}}{\cancel{x}} = 36 = L_r$

$\therefore \qquad L_r = 36$

Now; $\qquad V_r = \dfrac{V_p}{V_m} = \sqrt{L_r}$

$\Rightarrow \qquad V_p = V_m \cdot \sqrt{L_r}$

$\therefore \qquad V_p = 1.5 \times \sqrt{36}$

$$\boxed{\therefore \ V_p = 9 \text{ m/s}}$$

Now; $\qquad Q_r = \dfrac{Q_p}{Q_m} = \dfrac{A_p \cdot V_p}{A_m \cdot V_m}$

$\Rightarrow \qquad \left(\dfrac{L_p}{L_m}\right)^2 \cdot \left(\dfrac{V_p}{V_m}\right)^2 = L_r^2 \cdot \sqrt{L_r}$

$\therefore \qquad Q_p = Q_m \cdot L_r^2 \cdot \sqrt{L_r}$

$$= 2 \times (36)^2 \left(\sqrt{36}\right)$$
$$= 2 \times 6 \times 36 \times 36$$

$\therefore \qquad Q_p = 15552 \text{ m}^3/\text{sec. Ans.}$

(e) Scale ratio for pressure intensity

As, $\qquad P = \dfrac{\text{Force}}{\text{Area}} = \dfrac{\rho L^2 V^2}{L^2} = \rho V^2$

Pressure ratio, $P_r = \dfrac{P_p}{P_m} = \dfrac{\rho_p \cdot V_p^2}{\rho_m \cdot V_m^2}$

If Fluid is same; then: $\dfrac{P_\rho - P_m}{2} \ \rho_r = \dfrac{V_p^2}{V_m^2} = \left(\dfrac{V_p}{V_m}\right)^2 = L_r.$

$$\boxed{\therefore \ P_r = L_r}$$

(f) Scale ratio for work, energy, torque, moment, etc.

Torque = Force × Distance = $F \times L$

$\therefore$ Torque ratio, $T_r = \dfrac{T_p}{T_m} = \dfrac{(F \times L)_p}{(F \times L)_m} = F_r \times L_r = L_r^3 \times L_r = L_r^4$

$$\boxed{\therefore\ T_r = Lr^4}$$

(g) Scale ratio for power

As: Power = work per unit time.

$\Rightarrow$ Power = $F \times \dfrac{L}{T}$ $\qquad \left[\because\ P = F \times V \right]$

$\therefore$ Power ratio, $P_r = \dfrac{\rho_p}{\rho_m} = \dfrac{\dfrac{F_p \times L_p}{T_p}}{\dfrac{F_m \times L_m}{T_m}} = \dfrac{F_p}{F_m} \times \dfrac{L_p}{L_m} \times \dfrac{1}{\dfrac{T_p}{T_m}}$

$\Rightarrow P_r = F_r \cdot L_r \cdot \dfrac{1}{T_r} = L_r^3 \cdot L_r \cdot \dfrac{1}{\sqrt{L_r}}$

$$\boxed{\therefore\ P_r = L_r^{3.5}}$$

Example:

If $L_r = 10$ and $Q_p = 28.3$ m³/sec are given

Find the value of Q_m

We know: $L_r = \dfrac{L_p}{L_m} = 10$

$Q_r = \dfrac{Q_p}{Q_m} = \left(\dfrac{A_p}{A_m} \right) \left(\dfrac{V_p}{V_m} \right) = \left(\dfrac{L_p}{V_m} \right)^2 \left(\dfrac{V_p}{V_m} \right)$

$\Rightarrow \dfrac{Q_p}{Q_m} = L_r^2 \sqrt{L_r}$

$\therefore Q_m = \dfrac{Q_p}{L_r^{2.5}} = \dfrac{28.3}{(10)^{2.5}} = 0.0895$ m³/sec

(ii) $V_m = 2.4$ m/s

$V_p = ?$

$$V_r = \frac{V_p}{V_m} = \sqrt{L_r}$$

$$\therefore \qquad V_p = V_m \sqrt{L_r} = 2.4\sqrt{10}$$

$$\boxed{\therefore V_p = 7.589 \text{ m/sec}}$$

(iii) $H_m = 50$ mm

$H_p = ?$

Again; $L_r = \dfrac{H_p}{H_m}$

$$\therefore \ H_p = H_m \cdot L_r = 50 \times 10 = 500 \text{ mm}$$

(iv) $E_m = 4.5$ N $- m$

$E_p = ?$, $E_p \rightarrow$ Energy dissipated/s in prototype.

Now; $\qquad E_r = \dfrac{E_p}{E_m} = \left(\dfrac{F_p}{F_m}\right)\dfrac{L_p}{L_m}$

$$\frac{E_p}{E_m} = L_r^3 \cdot L_r = L_r^4$$

$$\therefore \qquad E_p = E_m \cdot L_r{}^4 = 4.5(10)^4$$

$$\therefore \qquad E_p = E_m \cdot L_r{}^4 = 3.5(10)^4$$

$$\therefore \qquad E_p = \frac{E_p}{E_m} = L_r^{3.5}$$

(v) $Q_m = \dfrac{1}{6}$ m^3/s

$L_r = \dfrac{L_p}{L_m} = 36$

$Q_p = ?$

We know; $\qquad Q_r = \dfrac{Q_p}{Q_m}\left(\dfrac{A_p}{A_m}\right)\left(\dfrac{V_p}{V_m}\right)$

$$\frac{Q_p}{Q_m} = \left(\frac{L_p}{L_r}\right)^2 \left(\frac{V_p}{V_m}\right) = L_r^2 \cdot \sqrt{L_r}$$

$$\therefore \qquad Q_p = Q_m \cdot L_r^{2.5}$$

$$= \frac{1}{6}(36)^{2.5}$$

$$\frac{q_p}{q_m} = \frac{Q_p/L_p}{Q_m/L_m} = \frac{Q_p}{Q_m} \times \frac{L_m}{L_p} = L_r^{2.5} \times \frac{1}{L_r} = L_r^{1.5}$$

$$q_p = q_m \times L_r^{1.5} = \frac{1}{6}(36)^{1.5}.$$

$$L_r = 50$$

width of model, $B_m = 600$ mm $= 0.6$ m

$$L_p = 15 \text{ m}$$
$$L_m = 1.5$$

4.10.4 Euler's Model Law

Eulers model law is the law in which the models are designed on Euler's number which means for dynamic similarity between the model and prototype. The Euler number for model and prototype should be equal. Euler's model law is applicable when the pressure forces are alone predominant in addition to the inertia force.

According to this law:

$$(E_u)_{\text{model}} = (E_u)_{\text{prototype}} \qquad \qquad \dots(1)$$

If: $\qquad\qquad V_m \rightarrow$ Velocity of fluid in model

$\qquad\qquad p_m \rightarrow$ Pressure of fluid in model

$\qquad\qquad P_m \rightarrow$ Density of fluid in model.

and $\qquad\qquad V_p, p_p, \rho \rightarrow$ Corresponding values in prototype; then:

Putting these values in equation we get:

$$\boxed{\frac{V_m}{\sqrt{\rho_m/\rho_m}} = \frac{V_p}{\sqrt{\rho_p/\rho_p}}} \qquad \qquad \dots(2)$$

If fluid is same in model and prototype, then equation 2 becomes as:

$$\boxed{\frac{V_m}{\sqrt{\rho_m}} = \frac{V_p}{\sqrt{\rho_p}}} \qquad \qquad \dots(3)$$

Thus, Euler's model law is applied for fluid flow problems where flow is taking place in a closed pipe in which case turbulence is fully developed so that viscous forces are negligible and gravity force and surface tension force is absent. This law is also used where the phenomenon of cavitation takes place.

4.10.5 Weber Model's Law

Weber model law is the law in which models are based on Weber's number, which is the ratio of the square root of inertia force to the surface tension force. Hence, where surface tension effects predominants in addition to inertia force, the dynamic similarity between the model and prototype is obtained by equating the weber number of the model and its prototype. Hence; according to this law:

$$(We)_{model} = (W_c)_{prototype}; \text{ where; } W_e \text{ is weber number} = \frac{V}{\sqrt{\sigma/\rho L}}$$

If $V_m \rightarrow$ Vet. of fluid in model; 6m $\rightarrow$ surface tensile force in model

$\rho_m \rightarrow$ density of fluid in model; $L_m \rightarrow$ Length of surface in model

and V_p, σ_p, P_p, $L_p \rightarrow$ Corresponding values of fluid in prototype.

Then; according to weber law; we have:

$$\frac{V_m}{\sqrt{\sigma_m/\rho_m L_m}} = \frac{V_p}{\sqrt{\sigma_p/\rho_p \cdot L_p}}$$

Weber mode law is applied in following cases:

1. Capillary rise in narrow passages
2. Capillary movement of water in soil
3. Capillary waves in channels
4. Flow over weirs for small heads.

4.10.6 Mach's Model Law

Mach Model law is the law in which models are designed on Mach number, which is the ratio of the square root of inertia force to elastic force of a fluid. Hence; where the forces due to elastic compression predominate in addition to inertia force, the dynamic similarity between the model and its prototype is obtained by equating the Mach number of the model and its prototype.

Hence; according to this law:

$$(M)_{model} = (M)_{prototype}$$

where; $$M = \text{Mach Number} = \frac{V}{\sqrt{K/\rho}}$$

If: $V_m \rightarrow$ Velocity of fluid in model

$K_m \rightarrow$ Elastic stress for model

$\rho_m \rightarrow$ Density of fluid in model

and V_p, K_p, $\rho_p \rightarrow$ Corresponding values for prototype.

Then; A/Q; Mach Law:

$$\frac{V_m}{\sqrt{K_m/\rho_m}} = \frac{V_p}{\sqrt{K_p/\rho_p}}$$

Mach model low is applied in the following cases:

1. Flow of aeroplane and projectile through air at supersonic speed, i.e. at a velocity more than the velocity of sound.
2. Aerodynamic testing.
3. Under water testing of torpedoes.
4. Water—hammer problems.

SOLVED PROBLEMS

1. Show that the resistance F to the motion of a sphere of diameter D moving with a uniform velocity V through a real fluid of density and viscosity μ is given by,

$$F = \rho D^2 V^2 f\left(\frac{\mu}{Vd\rho}\right)$$

Ans. The functional relationship is

$$F = f(D, V, \rho, \mu)$$

$$F = K\left(D^a \times V^b \times \rho^C \times \mu^d\right) \qquad \qquad ...(1)$$

where K is a dimensionless coefficient, using M, L, T system,

$$\left[MLT^{-2}\right] = [1]\left[L^a\right]\left[LT^{-1}\right]^b\left[ML^{-3}\right]^c\left[M^{-1}LT^{-1}\right]^d$$

For dimensional homogeneity, exponents of M, L and T are equated on both sides,

$$M : 1 = c + d; L : 1 = a + b - 3c - d,$$
$$B : -2 = -b - d$$

Three equations are there to solve 4 variables. So solution can be obtained in terms of one component which may be arbitrarily chosen.

Here out of experience components a, b and c are to be expressed in terms of d.

$$c = 1 - d, b = 2 - d, a = 2 - d$$

as D, V and ρ are recognized dimensionally groups appear very sound,

$$c = 1 - d, b = 2 - d, a = 2 - d$$

Substituting these values of exponents in expression (1)

$$F = k[D^{-d} \times v^{2-d} \times \rho^{-d} \times \mu^d]$$

Collecting like terms:

$$F = K\rho V^2 D^2 \left(\mu/VD\rho\right)^d$$

$$= \rho V^2 D^2 f\left(\mu/VD\rho\right)$$

2. The pressure drop ΔP in a pipe of a diameter D and length l depends on the density ρ and viscosity μ of the fluid flowing, mean velocity v of flow and the average height of protuberance t. Show that the pressure drop can be expressed

$$\Delta P = \rho V^2 f\left(\frac{l}{D}, \frac{\mu}{VD\rho}, \frac{t}{D}\right)$$

Ans. It may be denoted

$$\Delta P = f(D, l, \rho, \mu, v, t)$$

$$\Delta P = k(D^a \times l^b \times \rho^c \times \mu^d \times v^e \times t^f)$$

where k is non-dimensional constant, using $M - L - T$, system,

$$\left[ML^{-1}T^{-2} \right] = [1][L]^a [L]^b \left[ML^{-3} \right]^c \left[ML^{-1}T^{-1} \right]^d \left[LT^{-1} \right]^e [L]^f$$

Dimension of k, being a constant, is taken as unity, to prove homogeneity, the exponents of M, L and T on both sides,

$$M : 1 = c + d ; L : -1 = a + b - 3c - d + e + f,$$
$$T : -2 = -d - e$$

Substituting these values in expression

$$\Delta P = K \left[\rho V^2 \left(\frac{l}{D} \right)^b \left(\frac{\mu}{VD\rho} \right)^d \left(\frac{t}{D} \right)^t \right]$$

$$= \rho V^2 f \left(\frac{l}{D}, \frac{\mu}{VD\rho}, \frac{t}{D} \right)$$

Experimental investigation have shown ΔP in a linear function of $\dfrac{l}{D}$

$$\Delta P = \rho V^2 \frac{l}{D} f \left(\frac{\mu}{VD\rho}, \frac{t}{D} \right)$$

$$\frac{\Delta P}{w} = \frac{V^2}{g} \frac{l}{D} f \left(\frac{\mu}{VD\rho}, \frac{t}{D} \right)$$

$$\therefore \qquad \boxed{nf = \frac{4 flV^2}{2gD}}$$

3. Show by the use of Buckingham's *pi* theorem that the velocity through orifice is given by

$$V = \sqrt{2gH} \, f \left(\frac{D}{h}, \frac{\mu}{\rho VH}, \frac{\sigma}{\rho V^2 H} \right)$$

where H is lead causing flow, D is the diameter of the orifice, μ is the coefficient of viscosity, ρ is the mass density, σ is the surface tension and g is the gravitational acceleration.

Ans. It can be denoted that the functional relationship is,

$$f(v, D, H, g, \rho, \mu, \sigma) = 0 \qquad \qquad \qquad ...(1)$$

There are 7 physical quantities and 3 fundamental units, hence $(7 - 3) = 4\pi$ term. Choosing density ρ, viscosity v and head H as the repeating variables with unknown exponents, the non-dimensional π-terms.

(i) $\pi_1 = \rho^{a_1} V^{b_1} H^{c_1} D$

$$\left[M^0 L^0 T^0\right] = \left[ML^{-3}\right]^{a_1} \left[LT^{-1}\right]^{b_1} [L]^{C_1} [L]$$

Equating equation of M, L and T respectively one

$$0 = a_1;\ 0 = -3a_1 + b_1 + c_1 + 1;\ 0 = -b_1$$

Solution gives, $a_1 = 0;\ b_1 = 0;\ c_1 = -1$

$$\pi_1 = \rho^0 V^0 H^{-1} D = \frac{D}{H}$$

(ii) $\pi_2 = \rho^{a_2} V^{b_2} H^{c_2} g$

$$\left[M^0 L^0 T^0\right] = \left[ML^{-3}\right]^{a_2} \left[LT^{-1}\right]^{b_2} [L]^{c_1} \left[LT^{-2}\right]$$

Equating exponents of M, L, T respectively,

$$0 = a_1;\ 0 = -3a_1 + b_1 + c_1 + 1,\ 0 = -b_1$$

Solution gives, $a_1 = 0;\ b_1 = 0;\ c_1 = -1.$

$$\pi_1 = \rho^0 V^0 H^{-1} D = \frac{D}{H}$$

(ii) $\pi_2 = \rho^{a_2} V^{b_2} H^{c_2} g$

$$\left[M^0 L^0 T^0\right] = \left[ML^{-3}\right]^{a_2} \left[LT^{-1}\right]^{b_2} [L]^{c_1} \left[LT^{-2}\right]$$

Equating exponents of M, L, T respectively

$$a_2 = 0;\ 0 = -3a_2 + b_2 + c_2 + 1$$
$$0 = -b_2 - 2$$

Solution gives, $a_2 = 0;\ b_2 = -2,\ c_2 = 1,$

$$\pi_2 = \rho^0 V^{-2} H^1 g = \frac{gH}{V^2}$$

(iii) $\pi_3 = \rho^{a_3} V^{b_3} H^{c_3}$

$$\left[M^0 L^0 T^0\right] = \left[ML^{-3}\right]^{a_3} \left[LT^{-1}\right]^{b_3} [L]^{c_3} \left[ML^{-1}T^{-1}\right]$$

Equating components, M, L and T respectively

$$0 = a_3 + 1;\ 0 = -3a_3 + b_3 + c_3 - 1$$
$$0 = -b_3 - 1$$

which gives, $a_3 = -1,\ b_3 = -1,\ c_3 = -1$

$$\therefore \qquad \pi_3 = \rho^{-1} V^{-1} H^{-1} \mu$$

$$= \frac{\mu}{\rho V H}$$

(iv) $\pi_4 = \rho^{a_4} V^{b_4} H^{c_4} \sigma$

$$[M^0 L^0 T^0] = [ML^{-3}]^{a_4} [LT^{-1}]^{b_4} [L]^{c_4} [MT^{-2}]$$

Equating components of M, L and T respectively

$$0 = a_4 + 1, \ 0 = 3a_4 + b_4 + c_4, \ 0 = -b_4 - 2$$

Solution gives $\quad a_4 = -1, \ b_4 = -2, \ c_4 = -1,$

$$\pi_4 = \rho^{-1} V^{-2} H^{-1} \sigma = \frac{\sigma}{\rho V^2 H}$$

The functional relationship can then be

$$\phi\left[\frac{D}{H}, \frac{gH}{V^2}, \frac{\mu}{\rho VH}, \frac{\sigma}{\rho V^2 H}\right] = 0$$

or,

$$\frac{V^2}{gH} = \phi\left[\frac{D}{H}, \frac{\mu}{\rho VH}, \frac{\sigma}{\rho V^2 H}\right]$$

$$V = \sqrt{gH} \ \phi\left[\frac{D}{H}, \frac{\mu}{\rho VH}, \frac{\sigma}{\rho V^2 H}\right]$$

$$= \sqrt{2gH} \ \phi\left[\frac{D}{H}, \frac{\mu}{\rho VH}, \frac{\sigma}{\rho V^2 H}\right].$$

EXERCISE

1. In a certain pipe, the pressure difference Δp across its ends is known to depend on the length of the pipe l, its diameter d; the density ρ and viscosity μ of the fluid, the flow velocity v; the roughness of the pipe well given in terms of length ε which is the mean height of the roughness elements on the inside of pipe. Show that Euler number $\Delta P/\rho v^2$, the Reynolds number $v \ D\rho/\mu$ and the length ratios l/ρ and ε/D are functionally related.

2. Obtain a complete set of independent non-dimensional quantities when the following variables and involved in a particular situation. Power P; typical length l; number of revolution per unit time N; linear velocity v, density ρ and viscosity μ of fluid.

3. A fluid coupling transmitting a tongue T has a mean diameter D and contains a volume of fluid v. The primary and secondary runners are driven at speeds N_p and N_s fluid has viscosity ρ. Find a functional relationship in terms of dimensionless parameter.

Part B

Part B

Fluid Dynamics

Introduction, equations of motion, Euler's equation of motion, Bernoulli's equation from Euler's equation, Bernoulli's equation for real fluids.

5.1 Introduction

This is the study of motion which involves forces of action and reaction.

5.2 Equations of Motion

According to Newton's 2nd law of motion, Net force F acting on the fluid element in any direction is equal to mass m, multiplied by acceleration a in same direction.

So, $$F = ma \qquad \qquad \ldots(1)$$

For any of fluid flow, forces present are gravity force (F_g), the pressure force (F_p), viscous force/force due to viscosity (F_v), turbulence force (F_t), force due to compressibility (F_c).

So, replacing ma by the addition of all forces, written above, the equation of motion on equation 1 can be rewritten as,

$$F = F_g + F_p + F_v + F_t + F_c \qquad \qquad \ldots(2)$$

(i) For incompressible flow, if force due to compressibility, F_c is negligible and omitted from equation 2. the resulting net force,

$$F = F_t + F_p + F_v + F_t \qquad \qquad \ldots(3)$$

Equation (3) is also the equation of motion and it is called as Reynold's equations of motion.

(ii) If turbulence force is not taken into consideration, the generalised equations of motion is known as Navier Strokes equation.

$$F = F_g + F_p + F_v + F_c \qquad \qquad \ldots(4)$$

(iii) For an ideal flow (no viscosity), viscous force is taken as zero and the specific equation is known as Euler's Equation of motion.

$$F = F_g + F_p + F_t + F_c$$

5.3 Euler Equation of Motion

5.3.1 Statement

Euler's equation of motion is described as that equation in which forces due to gravity and pressure are taken into consideration.

5.3.1.1 Assumptions

→ Fluid is irrotational
→ Fluid is steady
→ Fluid is incompressible
→ Fluid is non viscous

5.3.1.2 Proof

Euler's Equation is established by applying Newton's 2nd law of motion to a small element of fluid moving within a stream tube.

Motion of the element is influenced by:

(a) Force due to pressure

Let P and $(P + dP)$ be the pressure intensities at the upstream and down stream face respectively.

Net pressure force acting on the element in the direction of motion is given by

$$PdA - (P + dP)\,dA = -dPdA \qquad \dots(1)$$

(b) Tangential force due to viscous shear

If the fluid element has a perimeter dP, then the shear force on the element is,

$$dF_s = \tau dPds \qquad \dots(2)$$

where τ is the frictional surface per unit area acting on the walls of stream tube.

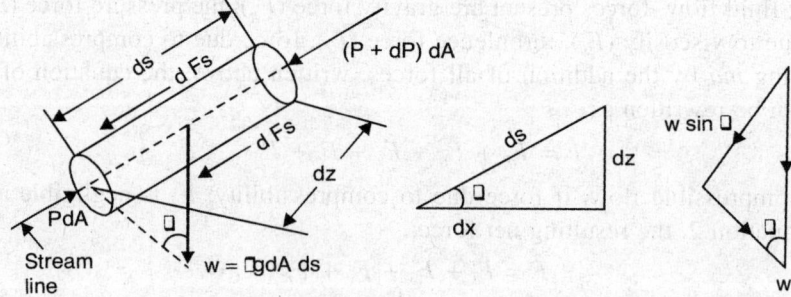

Fig. 5.1

(c) Gravity force acting in the directional field

If ρ is the density of the fluid, so the component of gravitational force in the direction of motion,

$$= \rho g dA ds\, \sin\theta$$

$$= \rho g dA dz \quad \left(\because \sin\theta = \frac{dz}{ds} \right) \qquad \dots(3)$$

As the resultant force in the direction of motion must be equal to the product of mass and acceleration in the direction.

$$-dpdA - \rho gdAdZ - \tau dpds = \rho dAds \, a_s \qquad \qquad ...(4)$$

As, velocity of an elementary fluid particle a long a stream line is a function of position and time only.

$$n = f(s, t) \qquad \qquad ...(5)$$

$$v = f(s, t)$$

$$dv = \frac{\partial v}{\partial s} \cdot ds + \frac{\partial v}{\partial t} \cdot dt$$

$$\frac{dv}{dt} = \frac{\partial v}{\partial s} \cdot \frac{\partial s}{\partial t} + \frac{\partial v}{\partial t}$$

$$\Rightarrow \qquad a_s = u\frac{\partial v}{\partial s} + \frac{\partial v}{\partial t} \left(\text{As } \frac{ds}{dt} = v \right) \qquad \qquad ...(6)$$

In a steady flow, $\dfrac{\partial v}{\partial t} = 0$ and the partial differential, becomes total differentials,

So, $$a_s = u\frac{dv}{ds} \qquad \qquad ...(7)$$

Substituting, these results of (7) in equation (4),

$$-dpdA - \rho gdAdz - \tau dpds = \rho dAvdv \qquad \qquad ...(8)$$

Dividing Equation (8) by $\rho dAds$ and rearranging,

$$v\frac{dv}{ds} + \frac{1}{\rho}\frac{dp}{ds} + g\frac{dz}{ds} = -\frac{\tau}{\rho}\frac{dp}{dA} \qquad \qquad ...(9)$$

(i) $v\dfrac{dv}{ds}$ convective acceleration

(ii) $\dfrac{1}{\rho}\dfrac{dp}{ds}$ force per unit mass caused by pressure distribution

(iii) $g\dfrac{dz}{ds}$ represents the force per unit mass resulting from gravitational pull.

(iv) $\dfrac{\tau}{\rho}\dfrac{dp}{dA}$ represents force per unit mass due to friction.

for ideal fluid $\tau = 0$,

$$vdv + \frac{dp}{\rho} + gdz = 0$$

$$\boxed{\frac{dp}{\rho} + gdz + vdv = 0} \qquad \qquad ...(10)$$

This equation is known as Euler's equation

The equation in honour of the Swiss mathematician Daniel Bernoulli (1700-1782) is known as Bernoulli's equation.

5.4 Bernoulli's Equation

5.4.1 Statement

The equation may be stated as 'the sum of the Kinetic energy (velocity head), the pressure energy (static head) and the potential energy (elevation head) of an ideal, in compressible fluid is constant along a streamline'.

5.4.2 Bernoulli's Equation from Euler's theorem

Bernoulli's equation relates velocity, pressure and elevation changes of a fluid in motion. This Equation may be obtained by integrating Euler's equation along the streamline for a constant density fluid.

Integrating Euler's equation (Equation 10)

Integration of $\dfrac{dP}{\rho} + gdz + vdv = 0$ gives

$$\int \frac{dP}{\rho} + \int gdz + \int vdv = \text{constant}$$

Assuming ρ to be constant.

$$\boxed{\frac{p}{\rho} + gz + \frac{v^2}{2} = \text{constant}}$$ ...(11)

Equation (11) is the Bernoulli's equation derived from Euler's equation (Equation 10) Bernoulli's equation for fluids

5.4.3 Assumptions

(a) flow is steady i.e., at a given point there is no variation of fluid properties with respect to time.

(b) fluid is ideal i.e., it does not exhibit any frictional effects due to fluid.

(c) flow is incompressible: fluid density is constant.

(d) flow is one dimensional.

(e) flow is imotional.

(f) flow is continuous and velocity is uniform over a section.

(g) only gravity and pressure forces are present and No energy in the form of Lect and Work is either added to or subtracted from the fluid.

5.4.4 Application of Bernoulli's theorem

Bernoulli's equation forms the basis for solving a wide variety of fluid flow such as (i) jets issuing from an orifice, jet trajectory, flow under a gate and over a weir (ii) flow meeting by obstruction meters, flow around submerged objects, (ii) flow associated with pumps and turbines.

5.4.5 Generalised Bernoulli's Equation for Fluids

Consider a perfect incompressible liquid, flowing through a non-uniform pipe.

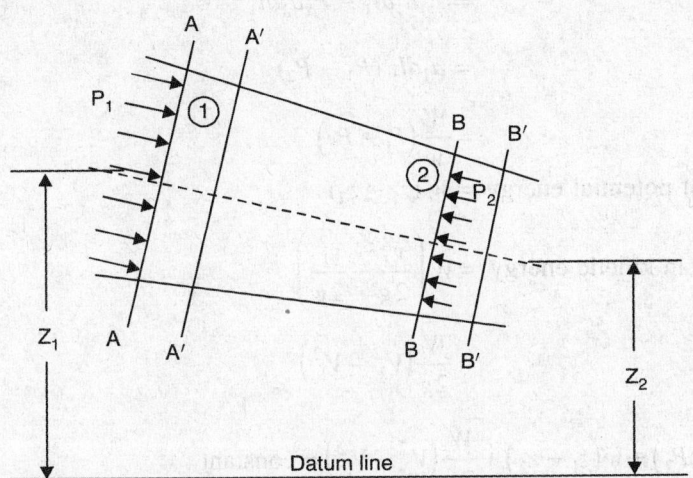

Fig. 5.2: Bernoulli's equation

Let us consider two section AA and BB of the pipe. Now let us assume that the pipe is running full and there is a continuity of flow between the two section.

z_1 = Height of AA above the datum.

P_1 = Pressure at AA

v_1 = Velocity of liquid at AA

a_1 = Cross-sectional area of the pipe at AA

Z_2, p_2, v_2, a_2 = Corresponding values at BB.

Let the liquid between the two section AA and BB move to $A'A'$ and $B'B'$. This movement of the liquid between AA and BB is equivalent to the movement of the liquid between AA and $A'A'$ to BB and $B'B'$.

Let W be the weight of the liquid between AA and $A'A'$.

$\therefore$
$$W = wa_1dl_1 = wa_2dl_2$$

$$a_1dl_1 = \frac{W}{w}$$

$$a_2dl_2 = \frac{W}{w}$$

$\therefore$
$$a_1dl_1 = a_2dl_2$$

Work done by pressure at AA in moving the liquid to $A'A'$.

$$= \text{Force} \times \text{distance}.$$

$$= P_1 \cdot a_1 \cdot dl_1$$

Similarly work done by pressure at BB in moving the liquid to $B'B'$

$$= -P_2 a_2 \cdot dl_2$$

Total work done by the pressure $= P_1 \, a_1 \, dl_1 - P_2 \, a_2 \cdot dl_2$

$$= P_1 a_1 dl_1 - P_2 a_1 dl_1 \qquad\qquad (\therefore a_1 dl_1 = a_2 dl_2)$$

$$= a_1 dl_1 \, (P_1 - P_2) \qquad\qquad \left(a_1 dl_1 = \frac{W}{w} \right)$$

$$= \frac{W}{w} \left(P_1 - P_2 \right) \qquad\qquad\qquad \ldots(1)$$

loss of potential energy $= w \, (z_1 - z_2)$ ...(2)

and gain in kinetic energy $= W \left(\dfrac{v_2^2}{2g} - \dfrac{v_1^2}{2g} \right)$

$$= \frac{W}{2g} \left(V_2^2 - V_1^2 \right) \qquad\qquad\qquad \ldots(3)$$

$$\frac{W}{w} \left(P_1 - P_2 \right) + w \left(z_1 - z_2 \right) + \frac{W}{2g} \left(V_2^2 - V_1^2 \right) = \text{constant}$$

$$\frac{P_1 - P_2}{w} + \left(Z_1 - Z_2 \right) + \frac{V_2^2 - V_1^2}{2g} = C$$

$$\frac{P}{w} + z + \frac{V^2}{2g} = \text{constant}$$

SOLVED PROBLEMS

1. A venturimeter with 200 mm at inlet and 100 mm throat is laid with axis horizontal and is used for measuring the flow of oil of specific gravity 0.8. The difference of levels in the U-tube differential manometer reads 180 mm of mercury whist 11.52×10^3 kg of oil is collected in 5 minutes. Find out the discharge coefficient for the meter (specific gravity of mercury is 13.6).

Ans. $A_1 = \dfrac{\pi}{4} (0.2)^2 = 0.0314 \text{ m}^2$, $A_2 = \dfrac{\pi}{4} (0.1)^2 = 0.00785 \text{ m}^2$

$$\frac{A_1}{A_2} = \frac{0.0314}{0.00785} = 4$$

Piezometric head $P_h = h \left(S_m - 1 \right) = 0.18 \left(\dfrac{13.6}{0.8} - 1 \right)$

$$= 2.88 \text{ m of oil}$$

Mass of oil = 11.52×10^3 kg in 5 minutes.

$\therefore$ Discharge of oil, $Q = \dfrac{11.52 \times 10^3}{5 \times 69} \times \left(\dfrac{1}{800}\right)$

$\qquad = 0.048$ m³/s

Substituting the given date,

$$Q = C_d \frac{A_1 A_2}{\sqrt{A_1^2 - A_2^2}} \sqrt{2gP_h} = C_d \frac{A_1}{\sqrt{\left(A_1/A_2\right)^2 - 1}} \sqrt{2gP_h} \quad 5/1$$

$$0.048 = C_d \frac{0.0314}{\sqrt{5^2 - 1}} \times \sqrt{2 \times 9.81 \times 2.88}$$

$$4.89 \times 0.048 = 0.236 \, C_d$$

$$C_d = 0.99$$

2. Find the discharge of water flowing through a pipe 30 cm in an inclined position a venturimeter with throat diameter 15 cm is inserted. The difference of pressure between the main and the throat is measured by a liquid of specific gravity 0.6 in an inverted U tube which gives a reducing of 30 cm. The loss of head between the main and the throat is 0.2 times of the kinetic head of the pipe.

$$A_1 = \frac{\pi}{4}(0.3)^2 = 0.07065 \text{ m}^2,$$

$$A_2 = \frac{\pi}{4}(0.15)^2 = 0.01766 \text{ m}^2$$

Applying Bernoulli's equation between the inlet and throat section of venturimeter.

$$\frac{P_1}{\rho g} + z_1 + \frac{V_1^2}{2g} = \frac{P_2}{\rho g} + z_2 + \frac{V_2^2}{2g} + \text{Lead loss}$$

$\Rightarrow \qquad \dfrac{V_2^2}{2g} - \dfrac{V_1^2}{2g} = \dfrac{P_1 - P_2}{\rho g} + \left(z_1 - z_2\right) - hf$

3. An orifice plate of orificemeter 10 cm has been fitted into a 25 cm diameter pipe that conveys oil of specific gravity 0.8. The pressure differential on the two sides of the orifice plate is measured by a mercury oil differential manometer. If the gauge shows a deflection of 80 cm of mercury, find the oil discharge in litres per second. Take (coefficient of discharge $C_d = 0.65$).

$$A_1 = \frac{\pi}{4}(0.25)^2 = 0.049 \text{ m}^2$$

$$A_2 = \frac{\pi}{4}(0.1)^2 = 0.00785 \text{ m}^2$$

$$\frac{A_1}{A_2} = \frac{0.0491}{0.00785} = 6.25.$$

Piezometric head, $P_h = h(S_m - 1)$

$$= 0.8(13.6/0.8 - 1) = 12.8 \text{ mcf or } 1$$

Again

$$Q = c_d \frac{A_1 A_2}{\sqrt{A_1^2 - A_2^2}} \sqrt{2gP_h}$$

$$= c_d \frac{A_1}{\sqrt{(A_1/A_2)^2 - 1}} \sqrt{2gP_h}$$

$$= 0.65 \times \frac{0.0491}{\sqrt{(6.25)^2 - 1}} \times \sqrt{2 \times 9.81 \times 12.8}$$

$$= 0.08197 \,\text{m}^3/s = 81.97 \text{ litres/sec}$$

For an inverted U-tube manometer, the piezometric head P_h is,

$$P_h = \frac{P_1 - P_2}{\rho g} + (z_1 - z_2)$$

$$= h(1 - S_m)$$

$$= 0.3(1 - 0.6) = 0.12 \text{ m of water}$$

Head loss $hf = 0.2 \times \dfrac{V_1^2}{2g}$

$$\frac{V_2^2}{2g} - \frac{V_1^2}{2g} = 0.12 - 0.2\frac{V_1^2}{2g}$$

or, $\dfrac{V_2^2}{2g} - \dfrac{0.8V_1^2}{2g} = 0.12$

From continuity equation

$$V_1 = \frac{A_2 V_2}{A_1} = \frac{0.01766}{0.07965}V_2 = \frac{V_2}{4}$$

$$\frac{V_2^2}{2g} - \frac{0.8}{2g}(V_2/4)^2 = 0.12$$

$$\frac{0.95V_2^2}{2g} = 0.12$$

Velocity at the throat section

$$V_2 = \frac{\sqrt{0.12 \times 2 \times 9.81}}{0.95} = 1.574 \ \text{m/s}$$

$$\text{Discharge } Q = 0.01766 \times 1.574$$

$$= 0.0278 \ \text{m}^3/\text{s}$$

4. A pitot-static tube is calibrated against a venturimeter which has a throat diameter of 75 mm and a discharge coefficient of 0.97. Air flows through the 128 mm diameter horizontal duct in which the venturimeter is installed and a differential pressure reading of 60 mm is obtained using a U-tube water manometer. The pitot-static tube is placed downstream of the venturimeter and U-tube water manometer is giving reading of 8 mm. Find flow velocity in the duct and velocity coefficient for pitot-static probe (density of air = 1.255 kg/m^3 and density of water = 1000 kg/m^3).

Ans. $A_1 = \dfrac{\pi}{4}(0.125)^2 = 0.01226 \ \text{m}^2,$

$$A_2 = \frac{\pi}{4}(0.075)^2 = 0.004415 \ \text{m}^2$$

$$\frac{A_1}{A_2} = \frac{0.01226}{0.004415} = 2.78$$

Piezometric head, $P_h = h(S_m - 1)$

$$= 0.06(1000/1.225 - 1)$$

$$= 48.92 \ \text{m of air}$$

Again $\qquad Q = C_d \dfrac{A_1 A_2}{\sqrt{A_1^2 - A_2^2}} \sqrt{2g \, P_h}$

$$= C_d \frac{A_1}{\sqrt{(A_1/A_2)^2 - 1}} \sqrt{2g P_h}$$

$$= 0.97 \times \frac{0.01226}{\sqrt{(2.78)^2 - 1}} \times \sqrt{2 \times 9.81 \times 48.92}$$

$$= 0.1337 \ \text{m}^3/\text{s}$$

Mean flow velocity through the duct, $V = \dfrac{0.1337}{0.01226} = 10.9 \ \text{m/s}.$

For the pitot-static tube,

Manometric deflection = 8 mm of water

$$h_a = 0.008 \times (1000/1.225 - 1)$$

$$= 6.53 \ \text{m of air}$$

Flow velocity $V = C_V \sqrt{2gh_d}$

$$= C_V \sqrt{2 \times 9.81 \times 6.53}$$

$$= 11.32 \ C_V$$

$$11.32 \, C_V = 10.90$$

Velocity coefficient for thr pitot tube

$$C_V = \frac{10.90}{11.32} = 0.962.$$

EXERCISE

1. A venturimeter of throat diameter 6 cm has a discharge coefficient of 0.978 and with a flow of 0.025 m³/s, the pressure differential is 10 N/m². Find the flow rate when an orifice of 6 cm is installed in the same pipe. The discharge coefficient for the orifice is 0.62 and pressure differential is same. **[Ans.: 0.01598 m³/s]**

2. A pitot-static tube is used to measure the speed of an airplane. If the pressure difference shown by a U-tube manometer is equivalent to h mm of water, find the relation between the manometer reading h in mm and air plane velocity v in m/s. Take air density equal to 1.22 kg/m³.

   Find the airplane speed if h is 100 of water. Take velocity coefficient or the tube as 0.98 and neglect the compressibility effects in air. **[Ans.: 4.0 $\sqrt{h\omega}$, 39.2 m/s]**

3. Water flows at the rate of 0.15 m³/s through a 15 cm diameter orifice in a 30 cm diameter pipe. If the pressure gauge fitted upstream and downstream of the orifice indicates reading of orifice 2 bar and 1 bar respectively. Find out discharge coefficient of orificemeter.
   [Ans.: 0.586]

4. Calculate the pressure drop and the power required to maintain 0.05 m³/s of petrol (sp. gs 0.7) through a steel pipe 0.2 m diameter and 1000 m long. Take Darcy coefficient of friction $f = 0.0025$ **[Ans.: 2.21 kW]**

5. An oil of mass density 950 kg/m³ and dynamic viscosity 1.0 Ns/m² is carried at the rate of 0.14 m³/s through a 30 cm diameter pipe over 1000 m length. Due to the change in temperature, viscosity of oil decreases by a factor of 8. If the quantity of oil to be used is same, differentiate between the cost of pumping.

$$\text{Ans.: Flow velocity } V = \frac{0.14}{\frac{\pi}{4}(0.3)^2} = 1.98 \ \text{m/s}$$

[**Hints:** Before change in temperature,

$$\text{Reynolds number } Re_1 = \frac{vd\rho}{\mu} = \frac{1.98 \times 0.3 \times 950}{1.0}$$

$$= 564.3]$$

...(1)

Hence the flow is laminar and frictional loss

$$hf_1 = \frac{32\,uvl}{wd^2} = \frac{32 \times 1.0 \times 1.98 \times 1000}{(950 \times 9.81) \times 0.3^2}$$

$$= 75.54 \text{ m.}$$

Power required for pumping, $P_1 = (950 \times 9.81) \times 0.14 \times 75.54$

$$= 98559 \text{ W}$$

$$= 98.56 \text{ kW}$$

After change in temperature, viscosity will decrease,

$$\mu_2 = \frac{\mu_2}{8} = \frac{1.0}{8} = 0.125 \text{ Ns/m}^2$$

$$Re_2 = \frac{1.98 \times 0.3 \times 950}{0.125} = 4514.4 \qquad \ldots(2)$$

Reynolds number in equation (2) is showing flow in turbulent.

$$4f = \frac{0.316}{(Re_2)0.25} = \frac{0.316}{(4514.4)0.25} = 0.0385$$

$$hf_2 = \frac{4\,flV^2}{2gd} = \frac{0.0385 \times 1000 \times 1.98^2}{2 \times 9.81 \times 0.30}$$

$$= 25.64 \text{ m.}$$

$$P_2 = (950 \times 9.81) \times 0.14 \times 25.64$$

$$= 33453 \text{ W} = 33.45 \text{ kW}$$

$\therefore$ Ratio of pumping costs $= \dfrac{P_2}{P_1} = \dfrac{33.45}{98.56} = 0.34$

Fluid Flow Measurements

Introduction, venturimeter, orifice meter, Pitot tube, frictional loss in pipe flow, Darcy equation for loss of head due to friction in pipes, Chezy's equation for loss of head due to friction in pipes, hydraulic gradient and total energy line.

6.1 Introduction

Different fluid flow measurements techniques are used commercially and for different engineering application.

6.2 Venturimeter

Venturimeters are invented by Mr. Clemens Herstal (1887) and has been named in the honour of famous Italian engineer Venturi.

This simple device has extensive use for water flow measurement.

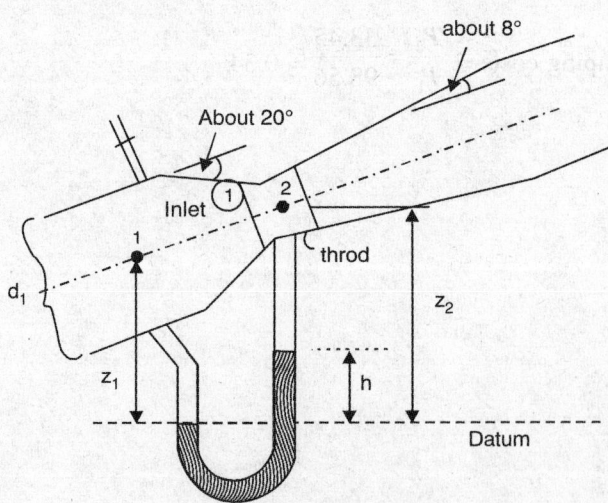

Fig. 6.1: A venturimeter

Construction

Venturimeter (obstruction meters)

The Venturimeter is a device for measuring discharge in a pipe. It consists of a rapidly converging section which increases the velocity of flow and hence reduces the pressure. It then returns to the original dimensions of the pipe by a gently diverging 'diffuser' section. By measuring the pressure differences the discharge can be calculated. This is a particularly accurate method of flow measurement as energy loss are very small.

Applying Bernoulli along the streamline from point 1 to point 2 in the narrow *throat* of the Venturi meter we have

$$\frac{p_1}{\rho g} + \frac{u_1^2}{2g} + z_1 = \frac{p_2}{\rho g} + \frac{u_2^2}{2g} + z_2$$

By the using the continuity equation we can eliminate the velocity u_2,

$$Q = V_1 A_1 = V_2 A_2$$

$$V_2 = \frac{V_1 A_1}{A_2}$$

Substituting this into and rearranging the Bernoulli equation we get

$$\frac{p_1 - p_2}{\rho g} + z_1 - z_2 = \frac{V_1^2}{2g}\left[\left(\frac{A_1}{A_2}\right)^2 - 1\right]$$

$$V_1 = \sqrt{\frac{2g\left[\dfrac{p_1 - p_2}{\rho g} + z_1 - z_2\right]}{\left(\dfrac{A_1}{A_2}\right)^2 - 1}}$$

To get the theoretic discharge this is multiplied by the area. To get the actual discharge taking in to account the losses due to friction, we include a coefficient of discharge

$$Q_{ideal} = V_1 A_1$$
$$Q_{actual} = C_d\, Q_{ideal} = C_d V_1 A_1$$

$$Q_{actual} = C_d A_1 A_2 \sqrt{\frac{2g\left[\dfrac{p_1 - p_2}{\rho g} + z_1 - z_2\right]}{A_1^2 - A_2^2}}$$

C_d = coefficient of diameter.

This can also be expressed in terms of the manometer readings.

$$p_1 + \rho g z_1 = p_2 + \rho_{man} gh + \rho g\left(z_2 - h\right)$$

$$\frac{p_1 - p_2}{\rho g} + z_1 - z_2 = h\left(\frac{\rho_{man}}{\rho} - 1\right)$$

Thus the discharge can be expressed in terms of the manometer reading:

$$Q_{actual} = C_d A_1 A_2 \sqrt{\frac{2gh(\rho_{man} - 1)}{A_1^2 - A_2^2}}$$

Purpose

The purpose of the diffuser in a Venturimeter is to assure gradual and steady deceleration after the throat. This is designed to ensure that the pressure rises again to something near to the original value before the Venturimeter. The angle of the diffuser is usually between 6 and 8 degrees. Wider than this and the flow might separate from the walls resulting in increased friction and energy and pressure loss. If the angle is less than this the meter becomes very long and pressure losses again become significant. The efficiency of the diffuser of increasing pressure back to the original is rarely greater than 80%.

Advantage

1. Because of smooth surface, it is not affected by wear and tear.
2. Well established application.
3. Ideally suitable for large flow of water, process fluids, gases etc.
4. High pressure recovers is possible.
5. Coefficient of discharge is high.
6. Independent of elevation

Disadvantage

1. Quite expensive in installation and replacement.
2. Space requirements are more.

6.3 Orifices and Mouthpieces

An orifice is a geometric opening in the side or bottom of a thin walled tank on vessel.

The opening serves to discharge the liquid contained in the tank and is regarded an orifice only if top edge of the opening lies below the liquid surface in the tank.

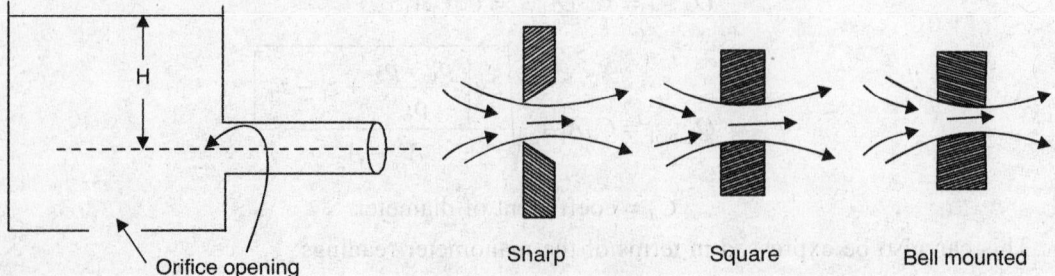

Fig. 6.2: Orifices and mouthpieces

Orifices can be classified on the basis of their shape, size, discharge.
(a) circular, rectangular or square
(b) usual shape can be circular with sharp, square or bell-mouthed edge at the entrance.
(c) Small or large

Flow through an orifice

The liquid particles approach the opening from all sides and move with acceleration along converging stream lines.

- Stream of liquid issuing from the orifice opening is called get.
- The section of least cross section and hence of maximum contraction is called vena-contracta.
- Beyond vena contracta, the stream lines become essentially parallel to each other.
- Gravitational effects, deflect the jet downwards in the form of parabolic trajectory.
- Section a-a where the *pr* is atmospheric and velocity can be assumed as zero.
- Section c-c where the jet is cylindrical and pressure is *Pc*.
- Since flow stream lines are very close together at a vena contracta, the flow velocity *V* is sensibly constant over this section.

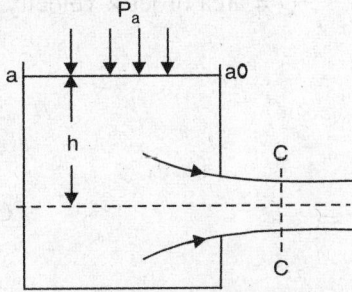

Fig. 6.3: Flow through an orifice

Its value can be determined by applying Bernoulli's equation to free surface of reason

$$\frac{V_a^2}{2g} + \frac{P_a}{w} + Z_a = \frac{V_c^2}{2g} + \frac{P_c}{w} + Z_c$$

Experiments indicate that when the orifice discharges freely into atmosphere the pressure across the jet at vena contracta is atmospheric with pressure datum as local atmospheric pressure $P_a = P_c = 0$ with elevation datum through the axis of orifice.

$$z_c = 0 \text{ and } z_a = h.$$

Velocity V_a at the surface of tank is practically zero compared to V_c. This stems from the fact that area of tank is very large as compared to the area of liquid jet.

$$O + O + h = \frac{V_c^2}{2g} + O + O$$

and hence the velocity from the orifice $v = \sqrt{2gh}$. This is known as Toricelli's theorem

6.3.1 Hydraulic coefficients

1. Coefficients of velocity $(C_v) = C_v = \dfrac{\text{actual velocity of jet at vena contracta}}{\text{Theoretical velocity}}$

$$= \frac{V}{\sqrt{2gh}} \Rightarrow \text{varies from 0.95 to 0.99}$$

2.

$$C_c = \frac{\text{area of jet at vena contracta}}{\text{area of orifice}}$$

$$= \frac{a_c}{a}$$

The value of coefficient of contraction varies 0.61 to 0.69.

3. Coefficient of discharge: C_d: Discharge through an orifice is

Q = area of jet × velocity of jet

$$= \left(C_c \times a\right) \times \left(C_v \times \sqrt{2gh}\right)$$

$$= C_c C_v \, a\sqrt{2gh}$$

$\Rightarrow$

$$C_d = \frac{\text{actual discharge}}{\text{theoretical discharge}}$$

$$= \frac{\text{actual velocity}}{\text{theoretical velocity}} \times \frac{\text{actual area}}{\text{theoretical area}}$$

$$= C_v \times C_c$$

6.3.2 Trajectory method to determine the coefficient of velocity.

x and y be the co-ordinates of the jet trajectory with original vena contracta. The horizontal distance travelled by the jet in the t is

$$n = Vt,$$

gravitational force attracts the jet down wards and during the same time interval t, the jet falls a vertical distance.

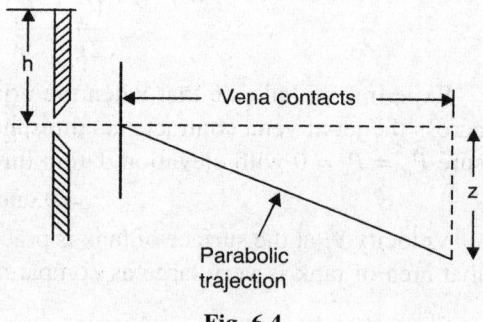

Vena contacts

Parabolic trajection

Fig. 6.4

$$Z = \frac{1}{z}gt^2$$

$$V = \sqrt{gx^2/2z}$$

$$V = C_v\sqrt{2gh}$$

$$C_d = \frac{\text{actual flow } Q}{\text{Ideal flow } Q}$$

$$= \frac{Q}{a\sqrt{2gh}}$$

$$H = \frac{V^2}{2g} + \frac{V^2}{2g}\left(\frac{V_c}{V} - 1\right)$$

Continuity equation.

$$\therefore \quad \frac{V_c}{V} = \frac{a}{a_c}$$

$$H = \frac{V^2}{2g} + \frac{V^2}{2g}\left(\frac{1}{C_c} - 1\right)^2$$

as

$$C_c = 0.62$$

$$H = 1.375\frac{V^2}{2g}$$

actual flow $V_C = 0.855\sqrt{2gH}$

theoretical flow $V + h = \sqrt{2gH}$

$$C_v = \frac{V_{hc}}{V_H} = \frac{0.855\sqrt{2gH}}{\sqrt{2gH}} = 0.855$$

$$C_d = C_c \times C_u = 1 \times 0.855 = 0.855$$

Discharge $Q = C_d \times \sqrt{2gH}$

$$= 0.855 \times \sqrt{2gH}$$

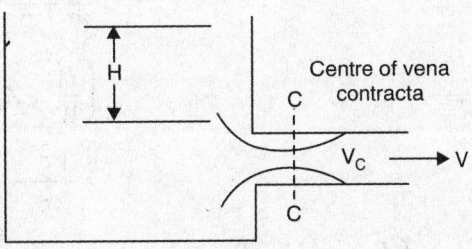

Fig. 6.5

6.4 Orifice Meter

Device to measure discharge of a fluid through pipe.

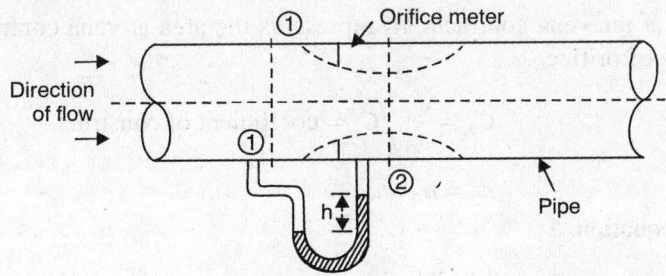

Fig. 6.6: Orifice meter

Advantages

1. Cheaper device,
2. C_d is smaller in orifice meter construction.

1. Flat circular plate which has a circular sharp edged hole called orifice, which is concentric with the pipe.
2. The orifice dia is kept 0.5 times of the dia of pipe, though it may vary from 0.4 to 0.8 times of pipe dia.

Working

A differential manometer is connected at section (1) which is at a distance of about 1.5 to 2.0 times the pipe dia upstream from the orifice plate, and at section (2), which is at a distance of about half the dia of orifice on the down stream side from orifice plate,

$(P_1 = P_2)$ at section 1
$V_1 = $ Velocity at section 1
$a_1 = $ area at section 1
$P_2, V_2, a_2 \rightarrow$ values at (2),

$$\frac{P_1}{W} + \frac{V_1^2}{2g} + Z_1 = \frac{P_2}{W} + \frac{V_2^2}{2g} + Z_2$$

$$\Rightarrow \qquad \left(\frac{P_1}{W} + Z_1\right) - \left(\frac{P_2}{W} + Z_2\right)$$

$$= \frac{V_2^2}{2g} - \frac{V_1^2}{2g}$$

But, $\left(\dfrac{P_1}{W} + Z_1\right) - \left(\dfrac{P_2}{W} + Z_2\right) = h = $ different head.

$$h = \frac{V_2^2}{2g} - \frac{V_1^2}{2g}$$

$$2gh = V_2^2 - V_1^2$$

$$V_2 = \sqrt{2gh + V_1^2} \qquad \qquad \text{...(1)}$$

at section 2, is at the vena contracta, a_2 represents the area at vena contracta, If a_0 is the area of orifice.

$$C_c = \frac{a_2}{a_0} \bigg/ C_c = \text{coefficient of construct.}$$

$$a_2 = a_0 \times c_c \qquad \qquad \text{...(2)}$$

By continuity equation,
Now,

$$a_1 v_1 = a_2 v_2$$

$$v_1 = \frac{a_2}{a_1} v_2$$

$$= \frac{a_0 c_c}{a_1} v_2 \quad \text{or,} \quad V_1 = \frac{a_0 c_c}{A_1} V_2 \qquad \qquad \text{...(3)}$$

Substituting the value of V_1 in equation (1)

$$V_2 = \sqrt{2gh + \frac{a_0^2 c_c^2 V_c^2}{a_1^2}}$$

$$V_2^2 = 2gh + \left(\frac{a_0}{a_1}\right)^2 c_c^2 V_c^2$$

$$V_2^2 = \left[1 - \left(\frac{a_0}{a_1}\right)^2 c_c^2\right] = 2gh$$

$$V_2 = \frac{\sqrt{2gh}}{\sqrt{1 - \left(\frac{a_0}{a_1}\right)^2 c_c^2}}$$

Discharge $Q = V_2 A_2 = V_2 \times a_0 c_c$

$$Q = \frac{a_0 c_c \sqrt{2gh}}{\sqrt{1 - \left(\frac{a_0}{a_1}\right)^2 c_c^2}} \qquad \qquad \dots(4)$$

and
$$C_d = C_c \frac{\sqrt{1 - \left(\frac{a_0}{a_1}\right)^2}}{\sqrt{1 - \left(\frac{a_0}{a_1}\right)^2 C_c^2}}$$

and
$$C_c = C_d \frac{\sqrt{1 - \left(\frac{a_0}{a_1}\right)^2 C_c^2}}{\sqrt{1 - \left(\frac{a_0}{a_1}\right)^2}}$$

$$\boxed{Q = \frac{C_d a_0 a_1 \sqrt{2gh}}{\sqrt{a_1^2 - a_0^2}}}.$$

Advantage and Disadvantage

Advantage

1. Low initial cost, simple in construction and replacement, as compared to venturimeter.

2. Can be used in wide range of pipe sizes (1.25 cm to 150 cm).

3. Very less space for installation.

Disadvantage

1. Pressure recovery is poor, and varies from 40 to 90% of differential pressure.

2. Coefficient of discharge has a low value.

3. Necessity of providing straightening vanes upstream.

4. Not so accurate because of erosion, conversion, scaling and different types of losses.

5. Tends to clog.

6.5 Pitot-Tube

Definition: It is a device used for measuring the velocity of flow at any point in a pipe or a channel. It is based on the principle that if the velocity of flow at a point becomes zero, the pressure there is increased due to the conversion of the kinetic energy into pressure energy. In its simplest form, the pitot-tube consists of a glass tube, bent at right angles as shown:

Working: The lower end, which is bent through 90° is directed in the upstream direction as shown the liquid rises up in the tube due to the conversion of kinetic energy into pressure energy. The velocity is determined by measuring the rise of liquid in the tube.

Consider the points (1) and (2) at the same level in such way that point (2) is just as the inlet of the pitot-tube and point (1) is far away from the tube:

Let

$p_1 \rightarrow$ Intensity of pressure at pt (1)

$v_1 \rightarrow$ Velocity of flow at (1)

$p_2 \rightarrow$ Pressure at point (2)

$v_2 \rightarrow$ Velocity at point (2); which is zero

$H \rightarrow$ Depth of tube in the liquid

$h \rightarrow$ Rise of liquid in the tube above the free surface.

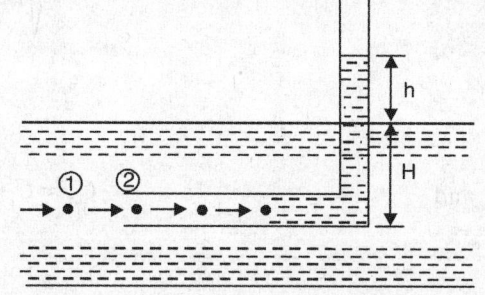

Fig. 6.7: Pitot-tube

Now; Applying Bernoulli's equation at points (1) and (2); we get:

$$\frac{p_1}{\rho g} + \frac{V_1^2}{2g} + Z_1 = \frac{p_2}{\rho g} + \frac{V_2^2}{2g} + Z_2$$

But; $z_1 = z_2$ [as points (1) and (2) are on the same line and also; $v_2 = 0$]

$\therefore$

$$\frac{p_1}{\rho g} = \text{Press head at (1)} = H$$

$$\frac{p_2}{\rho g} = \text{Press head at (2)} = (h + H)$$

Substituting these values, we get:

$$\therefore \qquad H + \frac{V_1^2}{2g} = (h + H)$$

$$\therefore h = \frac{V_1^2}{2g} \quad \text{or ;} \quad \boxed{V_1 = \sqrt{2gh}}$$

This is theoretical velocity. Actual velocity is given by: $(V_1)_{at} = C_V \cdot \sqrt{2gh}$. where; $C_v \rightarrow$ Co-efficient of pitot-tube.

$\therefore$ Velocity at any point, $(V_1)_{at} = C_V \cdot \sqrt{2gh}$.

Velocity of flow in a pipe by pitot-tube

For finding the velocity at any point in a pipe by pitot-tube, the following arrangements are adopted:

1. Pitot-tube along with a vertical piezometer tube as shown in Fig. (6.8)
2. Pitot-tube connected with piezometer tube as shown in Fig. (6.9)

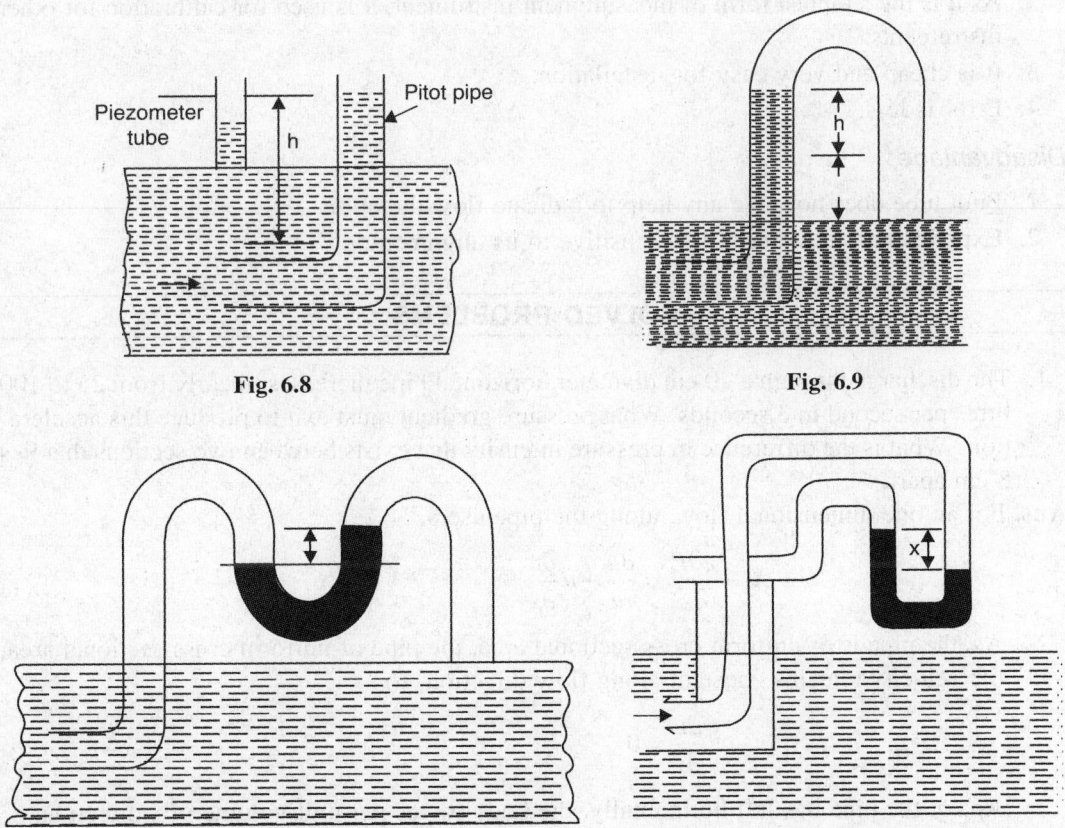

Fig. 6.8

Fig. 6.9

Fig. 6.10

Fig. 6.11

3. Pitot-tube and vertical piezometer tube connected with a differential U-tube manometer as shown in Fig. (6.10)
4. Pitot-static tube, which consists of two circular concentric tubes one inside the other with some annular space in between as shown in Fig. 6.11. The outlet of these two tubes are connected to the differential manometer where the difference of pressure head 'h' is measured by knowing the difference of the levels of the manometer liquid say x. Then;

$$h = x \left[\frac{S_g}{S_0} - 1 \right].$$

The section of least cross-section and maximum contraction is called vena contracta. Beyond vena-contracta, stream line becomes parallel to each other. Because, of gravitational effect, jet comes downward in the form of parabolic trajectory.

Advantage and Disadvantages

Advantage

1. It gives a good estimate of magnitude for determination of flow velocity.
2. As it is the simplest form of measurement instrument, it is used for calibration for other instruments.
3. It is cheap and very easy for installation.
4. Error is less.

Disadvantage

1. Pitot tube does not give any help to indicate flow direction.
2. Experiments show that it is insensitive to its alignment.

SOLVED PROBLEMS

---

1. The discharge through a 20 cm diameter horizontal pipe increases linearly from 25 to 100 litres per second in 3 seconds. What pressure gradient must exit to produce this acceleration? What is the difference in pressure intensity that exists between two sections that lies 8 cm apart.

Ans. For an one-dimensional flow, along the pipe users,

$$X - \frac{g \partial P}{\partial x} = \frac{\partial u}{\partial t} + u \frac{\partial u}{\partial x} \qquad \qquad ...(1)$$

As, the pipe is of uniform cross sectional area, the pipe of uniform cross-sectional area, the velocity remains constant along flow direction, so

$$\frac{\partial u}{\partial x} = 0 \qquad \qquad ...(2)$$

Again, the pipe has left horizontally, the body forces per unit volume in x direction,

$$X = 0,$$

Thus the expression (i) reduces to,

$$-\frac{1}{\rho}\frac{\partial P}{\partial x} = \frac{\partial u}{\partial t},$$

change in velocity as the flow changes from 25 to 100 litres per second is

$$\partial u = u_2 - u_1 = \frac{100 \times 10^{-3}}{\frac{\pi}{4}(0.2)^2} - \frac{25 \times 10^{-3}}{\frac{\pi}{4}(0.2)^3}$$

$$= 2.388 \text{ m/s}$$

This change occurs in 3 seconds,

$$\therefore \qquad \frac{\partial u}{\partial t} = \frac{2.388}{3} = 0.796 \text{ m/s}^2$$

and hence the pressure gradient,

$$\frac{\partial p}{\partial x} = -\rho\frac{\partial y}{\partial t}$$

$$= -100 \times 0.796 = -796 \text{ N/m}^2/\text{m}.$$

Difference in pressure between two sections that lie 8 m apart is

$$= \frac{\partial p}{\partial x} \times 8 = -796 \times 8 = -6368 \text{ N/m}^2$$

2. During the process of steam raising, it is required to supply 50 litres of water per minute into a boiler in which the pressure is 9.81 bar. Calculate the power expanded.

Volume of water = 50 litres per minute

$$= 50 \times 10^{-3} = \text{m}^3/\text{min}.$$

Weight of water supplied,

$$W = 9810 \times (50 \times 10^{-3}) = 490.5 \text{ N}$$

Pressure head in boiler, $\dfrac{P}{w} = \dfrac{9.81 \times 10^5}{9810} = 100 \text{ m}$

Energy expanded = 490.5 × 100 = 49050 Nm/min.

$$= 817.5 \text{ Watt}$$

3. A 60 cm diameter pipeline carries oil (sp.gr = 0.85) at 82500 m^3 per day. The friction head loss is 8.5 m per 1000 m of pipe run. It is planned to place pumping stations every 10 km along the pipe. Find the pressure drop in between pumping stations.

Ans. Applying Bernoulli's equation between two adjacent pumping stations.

$$\frac{P_1}{w} = \frac{V_1^2}{2g} + z_1 = \frac{P_2}{w} + \frac{V^2}{2g} + z_2 + hf$$

hf = head loss due to friction.

Since the pipeline is horizontal and of uniform cross sectional area,

$$z_1 = z_2 \text{ and } v_1 = v_2$$

$$\frac{P_1 - P_2}{w} = hf = \frac{8.5}{1000} \times (20 \times 1000) = 170 \text{ m.}$$

$$\therefore \qquad P_1 - P_2 = 170 \times (9810 \times 0.85)$$

$$= 1417545 \text{ N/m}^2 = 1417.55 \text{ kN/m}^2$$

4. Under a pressure of 5 MN/m², A water jet enters a 60 mm pipe and leaves by a 30 mm pipe with pressure of 4.5 MN/m². A vertical distance of 5 m separates the centres of the two pipes and exit pipe lies at a higher level. If discharge through the pipe lines is 42.4 × 10⁻⁴ m³/s. Find out the power to overcome frictional loss, where the specific weight of water = 9.8 kN/m².

Ans. Applying Bernoulli's equation at section 1 and 2

$$\frac{V_1^2}{2g} + \frac{P_1}{\rho g} + z_1 = \frac{V_2^2}{2g} + \frac{P_2}{\rho g} + z_2 + hf$$

Head loss $\qquad hf = \dfrac{V_1^2 - V_2^2}{2g} + \dfrac{P_1 - P_2}{\rho g} + (z_1 - z_2)$

$$\therefore \qquad V_1 = \frac{42.4 \times 10^{-4}}{\frac{\pi}{4}(0.06)^2} = 1.5 \text{ m/s}$$

$$V_2 = \frac{42.4 \times 10^{-4}}{\frac{\pi}{4}(0.03)^2} = 6 \text{ m/s.}$$

$$z_1 - z_2 = -5 \text{ m.}$$

$$hf = \frac{(1.5)^2 - (6)^2}{2 \times 9.81} + \frac{5 \times 10^6 - 4.5 \times 10^6}{9.8 \times 1000} - 50$$

$$= -1.72 + 51.02 - 5 = 44.3 \text{ m.}$$

$$\text{Power loss} = \rho g Q hf = (9.81 \times 1000) \times (42.4 \times 10^{-4}) \times 44.3$$

$$= 1842 \text{ watt} = 1.822 \text{ kW}$$

5. A 30 cm diameter pipe carries water under a head of 20 meters with a velocity of 3.5 m/s. If the axis of the pipe turns through 45°. What is the magnitude and direction of the resultant force on the band.

Ans.

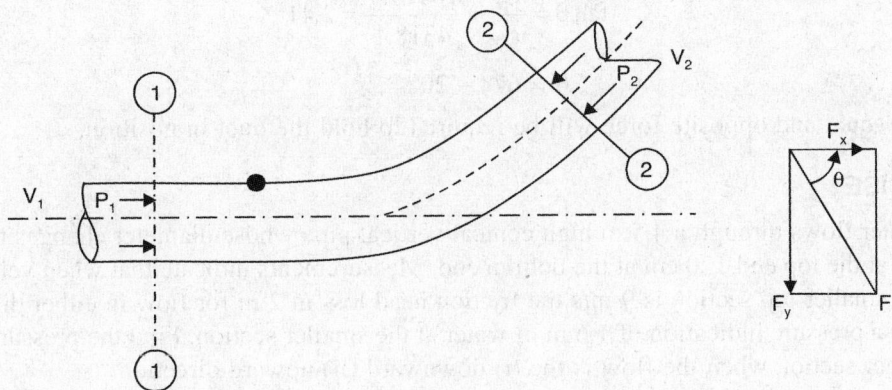

Fig. 6.12

The pipe is of uniform cross sectional area. Therefore, the flow velocities at 1–1 and 2–2 are taken to be same.

$$v_1 = v_2 = 3.5 \text{ m/s}$$

Flow rate,

$$Q = A_1 V_1 = \frac{\pi}{4}(0.3)^2 \times 3.5$$

$$= 0.247 \text{ m}^3/\text{s}.$$

Pressure intensity at sections 1 and 2 are taken to be same, at 2 sections.

$$P_1 = P_2 = \rho g h = 9810 \times 20 = 196200 \text{ N/m}^2$$

Force along the x-axis,

$$F_x = \text{Dynamic force} + \text{static force}$$

$$= \frac{wQ}{g}(V_1 - V_2 \cos 45°) + P_1 H_1 - P_2 A_2 \cos 45°$$

$$= \frac{wQ}{g}(V_1 - V_2 \cos 45°) + P_1 H_1 - P_2 A_2 \cos 45° \text{ S/m}$$

$$= \frac{9810 \times 0.247}{9.81}(0 - 3.5 \times 0.707) - 196200 \times \frac{\pi}{4}(0.3)^2 \times 0.707$$

$$= 612.4 - 98070 = -10419.4 \downarrow$$

The magnitude of resultant force acting on the bend is

$$F = \sqrt{F_x n_t F_y^2} = \sqrt{(4318.1)^2 + (10419.4)^2}$$

$$= 11278.5 \text{ N}.$$

The direction of resultant force with the x axis is:

$$\tan\theta = \frac{F_x}{F_y} = \frac{10419.4}{4318.1} = 2.41$$

$$\therefore \qquad \theta = 67° - 20'$$

An equal and opposite force will be required to hold the duct in position.

EXERCISE

1. Water flows through a 4.5 m high conical vertical pipe whose diameter changes from 40 cm at the top end 120 cm at the bottom end. Measurements indicate that when velocity at the smaller top section is 9 m/s the friction head loss in 2 m for flow in either direction. For a pressure indication of 1.6 m of water at the smaller section. Find the pressure at the larger section when the flow is the (i) downward (ii) upward direction.

   [Ans.: 18.68 m, 22.68 m]

2. Water enters a reducing pipe horizontally and comes out vertically in the downward direction. If the inlet velocity is 5 m/s and pressure is 80 kpa gauge whereas inlet and exit diameters are 30 cm and 20 cm respectively. Calculate the components of reacting on the pipe. **[Ans.: 14.56 kN, −8.07 kN]**

3. A 3 cm diameter nozzle of a fire base is held by the fireman 2 m off the ground. If discharge through the nozzle is 0.03 m³/s. What is the maximum diameter that the fireman can stand back and still get the water on a roof which is 4 cm above the ground. At what angle will holding the nozzle.

4. A 30 cm diameter pipe causing 0.045 m³/s of water has a pressure of 3.5 bar at a certain flow section. Find out the total energy relative to a datum 6 m below the pipe.

6.6 Flow through Pipes

A pipe is a closed conduit through which fluid flows under pressure. Pipe runs full when the fluid has no free surface.

When the fluid flows through piping system. Some of the potential energy is lost to overcome hydraulic resistance.

6.6.1 Loss of energy in pipe

The loss of energy is classified as

1. Major Losses

Friction Losses are considered as major losses. Darcy-Weisback formula and chezy's formula are two most important formula for these.

2. Minor losses

Minor losses or local resistance which result from flow disturbances caused by (a) Sudden enlargement of pipe (b) Sudden contraction of pipe (c) Any bend in pipe (d) pipe fitting (e) An obstruction in pipe.

6.6.2 *Friction Loss in Pipe Flow*

The frictional pressure drop associated with fluid flow is called Major Pipe Loss and construc-tions of pipe fittings are collectively referred to as Minor Pipe Loss.

While analysing the flow through long pipes, local losses due to change in cross-section can be neglected in comparison to the loss of head caused by viscous function. However, in short pipes, the minor losses becomes quite significant and sometimes as important as viscous fric-tion losses.

6.6.2.1 *Darcy equation for loss of head due to friction in pipes*

Experimental measurements on turbulent flow through pipes depicts that the frictional force effects fluid flow and for a steady flow, frictional effects associated with fluid are proportional to:

(i) Pipe length l (ii) Wetted Perimeter P (iii) V^n, where V is the average flow velocity, n is an index depending on the material and nature of pipe surface. Rougher the surface, greater in the value of n. With turbulent flow in the commercial piper, $n = 2$ considering a uniform horizontal pipe, having steady flow, let 1-1 and 2-2 are two sections of pipe are let P_1 = Pressure intensity at section 1-1.

V_1 = Velocity of flow at section 1-1.

L = Length of pipe between sections 1-1 and 2-2.

d = Diameter of pipe.

f = Frictional resistance per unit wetted area per unit velocity,

h_f = Loss of head due to friction.

and P_2, V_2 = Values of pressure intensity and velocity at section 2.

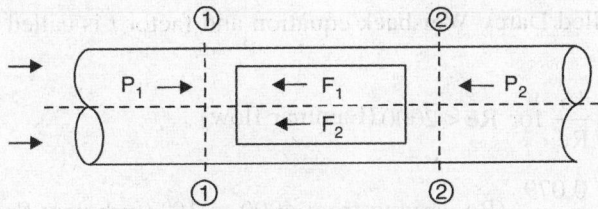

Fig. 6.13: Darcy equation

Forces acting on the fluids acting between sections 1 and 2 are:

(i) Propelling force = $(P_1 - P_2)A$ ...(1)

where A is cross-sectional area of pipe line.

(ii) Frictional Resistance force = $f'PlV^2$ ...(2)

Under equilibrium condition, propelling force = frictional resisting force

$$(P_1 - P_2)A = f'plv^2$$

$$\frac{P_1 - P_2}{W} = \frac{f'}{W}\frac{P}{A}lV^2 \qquad \text{(dividing both side by } W)$$

$$\Rightarrow \qquad h_f = \frac{2gf'}{W}\left(\frac{P}{A}\right)\frac{lV^2}{2g} \qquad \text{(multiplying and dividing by } 2g)$$

$$\Rightarrow \qquad h_f = \frac{2gf'}{W} \times \frac{l}{y_m} \times \frac{V^2}{2g} \qquad \qquad \dots(3)$$

Ratio $\dfrac{A}{P}$ is called hydraulic mean depth and denoted by 7m.

$$h_f = f \times \frac{l}{y_m} \times \frac{V^2}{2g} \qquad \left(f = \frac{2gf}{W} = \text{constant} \right) \qquad \dots(4)$$

for a circular pipe running full

$$A = \frac{\pi}{4} d^2 \text{ and perimeter } P = \pi d$$

$$y_m = \frac{A}{P} = \frac{\frac{\pi}{4} d^2}{\pi d} = \frac{d}{4} \qquad \qquad \dots(5)$$

∴ In equation (4), putting the value of equation (5).

$$h_f = \frac{4 f l v^2}{2gd} \quad \dots(6)$$

On $\qquad \qquad h_f = \frac{\lambda l v^2}{d2g} \qquad \qquad \left(\lambda = 4f = \text{friction factor} \right)$

Equation (6) is called Darcy-Weisback equation and factor f is called Darcy coefficient of friction.

$$f = \frac{16}{\text{Re}} \text{ for Re} < 2000 \text{ (Laminar flow)}$$

$$= \frac{0.079}{\text{Re}^{14}} \text{ (Re varying from 4000 to } 10^6 \text{ (turbulent flow)}.$$

Chezy equation for loss of head due to friction in pipes.
Equating propelling force due to pressure difference and frictional resistance force

$$(P_1 - P_2)A = f' P l V^2$$

$$\frac{P_1 - P_2}{W} = \frac{f'}{W} \frac{P}{A} l v^2$$

Mean flow velocity $v = \sqrt{w/f'} \times \sqrt{A/P \times \dfrac{h_f}{l}}$ the factor $\sqrt{w/f'}$ is called Chezy constant C,

∴ Mean flow velocity $V = c \sqrt{y_{mi}}$.

This is called chezy formula.

Hydraulic gradient and total energy line. The energy gradient line (EGL) and hydraulic gradient line (HGL) are the graphical representations of the longitudinal variation in total lead and piezo metric lead at salient points of a pipe line.

Hydraulic Gradient Line: It is defined as the line which gives the sum of pressure Lead $\left(\dfrac{P}{W}\right)$ and detum lead z of a flowing fluid with reference to some reference line.

Total Energy Line: It is defined as the line which gives the sum of pressure lead, datum lead and kinetic lead of a flowing fluid in a pipe with respect to some reference line.

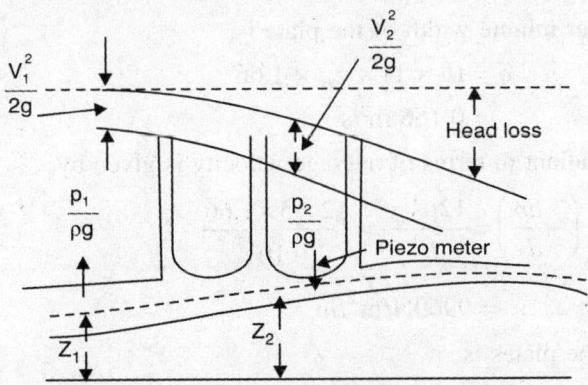

Fig. 6.14: Hydraulic gradient and total energy lines

SOLVED PROBLEMS ON FLOW THROUGH PIPES

---

1. An oil of specific gravity 0.87 and viscosity 4.8 poise flows in 10 cm diameter horizontal pipe at the rate of 5 litre per second. Find out the flow is laminar or turbulent.

Ans. As,

Flow will be laminar if Reynolds number, $Re = \dfrac{Vd\rho}{\mu} < 2000,$

Given, $d = 0.1$ m, $V = \dfrac{5 \times 10^{-3}}{\dfrac{\pi}{4}(0.1)^2} = \dfrac{5 \times 10^{-3}}{0.000785}$

$$= 6.369 \text{ m/s}$$

$\rho = 870 \text{ kg/m}^3$, $\mu = 4.8$ poise $= 4.8 \times 0.1$

$$= 0.48 \text{ Ns/m}^2$$

$$Re = \frac{6.369 \times 0.1 \times 870}{0.48} = 1154.38 < 2000$$

So the flow is laminar.

2. Two fixed parallel plates kept 10 cm apart, have laminar flow of 0.1 between them with a maximum velocity 2.5 m/s. Taking dynamic viscosity of oil to be $\mu = 53.0$ Ns/m². Find out

(i) the discharge per meter width (ii) the shear stress at the plates (iii) the pressure differences between two points 50 m apart (iv) velocity at 5 cm from the plate (v) the velocity gradient at the plates end.

Ans. As, the average velocity is two thirds of maximum velocity,

$$V_{av} = \frac{2}{3} \times 2.5 = 1.67 \, \text{m/s}$$

$$V_{av} = \frac{2}{3} \times 2.5 = 1.66 \, \text{m/s}$$

(i) The discharge per minute width of the plate is,

$$q = (b \times 1) \times v_{av} \times 1.66$$
$$= 0.166 \, \text{m}^3/\text{s}$$

(ii) The pressure gradient in terms of average velocity is given by,

$$\left(-\frac{dp}{dx}\right) = \frac{12\mu V_{av}}{b^2} = \frac{12 \times 5 \times 1.66}{(0.10)^2}$$

$$= 9960 \, \text{N/m}^2/\text{m}$$

Shear stress at the plates is:

$$\tau = \left(-\frac{dp}{dx}\right)\frac{b}{2} = \frac{9960 \times 1.10}{2} = 498 \, \text{N/m}^2$$

(iii) For laminar flow between two parallel plates,

$$P_1 - P_2 = \frac{12\mu V_{av} l}{b^2} = \frac{12 \times 5 \times 1.66 \times 50}{(0.10)^2}$$

$$= 498000 \, \text{N/m}^2$$

(iv) The velocity distribution across the passage between two stationary plate surfaces is

$$u = \frac{1}{2u}\left(-\frac{dp}{dx}\right)\left(by - y^2\right)$$

$$(u)_{y=5} = \frac{1}{2\mu}\left(-\frac{dp}{dx}\right)\left(by - y^2\right)$$

$$= \frac{1}{2 \times 5} \times 9960 \times \left\{0.10 \times 0.05 - (0.05) \times (0.05)\right\}$$

$$= 996(0.005 - 0.0025)$$

$$= 996 \times 0.0025$$

$$= 2.49 \, \text{m/s}.$$

(v) Velocity gradient

$$\frac{du}{dy} = \frac{d}{dy}\left[\frac{1}{2\mu}\left(-\frac{dp}{dx}\right)(by - y^2)\right]$$

$$= \frac{1}{2\mu}\left(-\frac{dp}{dx}\right)(b - 2y)$$

∴ Velocity gradient at the surface of plates

$$\left(\frac{du}{dy}\right)_{y=0} = \frac{1}{2\mu}\left(-\frac{dp}{dx}\right)b$$

$$= \frac{1}{2\times 5}\times 9960\times 0.10$$

$$= 99.6 \text{ m/s/m.}$$

3. Show that the discharge per unit width between two parallel plates distance b apart, when one plate is moving at velocity v while the other one is held stationary, for the condition of zero shear stress at the fixed plate: $q = bv/3$.

Ans. For Couette flow, velocity distribution,

$$u = \frac{V_y}{b} + \frac{1}{2\mu}\left(-\frac{dp}{dx}\right)(by - y^2)$$

Discharge per unit width

$$q = \int_0^b u\, dy = \int_0^b \left\{\frac{Vy}{b} + \frac{1}{2\mu}\left(-\frac{dp}{dx}\right)(by - y^2)\right\} dy$$

$$= \left[\frac{Vy^2}{2b} + \frac{1}{2\mu}\left(-\frac{dp}{dx}\right)\left(\frac{by^2}{2} - \frac{y^3}{3}\right)\right]_0^{b_1}$$

$$= \frac{Vb^2}{2b} + \frac{1}{2\mu}\left(-\frac{dp}{dx}\right)\left(\frac{b^3}{2} - \frac{b^3}{3}\right)$$

$$= \frac{Vb}{2} + \frac{1}{2\mu}\left(-\frac{dp}{dx}\right)\frac{b^3}{6} \qquad \ldots(1)$$

Shear stress

$$\tau = \mu\frac{du}{dy} = \mu\frac{d}{dy}\left[\frac{V^4}{b} + \frac{1}{2\mu}\left(-\frac{dp}{dx}\right)(by - y^2)\right]$$

$$= \mu\left[\frac{V}{b} + \frac{1}{2\mu}\left(-\frac{dp}{dx}\right)(b - 2y)\right]$$

Shear stress at the surface of fixed plate = 0

$$0 = \mu \left[\frac{V}{b} + \frac{1}{2\mu} \left(-\frac{dp}{dx} \right) b \right]$$

$$\frac{V}{b} + \frac{1}{2\mu} \left(-\frac{dp}{dx} \right) b = 0,$$

or,
$$\frac{1}{2u} \left(-\frac{dp}{dx} \right) = \frac{V}{b^2} \qquad \qquad ...(2)$$

putting the value of equation (2) in equation (1),

$$q = \frac{Vb}{2} - \frac{V}{b^2} \times \frac{b^3}{6}$$

$$= \frac{Vb}{2} - \frac{Vb}{6} = \frac{Vb}{3}$$

$$\boxed{q = \frac{Vb}{3}}.$$

4. A horizontal pipe of 10 cm diameter conveys an oil of specific gravity 0.9 and dynamic viscosity 0.8 kg/ms. A pressure drop os 50 kN/m² per meter of pipe length is measured. Find (i) flow rate of oil and centre line velocity (ii) wall shear stress and the frictional drag over 200 m of pipe length (iii) power of pump if overall efficiency is 60%.

Ans. The pressure loss for laminar flow through a pipe is given by:

$$P_1 - P_2 = \frac{32 \mu V l}{d^2}$$

$$P_1 - P_2 = \frac{32 \mu V l}{d^2}$$

$$0.5 = 51.2 \, V$$

$$v = 9.76 \, \text{m/s}$$

EXERCISE

1. A pipe 10 cm in diameter and 1000 cm long is used to pump oil of viscosity 3.5 poise and specific gravity 0.92 at the rate of 1200 lit/min. The first 300 m of the pipe is laid along the ground sloping upwards at 15° to the horizontal. Find out the nature of the flow. Determine the pressure required for the pump and power if the efficiency of the pimp as 60%.
   [**Ans.:** Laminar, p_r = 100125 m of oil power = 301.2 kW]
2. A 5 cm diameter takes off abruptly from a large tank and runs 45 m, and next discharge directly into open air with a velocity of 1.5 m/s compute the necessary height of water surface above point discharge.   [**Ans.:** Friction coefficient f = 0.0065) (11.01 m/s]

3. A pipeline, 40 m long, is connected to a water tank at one end and discharges freely into the atmosphere at the other end. For the 1st 25 m of its length from the tank, the pipe is 15 cm diameter and then its diameter is suddenly enlarged to 30 cm. The height of water level in the tank is 8 above the centre of pipe. Find out flow rate if pipe friction factor 0.04.

(i) Reynolds number

$$R_e = \frac{Vd\rho}{\mu}$$

$$= \frac{9.76 \times 0.1 \times 900}{0.8}$$

$$= 878.4$$

As the Reynolds number is less than zero, hence the flow is laminar flow.

Discharge $Q = AV_1 = \frac{\pi}{4}(0.1)^2 \times 4.76$

$$= 0.0766 \text{ m}^3/\text{s}$$

Centre line velocity $V = 2V_1 = 2 \times 9.76$

$$= 19.52 \text{ m/s}$$

(ii) Wall shear stress

$$\tau_0 = \frac{R}{2}\left(-\frac{dp}{dx}\right) = \frac{0.05}{2}\left[\left(50 \times 10^3\right)/2\right]$$

$$= 625 \text{ N/m}^2$$

Frictional drag for 200 m length of pipe,

$$F_D = \tau_0(\pi dl) = 62.5(\pi \times 0.1 \times 200)$$

$$= 39250 \text{ N}$$

(iii) Power $p = F_D \times v_1 = 383080$ N m/s $= 383$ kW

Unit
7

Laminar Flow and Viscous Effects

> Reynold's number, critical Reynold's number, laminar flow through circular pipe (Hagen-Poiseulle's equation), laminar flow between parallel stationary plates.

Introduction

All real fluids are viscous. In this chapter, laminar flow fluids will be discussed with viscous effects.

7.1 Reynolds Number

As fluid is classified as laminar and turbulent flow.

In laminar flow, fluid particles move in flat or curved layers (unmixing) or streams with smooth continuous path. This is only possible only at low velocities (highly viscous).

In turbulent flow, motion of the fluid particles are considered as irregular if the flow velocities are increased and/or fluid is less viscous.

It was Osborne Reynolds, an English scientist, who experimentally confirmed the existence of two distinct layers with different properties, and under some certain conditions,

Definition

The dimensionless number named by his name is actually the deciding factor for flow type. The quantity $\dfrac{\rho v d}{\mu}$ is called Reynolds number (Re).

$$= \frac{\text{inertia force}}{\text{viscous force}}$$

μ = Kinematic viscosity

ρ = Density of flow

v = Velocity of flow

d = Diameter of the pipe

If Re < 2000 for circular pipe, flow is known as laminar flow.

If Re > 4000, flow is said to be turbulent.

⇒ Re lies between 2000 to 4000, flow is transient flaw.

7.2 Critical Reynolds Number

The laminar flow at higher Reynolds number is unstable and even if slightest disturbance transforms it into turbulent flow instantly.

Critical Reynolds number is that number which decides the property and phase of flow.

There are two kinds of Reynolds number.

(i) Lower critical number (ii) Upper critical number.

(i) Lower critical number is that number below which disturbances of any magnitude are eventually viscous action. The approximate values of lower critical reynolds number are:

> (i) 2320 for circular pipes and tubes.
>
> (ii) 1000 for parallel walls.
>
> (iii) 500 for open channel flows.
>
> (iv) 1 for flow around a sphere.

(ii) The higher critical Reynolds number is really indeterminate. This critical Reynolds number depends upon the care taken to prevent any initial disturbance from affecting the flow. This critical velocity for various pipe diameters has the value.

d cm	0.1	10.0	100
V_{cr} cm/s	288	2.88	0.288

7.3 Laminar Flow through Circular Pipe

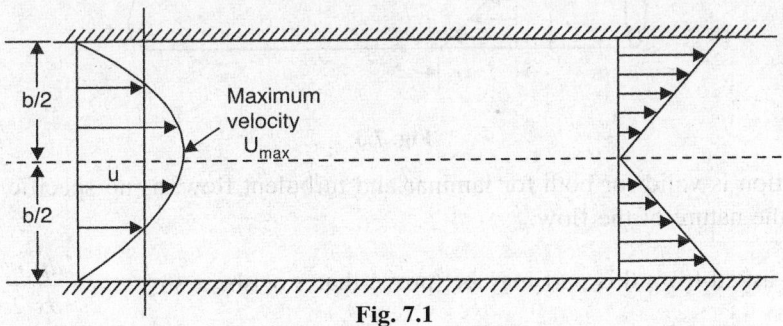

Fig. 7.1

7.3.1 Shear distribution across a section

For analysing laminar flow of an incompressible fluid through a circular pipe of radius R, the equilibrium of forces on a small concentric cylindrical fluid element of radius R, length dx and located at a distance y from the bottom internal surface of pipe.

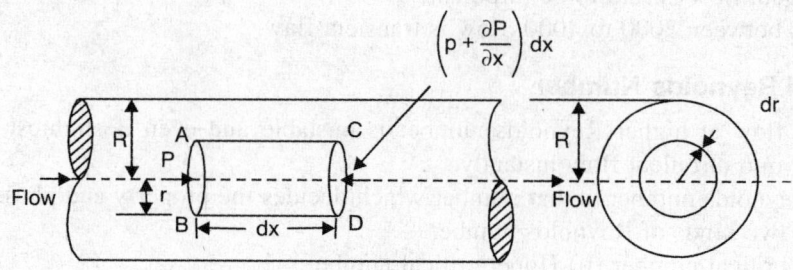

Fig. 7.2

$$\tau = -\frac{\partial P}{\partial x} \cdot \frac{r}{2} \qquad \ldots(2)$$

or in the limits,

$$\frac{dp}{dx} = -\frac{2\tau}{r}$$

or,

$$\tau = -\frac{dp}{dx}\frac{r}{2} \qquad \ldots(3)$$

$$\tau_{max} = \frac{1}{2}\left(-\frac{dp}{dx}\right)R$$

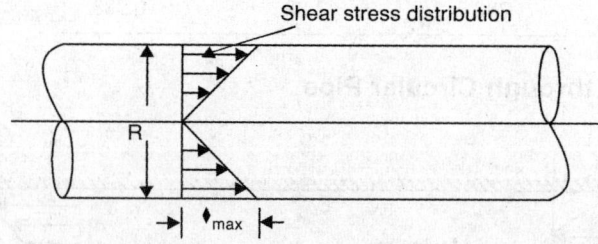

Fig. 7.3

This equation is valid for both for laminar and turbulent flow, as no specific assumptions made about the nature of the flow.

[Question asked from this part may be: prove that τ or shear stress $= -\frac{dp}{dx}\frac{r}{2}$, '–' ve sign proves that radius r is measured in the direction opposite to flow direction]

7.3.2 Velocity distribution across the pipe

From Newton's law of viscosity for laminar flow,

$$\text{Shear stress } \tau = \mu\frac{du}{dy} \qquad \ldots(4)$$

But from equation (3),

$$\tau = -\frac{dp}{dx}\frac{r}{2} \qquad \qquad ...(3)$$

For equilibrium, the summation of the pressure and viscous forces in x direction must be equal to zero.

Then the forces acting on the fluid element are:

1. The pressure force $(P \times \pi r^2)$ on face AB.

2. The pressure force $\left(p + \frac{\partial p}{\partial x} \cdot dx\right)\pi r^2$ on face CD.

3. The shear force $\tau \times 2\pi r dx$ on the surface of fluid element.

For equilibrium, the summation of pressure and viscous forces in direction of flow must be equal to zero, as there is no acceleration.

$$p\pi r^2 - \left[\left(p + \frac{\partial p}{\partial x}dx\right)\right]\pi r^2 - 2\pi r dx\,\tau = 0 \qquad \qquad ...(1)$$

$$\Rightarrow \qquad -\frac{\partial p}{\partial x} \cdot dx\,\pi r^2 - \tau \times 2\pi r dx = 0$$

$$\Rightarrow \qquad -\frac{\partial p}{\partial x} \cdot r - 2\tau = 0$$

$$\tau = -\frac{\partial p}{\partial x} \cdot \frac{r}{2}$$

But, if y is measured from pipe wall,

$$y = R - r,\ dy = -dr.$$

$$\therefore \qquad \tau = -\frac{du}{dr} \qquad \text{(Putting } dy = -dr \text{ in equation 4)} \qquad ...(5)$$

Equating equation (3) and (5)

$$-\frac{dp}{dx}\frac{r}{2} = -\frac{du}{dr}$$

As, $\frac{dp}{dx}$ is not a function of r, integration with respect to r, yields the following expression for velocity distribution,

$$u = \frac{1}{4\mu}\left(\frac{dp}{dx}\right)r^2 + C \qquad \qquad ...(6)$$

The constant of integration 'C' can be evaluated from boundary condition: $u = 0$ at $r = R$,

So, $$C = -\frac{1}{4\mu}\left(\frac{dp}{dx}\right)R^2 \qquad \qquad ...(7)$$

Putting the values of (7) in equation (6),

$$u = \frac{1}{4\mu}\left(-\frac{dp}{dx}\right)\left(R^2 - r^2\right) \qquad \ldots(8)$$

In equation (8), values of μ, $\frac{dp}{dx}$ and R are constant, which means the velocity u varies with the square of r.

[Equation (8), shows velocity distribution across the section of pipe is parabolic]

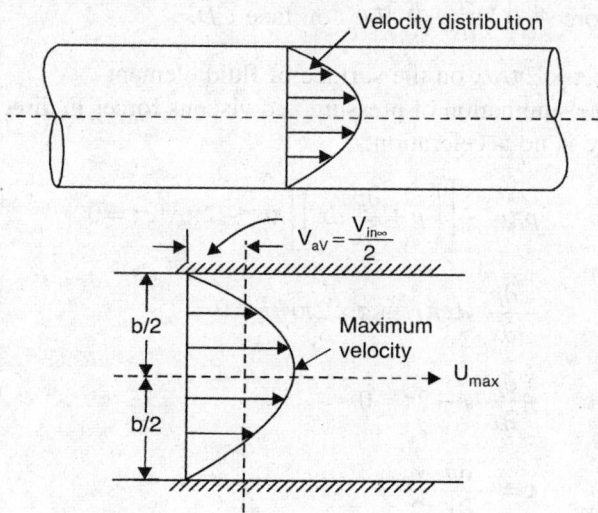

Fig. 7.4

Velocity distribution across a P section in a circular pipe.

7.3.3 Ratio of Maximum velocity to average velocity (Prove that ratio of maximum velocity to average velocity is 2)

The velocity is maximum, when $r = 0$ and respective velocity U_{max}

$$U_{max} = -\frac{1}{4\mu}\frac{dp}{dx}R^2 \qquad \ldots(9)$$

Considering the flow through a circular ring element of radius r and thickness dr so, the discharge of flowing per second through this elementary ring.

$$dQ = \text{Velocity at a radius} \times \text{area of ring element}$$
$$= u \times 2\pi\, rdr \qquad \ldots(10)$$

From equation (8), putting the value of u in equation (10),

$$dQ = -\frac{1}{4\mu}\frac{dp}{dx}\left[R^2 - r^2\right] \times 2\pi r\, dr \qquad \ldots(11)$$

Total discharge Q is obtained by considering the flow through a ring and by taking integration of equation (11).

$$Q = \int_0^R dQ = \int_0^R -\frac{1}{4\mu}\frac{dp}{dx}\left(R^2 - r^2\right) \times 2\pi r\, dr$$

$$= \frac{1}{4\mu}\left(-\frac{dp}{dx}\right) \times 2\pi \int_0^R \left(R^2 - r^2\right) r\, dr$$

$$= \frac{1}{4\mu}\left(-\frac{dp}{dx}\right) \times 2\pi \int_0^R \left(R_r^2 - r^3\right) dr$$

$$= \frac{1}{4\mu}\left(-\frac{dp}{dx}\right) \times 2\pi \left[\frac{R^2 r^2}{2} - \frac{r^4}{4}\right]_0^R$$

$$= \frac{1}{4\mu}\left(-\frac{dp}{dx}\right) \times 2\pi \left[\frac{R^4}{2} - \frac{R^4}{4}\right]$$

$$= \frac{1}{4\mu}\left(-\frac{dp}{dx}\right) \times 2\pi \times \frac{R^4}{4}$$

$$= \frac{\pi}{8\mu}\left(-\frac{dp}{dx}\right) R^4$$

The average, $\bar{u}$, is obtained by dividing the discharge of the fluid across the section by the area of the pipe πR^2.

$$\bar{u} = \frac{Q}{\text{area}} = \frac{\dfrac{\pi}{8\mu}\left(-\dfrac{dp}{dx}\right) R^4}{\pi R^2}$$

$$\bar{u} = \frac{1}{8\mu}\left(-\frac{dp}{dx}\right) R^2 \qquad\qquad\qquad \dots(12)$$

Dividing the equation (12) by equation (9)

$$\frac{U_{max}}{\bar{u}} = \frac{\dfrac{1}{4\mu}\dfrac{dp}{dx} R^2}{\dfrac{1}{8\mu}\left(-\dfrac{dp}{dx}\right) R^2} = 2,$$

average flow occur at a radial distance $0.707\,R$ from centre of pipe.

7.3.4 *Laminar Flow through Circular Pipe (Hagen-Poiseuille equation)*

Drop of pressure for a given length (L) of a pipe

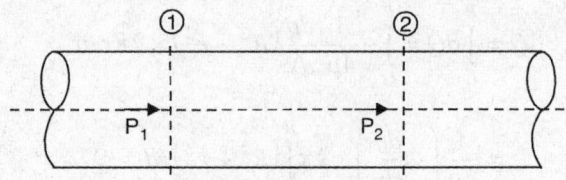

Fig. 7.5

From equation 12,

$$\bar{u} = \frac{1}{8\mu}\left(-\frac{dp}{dx}\right)R^2$$

or,

$$\left(-\frac{dp}{dx}\right) = \frac{8\mu\bar{u}}{R^2} \qquad\qquad ...(13)$$

Integrating the above w.r.t. x,

$$-\int_2^1 dp = \int_2^1 \frac{8\mu\bar{u}}{R^2}\,dx$$

$$\Rightarrow \qquad -\big[P_1 - P_2\big] = -\frac{8\mu\bar{u}}{R^2}\big[x_1 - x_2\big]$$

As $\qquad x_2 - x_1 = L,$

So $\qquad -\big[P_1 - P_2\big] = -\frac{8\mu\bar{u}}{R^2}\big(x_1 - x_2\big)$

$\therefore \qquad\qquad P_1 - P_2 = \frac{8\mu\bar{u}}{R^2}L$

As, $\qquad\qquad R = \frac{D_0}{2}$

$\therefore \qquad\qquad P_1 - P_2 = \frac{32\mu\bar{u}L}{D^2}$

$$\Delta P = \frac{32\mu\bar{u}L}{R^2}, \ \Delta P = \frac{32\mu\bar{u}L}{R^2}$$

$\therefore \qquad$ Loss of Pressure Lead $= \dfrac{P_1 - P_2}{\rho g}$

$$\frac{P_1 - P_2}{\rho g} = \frac{32\mu\bar{u}L}{\rho g D^2}$$

This equation is called Hagen-Poiseuille formula.

Considering a viscous fluid flowing between two fixed plates from left to right. If a fluid element of length dx and thickness dy at a distance y from the lower plate.

(i) If P is the intensity of pressure on the face AB of the fluid element then intensity of pressure on the face CD will be $\left(p+\dfrac{\partial p}{\partial x}dx\right)$, let τ is the shear stress acting on the BC, then the shear stress on the face AD, will be $\left(\tau+\dfrac{\partial \tau}{\partial y}\cdot dy\right)$.

(ii) If the width of the element in the direction perpendicular to the paper is taken as one, then the forces acting on the fluid element are:

(a) The pressure force, $p \times dy \times 1$ on face AB

(b) The pressure force, $\left(p+\dfrac{\partial p}{\partial x}\cdot dx\right)dy \times 1$ on face CD.

(c) Shear force, $\tau \times dx \times 1$ on face BC.

(d) The shear force $\left(\tau+\dfrac{\partial \tau}{\partial y}\cdot dy\right)dx\times 1$ on face AD.

7.4 Laminar Flow between Parallel Stationary Plates

If laminar flow between two parallel stationary flat plates separated by a small gap b, the flow analysis shall be made here by assuming.

Assumptions

1. There is no variation in the fluid properties in the y direction, $\dfrac{d}{dy}(\ )=0$

2. The plates are of infinite extent in z direction for the flow to be two-dimensional and long enough in the x direction for the flow to remain parallel.

3. The flow has a zero velocity relative to an adjacent solid surface giving the bound any conditions:

$$u = 0 \text{ at } y = 0;\ u = 0 \text{ at } y = b$$

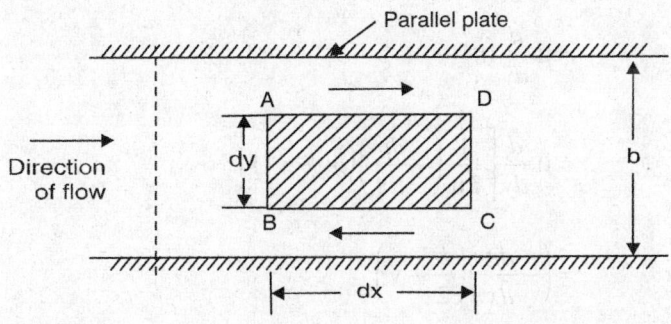

Fig. 7.6

(i) Shear stress is zero, at $y = b/2$, i.e. at the centre of flow passage.

(ii) Shear stress is maximum, at $y = 0$ i.e. at the surface of lower plate.

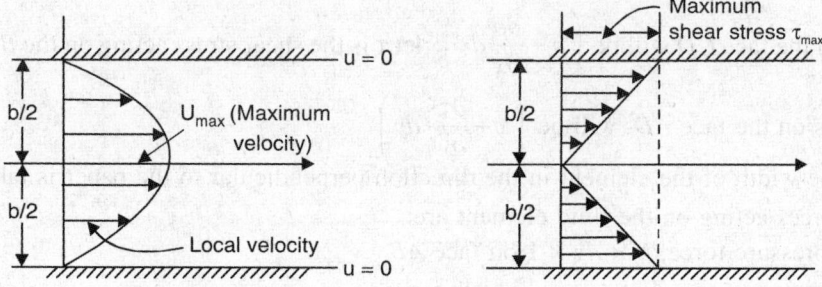

Fig. 7.7

Again, shear stress is maximum at $y = b$, at the surface of the upper plate.

$$\tau_{max} = \left(-\frac{dp}{dx}\right)\left(\pm\frac{b}{2}\right)$$

Thus, shear stress distribution has a linear variation from a zero value at midway between the plates to a maximum value at surface boundaries.

Substituting these values in equation 5,

$$u = \frac{1}{2u}\left(-\frac{dp}{dx}\right)\left(by - y^2\right) \qquad \ldots(6)$$

'−ve' sign for pressure gradient indicates that there is drop in pressure in flow direction.

Hence, the velocity profile is parabolic with its vertex at the centre of the flow passage. This type of flow is known as Plane Poiseuille flow.

(a) Shear stress distribution

Shear stress distribution can be obtained from equation, (2)

$$\tau = \mu\frac{du}{dy}$$

$$= \mu\frac{d}{dy}(u)$$

$$= \mu\frac{d}{dy}\left[\frac{1}{2\mu}\left(-\frac{dp}{dx}\right)\left(by - y^2\right)\right]$$

$$= \left(-\frac{dp}{dx}\right)\left(\frac{b}{2} - y\right)$$

As there is no acceleration, resultant force in the direction of flow is zero,

$\Rightarrow$ $$pdy \times 1 - \left(p + \frac{\partial p}{\partial x} dx\right) dy \times 1$$

$\Rightarrow$ $$-\tau dx \times 1 + \left(\tau + \frac{\partial \tau}{\partial y} dy\right) dx \times 1 = 0$$

$\Rightarrow$ $$-\frac{\partial p}{\partial x} dx\, dy + \frac{\partial \tau}{\partial x} dy\, dx = 0$$

dividing by $dx\, dy$, we get,

$$-\frac{dp}{dx} + \frac{\partial \tau}{\partial y} = 0$$

or, $$\frac{dp}{dx} = \frac{d\tau}{dy} \qquad \qquad \dots(1)$$

(b) Velocity distribution

From Newton's law of viscosity,

$$\tau = \mu \frac{du}{dy} \qquad \qquad \dots(2)$$

By, putting the value of equation 2 in equation (1)

$$\frac{dp}{dx} = \frac{d}{dy}\left(\mu \frac{du}{dy}\right)$$

$$= \mu \frac{d^2 u}{dy^2}$$

$$\frac{d^2 u}{dy^2} = \frac{1}{\mu} \frac{dp}{dx} \qquad \qquad \dots(3)$$

Integrating the above equation, with respect to y,

$$\frac{du}{dy} = \frac{1}{\mu} \frac{dp}{dx} y + c_1 \qquad \qquad \dots(4)$$

$$\left(\text{as } \frac{dp}{dx} \text{ is independent of } y\right)$$

Integrating equation (4),

$$u = \frac{1}{\mu} \frac{dp}{dx} \frac{y^3}{2} + c_1 y + y c_2 \qquad \qquad \dots(5)$$

c_1 and c_2 are the constant of integration.

By putting two boundary conditions.

(i) at $y = 0$, $u = 0$ (ii) at $y = b$, $u = 0$.

(i) at $y = 0$, $u = 0$ and putting these value in equation (5)

$$0 = 0 + c_1 \times 0 + c_2$$

or $\qquad c_2 = 0$

(ii) at $y = b$, $u = 0$

as the flow divided by

$$V_{max} = \frac{Q}{b \times 1} = \frac{b^3}{12\mu}\left(-\frac{dp}{dx}\right)$$

Prove that ratio between average to maximum velocity is $\dfrac{V_{av}}{V_{max}} = \dfrac{2}{3}$

(c) Pressure drop in terms of average velocity

$$-dp = \frac{12\mu V_{av}}{b^2} dx$$

if for a finite length of plates between $x = x_1$ and $x = x_2$

$$p_1 - p_2 = \frac{12\mu V_{av}}{b^2}(x_2 - x_1)$$

$$= \frac{12\mu V_{av} l}{b_2}$$

$$\frac{p_1 - p_2}{w} = \text{pressure head} = \frac{12\mu V_{av} l}{wb}$$

(d) Maximum Velocity

For velocity to be maximum,

$$\frac{du}{dy} = 0$$

i.e. $\qquad \dfrac{du}{dy} = \dfrac{1}{2\mu}\left(-\dfrac{dp}{dx}\right)(b - 2y) = 0$

i.e. Maximum velocity occurs at the centre of flow passage.

$$U_{max} = \frac{1}{2\mu}\left(-\frac{dp}{dx}\right)\left(\frac{b^2}{2} - \frac{b^2}{4}\right)$$

$$= \frac{b^2}{8\mu}\left(-\frac{dp}{dx}\right) \qquad\qquad ...(7)$$

(e) *Average velocity*

Let Q be discharge per unit width of flow passage,

$$Q = \int_0^b \frac{1}{2\mu}\left(-\frac{dp}{dx}\right)(by - y^2)\,dy$$

$$\boxed{Q = \frac{b^3}{12\mu}\left(-\frac{dp}{dx}\right)}$$

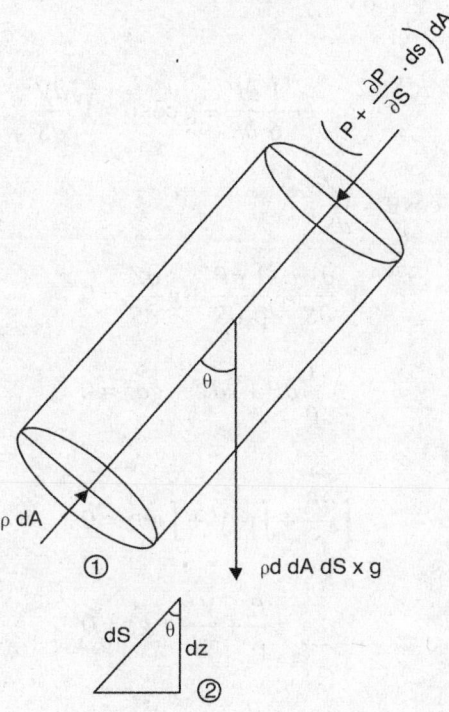

Fig. 7.8

Let us consider only gravitational and pressure related forces. Considering a stream flow in the direction S, we have 3 different forces.

(i) $Pd\,A$ acting on the same direction as S

(ii) $\left(p + \dfrac{\partial p}{\partial S} \cdot dS\right) dA$ acting opposite direction.

(iii) Component of weight $SdAds + g$ {wt = Density × Area × length × g}
Balancing the forces,

$$PdA - \left(P + \frac{\partial P}{\partial S} \cdot dS\right)dA - \rho dA dS\, g \cos\theta$$

$$= \text{Mass of fluid} \times \text{acceleration} \qquad \ldots(1)$$

$$\text{acceleration} = \frac{dV}{dt} = \frac{dV}{dS} \cdot \frac{dS}{dt} + \frac{dV}{dt}$$

for steady flow $\dfrac{\partial V}{\partial t} = 0$

$$a = V \frac{\partial V}{\partial S} \qquad\qquad \left\{ \because \ \frac{\partial S}{\partial t} = V \right\}$$

$$pd\Delta - Pd\Delta - \frac{\partial D}{\partial S} \cdot dS \cdot dA - \rho dAds g \cos\theta = \int dAdS \times \frac{vdV}{ds}$$

Dividing by $\rho dAds$

$$-\frac{1}{\rho} \frac{\partial P}{\partial S} - g \cos\theta = V \frac{dV}{dS}$$

But turn diagram (2), $\cos\theta = \dfrac{dZ}{dS}$

$$\therefore \qquad\qquad V \frac{\partial V}{\partial S} + \frac{1}{\rho} \frac{\partial P}{\partial S} + g \frac{\partial Z}{\partial S} = 0$$

or $\qquad\qquad\qquad \dfrac{1}{\rho} \partial P + v dV + g dz = 0$ \hfill …(2)

Integrating (2), we get

$$\int \frac{\partial P}{\rho} + \int v \partial V + \int g \partial z = 0$$

$$\frac{P}{\rho} + \frac{V^2}{d} + gz = 0$$

$$\frac{P}{\rho g} + Z + \frac{V^2}{dg} = 0 \hspace{3cm} …(3)$$

Dividing by g

$$\frac{P}{\rho g} = \text{Pressure head}$$

$$\frac{V^2}{\rho g} = \text{Kinetic head}$$

$$Z = \text{Potential head}$$

(2) is called Euler's equation of motion.

and (3) is called Bernoulli's Equation of motion.

Bernoulli's equation of motion states that the total head of an irrotational, steady and in compressible fluid is sum of its pressure head, kinetic head and potential head.

Assumptions made:

(i) fluid is steady

(ii) fluid is ideal {viscosity = 0}

(iii) fluid is Irrotational

(iv) fluid is Incompressible {S = constant}

$$Q = 70 \text{ Litres/Sec} = \frac{70}{1000} = 0.07 \text{ m}^3/\text{sec.}$$

$$d_1 = 250 \text{ mm} = 0.25 \text{ m}$$

$$A_1 = \frac{\pi}{4} d_1^2 = 0.049 \text{ m}^2$$

$$d_2 = 500 \text{ mm} = 0.5 \text{ m}$$

$$A_2 = \frac{\pi}{4} d_2^2 = 0.19625 \text{ m}^2$$

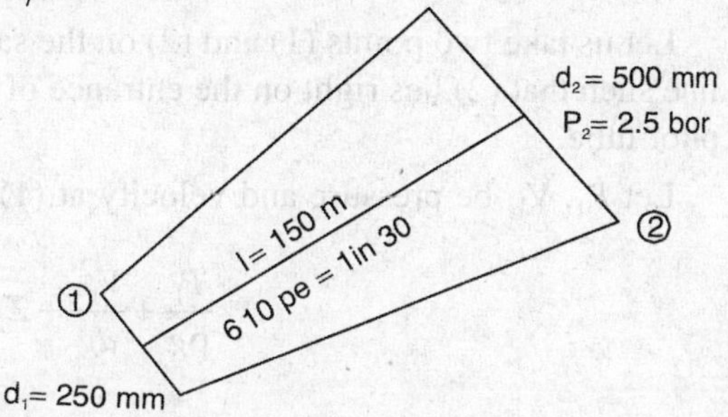

$$V_1 = \frac{Q}{A_1} = \frac{0.07}{0.049} = 1.4285 \text{ } \mu/\text{sec}$$

Fig. 7.9

$$V_2 = \frac{Q}{A_2} = \frac{0.07}{0.19625} = 0.3966 \text{ } \mu/\text{sec}$$

$$Z_1 = 0$$

Slope = line 30.

$$\therefore \quad Z_2 = \frac{1}{30} \times (150) = 5 \text{ m.}$$

$$P_2 = 2.5 \text{ bar} = 2.5 \times 10^2 = 250 \text{ } kPa = 25000 \text{ N/m}^2$$

From Bernoulli's Equation, we have,

$$\frac{P_1}{\rho g} + \frac{V_1^2}{dg} + Z_1 = \frac{P_2}{\rho g} + \frac{V_2^2}{dg} + Z_2$$

$$\frac{P_1}{\rho g} = \frac{P_2}{\rho g} + \frac{V_2^2}{dg} + Z_z - \frac{V_1^2}{dS} - 21$$

$$\frac{P_1}{1000 \times 9.81} = \frac{25000}{1000 \times 9.81} + \frac{(0.3566)^2}{2 \times 9.81} + 5 = \frac{(1.4286)^2}{2 \times 9.81} - 0$$

$$= 2.548 + 6.4 \times 10^{-3} + 5 - 0.10402$$

$$P = (7.450) \text{ } 9.81 \times 1000$$

$$= 73.088 \times 10^3 \text{ N/m}^2$$

$$= 73.088 \text{ bar}$$

Pitot-Static tube watts on bernoulli's principle. When the velocity of point becomes zero, the increase in pressure is caused due to the change of kinetic energy to pressure energy at that point. In its simplest form a pitot tube consists of a glass tube, which has been bent at right angle. It is dipped inside the flowing fluid as shown in figure. Thus the upstreams caused the fluid to rise inside the pitot tube.

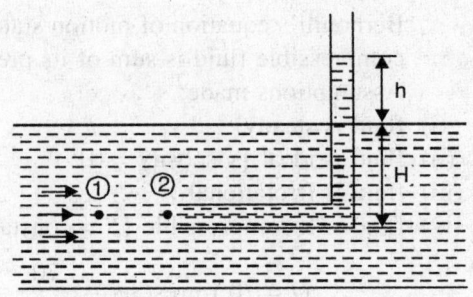

Fig. 7.10

Let us take two points (1) and (2) on the same line such that (2) lies right on the entrance of the pitot tube.

Let P_1, V_1, be pressure and velocity at (1) and P_2, V_2 corresponding values at (2)

$$\frac{P_1}{\rho g} + \frac{V_1^2}{dg} + Z_2 = \frac{P_2}{\rho g} + \frac{V_2^2}{dg} + Z_2$$

but

$$Z_1 = Z_2 = 0$$

$$V_2 = 0$$

$\therefore$

$$\frac{P_d}{\rho g} - \frac{P_{201}}{\delta g} = \frac{V_{21}^2}{dg}$$

$$\frac{P_1}{\rho g} = H \ \{\text{pressure head at } (1)\}$$

$$\frac{P_2}{\rho g} = H + h \ \{\text{pressure head at } (2)\}$$

$$\frac{V_1^2}{dg} = H + h - H = h$$

$$V_1^2 = 2gh$$

$$V_1 = \sqrt{2gh}$$

$$V_{\text{actual}} = C_v \sqrt{2gh}$$

$$C_v = \text{Coef of Pitot tube.}$$

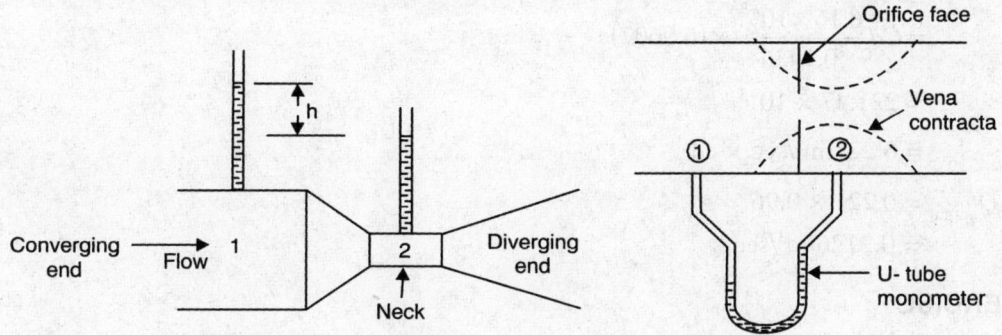

Fig. 7.11: Venturimeter

Venturimeter	Orifice plate
1. The setup is more expensive	The instrument is cheaper
2. Higher accuracy	Lower accuracy
3. Not easy to apply	Easier application

(c) d_1 = 50 cm = 0.5 m

d_2 = 20 cm = 0.2 m

P_1 = 1.8 bar = 180 kpa = 18×10^3 N/m^2

P_2 = 30 cm of Hg; $\dfrac{P_2}{\rho g} = \dfrac{-30 \times 13.6}{100} = 4.08$

h_f = 10% n

$n = \dfrac{P_2 - P_1}{\rho g} = 4.08 - \dfrac{18 \times 10^3}{1000 \times 9.81} = 2.245$

$h_f = \dfrac{10}{100} \times 2.245 = 0.2245\ \mu$

Cd = 0.96

$Q = \dfrac{a_1 a_2 \sqrt{2gh}}{\sqrt{a_1^2 - a_2^2}} \times C_d$

$a_1 = \dfrac{\pi}{4}(d_1)^2 = 0.19625$

$a_2 = \dfrac{\pi}{4}(d_2)^2 = 0.0314$

h = h + hf

$Q = Cd \dfrac{0.19625 \times 0.0314}{\sqrt{(0.19625)^2 - (0.0314)^2}} \times \sqrt{2 \times 9.81 \times (2.245 + 0.2245)}$

$= Cd \dfrac{6.16 \times 10^{-3}}{\sqrt{0.0385 - 9.859 \times 10^{-4}}} \times \sqrt{2 \times 9.81 \times 2.4645}$

$$= Cd \frac{6.16 \times 10^{-3}}{0.1936} \times (6.9607)$$

$$= 221.47 \times 10^{-3}$$

$$= 0.221 \text{ m}^3/\text{sec} \times \text{cd}$$

$$Q_{\text{actual}} = 0.221 \times 0.96$$

$$= 0.2126 \text{ m}^3/\text{sec}$$

EXERCISE

1. Determine the direction and amount of flow per metre width between two parallel plates when one is moving relative to the other with a velocity of 3 m/s in the negative x direction, ($\frac{dP}{dx}$ is given as -100 N/cm^2/cm), u is given as 0.4 poise, and distance between the plates is 1 mm. **[Ans.:** 0.207 m^3/s and in positive x direction]

2. An oil having viscosity of 7 poise and specific gravity 0.85 flows through a horizontal 50 mm diameter pipe with a pressure drop of 18 kN/m^2 per metre.

Flow Past Immersed Bodies

> Drag, lift, expression for lift and drag, pressure drag and friction drag, boundary layer concept, displacement thickness, momentum thickness and energy thickness, velocity of sound in a fluid, Mach number, propagation of pressure waves in a compressible fluid.

8.1 Drag Force

When the fluid flows past a solid body, the fluid encounters resistance. This resistance force remains the same, whether the body moves through the fluid or around the body.

Thus when an aeroplane flies, a car moves, or submarine moves through water, the propulsion unit has to exert a force which is sufficient enough to balance resistance offered by a fluid. The force which is behind the balancing of resistance is known as drag force.

8.2 Lift Force

When a body is either unsymmetrical body or its axis has not aligned with the flow direction, both the pressure and shear forces yields a combined resultant force F. Now lift force is known as the component of resultant force which acts normal to free stream direction. Let p and τ represent the pressure intensity and shear stress acting on imaginary elementary are dA and a normal to this area is inclined at an angle θ with the vertical axis. The element of drag SFD due to p and τ is

$$\delta F_D = \quad p\, dA \sin\theta \quad + \quad \tau dA \cos\theta$$
$$\text{(pressure drag)} \quad \text{(friction drag)}$$

and its integration round the whole area yields the total drag F_D:

$$F_D = \int_A (p \sin\theta + \tau \cos\theta)\, dA \qquad \qquad \dots(8.1)$$

The relative magnitude of the two components of the total drag, i.e. pressure drag and friction drag depends on the shape and orientation of the body. When a thin plate is set with

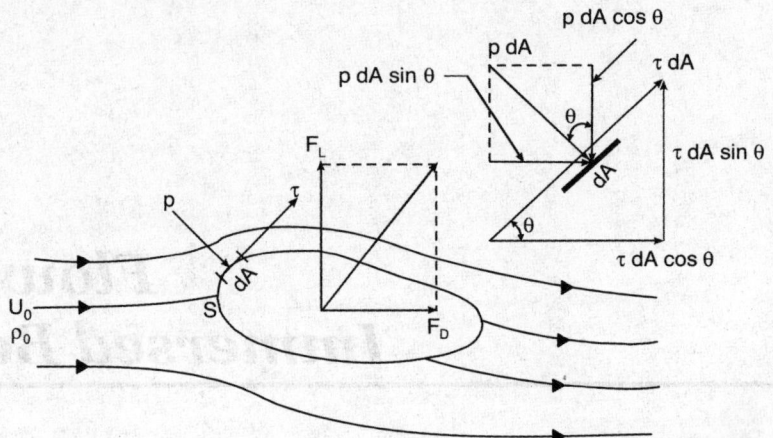

Fig. 8.1: Pressure and frictional forces on an elementary surface of an immersed body

its axis along the direction of flow (Fig. 8.2a), the pressure drag $\int p \, dA \sin\theta$ is essentially zero and the total drag is entirely due to shear stresses. However, when the same plate is held with its axis normal to flow direction (Fig. 8.2b), the flow separates at the edges forming a turbulent wake behind the plate. The friction drag $\int \tau \, dA \cos\theta$ is then zero, and the pressure force contributes totally to drag the body.

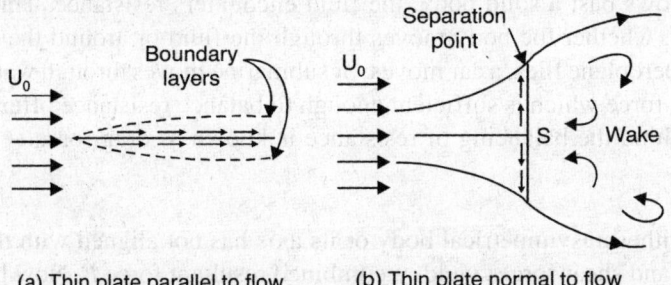

(a) Thin plate parallel to flow   (b) Thin plate normal to flow

Fig. 8.2: Thin flat plate held parallel and perpendicular to flow

Likewise, the differential lift force is,

$$\delta F_L = p \, dA \cos\theta + \tau \, dA \sin\theta$$

and the total lift force is,

$$F_L = \int_A (p \cos\theta + \tau \sin\theta) \, dA \qquad \qquad \dots(2)$$

Both the pressure and shear stress distribution, and hence the lift and drag depend on the geometrical configuration of the body, the fluid density, the fluid viscosity, the elastic property of the fluid, and the flow velocity. General functional equation for lift and drag can, thus, be expressed as:

$$F = f(U_0, d, \rho, \mu, E, g)$$

By employing the method of dimensional analysis and choosing U_0, d and ρ as repeated variables, this functional relationship can be recast in the dimensionless form:

$$\frac{F}{\rho U_0^2, d^2} = f\left(\frac{U_0 d\,\rho}{\mu}, \frac{U_0}{\sqrt{E/\rho}}, \frac{U_0}{\sqrt{gd}}\right) = f\left(R_e, M, F_r\right) \qquad \ldots(3)$$

At low speeds (Mach number $M < 0.3$), change in the density of the fluid is insignificant and so the effect of Mach number can be neglected.

Further, the Froude number $U_0/\sqrt{gd}$ will affect the value of F only when interface of two liquids is involved as in the motion of a ship on the sea surface. For a fully-submerged body, Froude number has no effect. Therefore, with the stipulations of flow at low velocity and with the body fully-submerged in the surrounding fluid, equation (3) can be rewritten as:

$$\frac{F}{\rho\, U_0^2, d^2} = f\left(R_e\right) \qquad \ldots(4)$$

The parameter d^2 is proportional to the area of the body, and parameter $\rho U_0^2/2$ represents the dynamic pressure of the undisturbed flow stream. Further, equation (4) applies equally to both lift and drag which can thus be expressed in dimensionless terms by the definition of lift and drag coefficients:

Lift coefficient $C_L = \dfrac{F_L}{(1/2)\,\rho\, U_0^2 \times A}$;

Drag coefficient $C_D = \dfrac{F_D}{(1/2)\,\rho\, U_0^2 \times A}$ $\qquad \ldots(5)$

Evidently the lift and drag coefficients depend upon the flow Reynolds number and characteristic area A of the body; usually this area means the largest projected area of the immersed body or the projected area perpendicular to the direction of flow. A projected area refers to the area that would be seen by a person looking at the body with head on. For example, the projected area of a cylinder (diameter d and length l) with its axis normal to flow is dl. For an air foil, A refers to the product of span and mean chord.

The coefficients C_L and C_D are of prime importance and are invariably used for correlating aerodynamic lift forces.

8.3 Pressure Drag and Friction Drag

8.3.1 Pressure Drag

It is the force due to pressure in the direction of fluid motion.

$$\text{Pressure drag} = \int P\cos\theta\, dA$$

It is also known as form drag.

8.3.2 Skin Friction Drag

It is the force due to shear stress in the direction of fluid motion. It is also known as friction, skin or shear drag.

$$\text{Skin friction drag} = \int \tau_0 \sin\theta\, dA$$

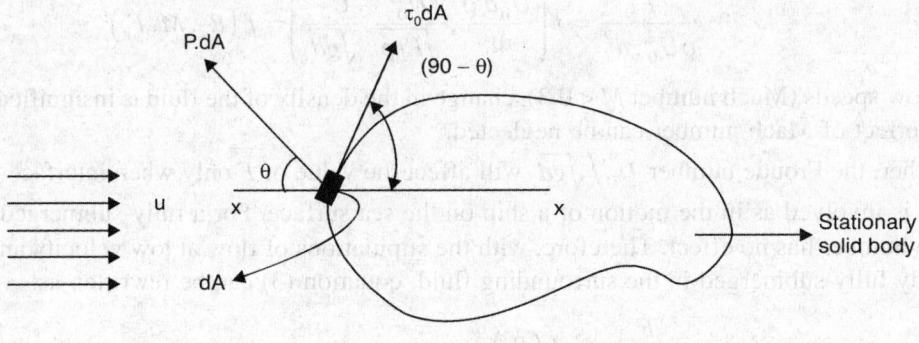

Fig. 8.3

When θ = 90°, pressure drag will be zero and total drag will be only friction drag.

When θ = 0 or, area parallel to U, skin friction drag will be zero and total drag due to the pressure difference between the upstream and down stream side of the plate. If the plate is held at an angle with the direction of flow, both the drag will exist and total drag will be sum of these two drags.

8.4 Boundary Layer

When a real fluid flows past a solid body, the fluid particles adhere to the boundary and condition of no slip occurs. That is velocity of fluid close to the boundary of solid body will be same as that of boundary. If boundary is stationary, velocity of fluid at the boundary will be zero

and velocity will be higher away from the boundary and velocity gradient $\dfrac{du}{dy}$ will exist.

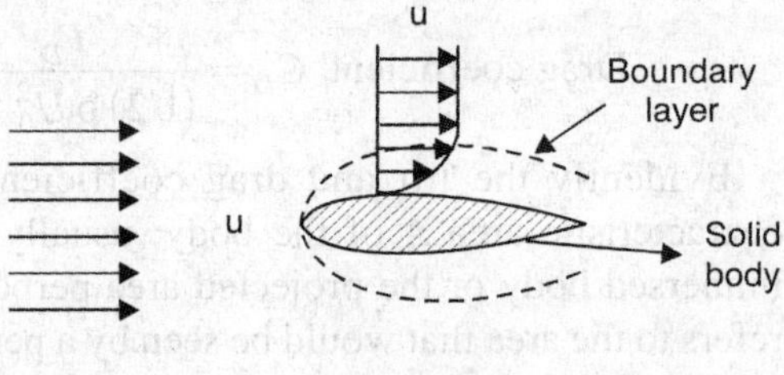

Fig. 8.4

The velocity of fluid increases from zero velocity on the stationary boundary to free stream velocity u of the fluid in the direction normal to the boundary.

Narrow region in the vicinity of solid boundary where variation of velocity from zero to u takes place, is called boundary layer.

8.5 Displacement Thickness (δ)

It is defined as the distance, measured perpendicular to the boundary of the solid body, by which the boundary should be displaced to compensate for the reduction in flow rate on account of boundary layer formation. It is denoted by δ. It is also defined as "the distance, perpendicular to the boundary by which the free stream is displaced due to the

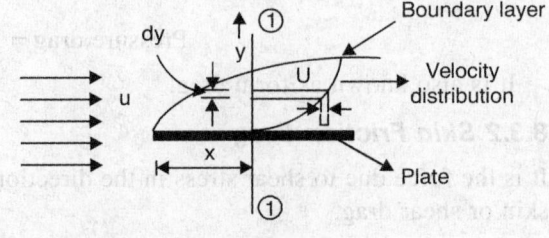

Fig. 8.5

formation of boundary layer".

$$\delta = \int_0^\delta \left(1 - \frac{u}{U}\right) dy$$

8.6 Momentum Thickness (θ)

Momentum thickness is defined as the distance, measured perpendicular to the boundary of the solid body by which the boundary should be displaced to compensate for the reduction in momentum of the flowing fluid on account of boundary layer formation. It is denoted by θ.

$$\theta = \int_0^\delta \frac{u}{U}\left[1 - \frac{u}{U}\right] dy$$

8.7 Energy Thickness (δ″)

It is defined as the distance measured perpendicular to the boundary of the solid body, by which the boundary should be displaced to compensate for the reduction in kinetic energy of the flowing fluid on account of boundary layer formation.

It is denoted by δ″

$$\delta'' = \int_0^\delta \frac{u}{U}\left[1 - \frac{u^2}{U^2}\right] dy$$

8.8 Velocity of Sound in a Fluid, Mach Number

From differential form of continuity,

$$\frac{d\rho}{\rho} + \frac{dA}{A} + \frac{dV}{V} = 0$$

For incompressible flow, there is no change in density and $\frac{d\rho}{\rho} = 0,$

$$\therefore \qquad \frac{dA}{A} = -\frac{dV}{V}$$

For compressible flow,

$$\frac{d\rho}{\rho} + V\, dV = 0$$

$$\frac{dp}{d\rho} + \frac{d\rho}{\rho} + V\, dV = 0$$

$$\frac{a^2\, d\rho}{\rho} + V\, dV = 0$$

Substituting for $\dfrac{d\rho}{\rho}$ from continuity equation,

$$a^2\left(\frac{dA}{A}+\frac{dV}{V}\right)V\,dV,$$

or,

$$\frac{dA}{A}=\frac{V\,dV}{a^2}-\frac{dV}{V}$$

$$=\frac{dV}{V}\left(\frac{V^2}{a^2}-1\right)$$

$$=\frac{dV}{V}\left(M^2-1\right).$$

where M is Mach number $=\dfrac{V}{a}$.

where V is the velocity of fluid and a is the velocity of sound.

The ratio $\dfrac{dA}{A}$ can be negative or positive depending on the Mach number of the flow.

There are three kinds of Mach number and three different flow depending on that:

(i) Subsonic Flow: $M < A$, here in this case, dA and dV are opposite in sign, i.e. an increase of cross-sectional area causes a reduction of velocity and vice versa.

(ii) Supersonic Flow: ($M > 1$), dA and dV are of same sign. An increase of cross-sectional area causes an increase of velocity and vice versa.

(iii) Sonic Flow ($M = 1$), dA must be zero, since the 2nd derivative is positive.

8.9 Propagation of Pressure Waves in a Compressible Fluid

When a flow changes from supersonic to subsonic, flow conditions take place through the setting up of a pressure wave, called the shock wave.

Whereas the Mach Number and velocity drop across a shock, there occurs a sudden rise (discontinuity) other flow parameters namely the pressure, density, temperature and entropy.

A shock appears in two form:

(i) Normal shock wherein the plane of shock is substantially perpendicular to the direction of flow.

(ii) Oblique shock wherein the plane of shock is inclined to the flow direction.

A control surface drawn around the wave ($M_1 > 1$) upstream of the shock and the associated flow variables are velocity V_1, pressure P_1, density ρ_1 and temperature T_1. On passing through the shock, the flow becomes subsonic ($M_2 < 1$) with velocity V_2, pressure P_2, density ρ_2 and temperature T_2. These changes in the flow variables occurring a shock are not reversible and so not isentropic.

SOLVED PROBLEMS

---

1. A truck having a projected area of 6.5 m² travelling at 70 km/hr has a total resistance of

2000 N, of this 20% is due to rolling friction and 10% due to surface friction. The rest is due to form drag. Find coefficient of form drag. Take $\rho = 1.22$ kg/m^3 for air.

Ans. Speed of Truck = 76 km/hr

$$= \frac{76 \times 1000}{3600} = 19.44 \text{ m/s}$$

Resistance due to rolling friction $= \dfrac{20 \times 2000}{100} = 400$ N.

Resistance due to surface friction $= \dfrac{10 \times 2000}{100} = 200$ N.

$\therefore \qquad$ Resistance due to form drag $= 2000 - 400 - 200 = 1400$ N.

Form Drag $= C_D \times \dfrac{1}{2}\rho U_0^2 \times A$

$$1400 = C_D \times \frac{1}{2}\rho U_0^2 \times A$$

$$= C_D \times \left(\frac{1}{2} \times 1.22 \times 19.44^2\right) \times 6.5$$

$\therefore$ Coefficient of form drag $C_D = \dfrac{1400 \times 2}{1.22 \times 19.44^2 \times 6.5} = 0.934$.

2. A jet plane having a wing area of 20 m^2 and weighing 25 KN flies at 950 km/hr speed. If engine develops 8500 kw and has a mechanical efficiency of 60%. Determine the lift and drag coefficients for the wind. Take specific weight of air = 12 N/m^3.

Ans. Speed of jet plane,

$$U_0 = 950 \text{ km/hr}$$

$$= \frac{950 \times 1000}{3600} = 264 \text{ m/s}.$$

(i) When the jet flies, its weight must be balanced by the lift force.

$$w = C_L \times \left(\frac{1}{2}\rho U_0^2\right) \times A$$

$$\Rightarrow \quad 25 \times 10^3 = C_L \times \left(\frac{1}{2} \times \frac{12}{9.81} \times 264^2\right) \times 20 = 852550 \, C_L$$

$\therefore \quad$ Coefficient of lift $C_L = \dfrac{25 \times 10^3}{852550} = 0.0293$

(ii) Power available to drag resistance = 8500 × 0.6 = 5100 kw

$$P = F_D \times U_0$$

$$\therefore \quad F_D = \frac{P}{U_0}$$

$$\therefore \quad \text{Drag force} \ = \frac{5100 \times 10^3}{264} = 19318$$

$$\text{Drag force } F_D = C_D \left(\frac{1}{2} \rho U_0^2 \right) \times A$$

$$19318 = C_D \left(\frac{1}{2} \times \frac{12}{9.81} \times 264^2 \right) \times 20$$

$$= 852550 \text{ CD}$$

$$\therefore \quad \text{Coefficient of drag } \ C_D = \frac{19318}{852550} = 0.226$$

3. An aeroplane weighing 40 KN is flying in a horizontal direction at 360 km/hr. The plane spans 15 m and has a wing surface area of 35 m^2. Determine lift coefficient and the power required to drive the plane. Drag coefficient $C_D = 0.03$ and for air $\rho = 1.20$ kg/m^3.

$$U_0 = 360 \text{ Km/hr}$$

$$= 360 \times \frac{1000}{3600} = 100 \text{ m/s.}$$

For equilibrium in vertical direction, lift equals weight of aeroplane.

$$\text{Weight} = C_L \times \frac{1}{2} \rho U_0^2 \times A$$

$$40 \times 10^3 = C_L \left(\frac{1}{2} \times 1.20 \times 100^2 \right) \times 35$$

$\therefore$ Lift coefficient

$$C_L = 0.19$$

$\therefore \qquad \text{Drag force } \ F_D = C_D \times \frac{1}{2} \rho U_0^2 \times A$

$$= 0.03 \times \left(\frac{1}{2} \times 1.20 \times 100^2 \right) \times 35 \ = 6300 \text{ N}$$

$\therefore$ Power required $= F_D \times U_0 = 6300 \times 100 = 630 \times 10^3$ w.

(i) Lift force per unit span $= \dfrac{40 \times 10^3}{15} = 2667$ N.

if lift $= \rho U_0 \, \Gamma$

Circulation $\Gamma = \dfrac{2667}{1.22 \times 100} = 21.86$ m^2/s

4. A rotational mixing device is constructed with two circular disks each 10 cm in diameter. These disks are spaced 1 m apart on the two end of a horizontal rod whose centre has a vertical shaft attached to it. Estimate the power required for rotation when this mixer turns 60 resolutions per minute in a solution having density 950 kg/m^3 $\mu = 0.08075$ kg m/s. For $3000 < R_e < 5000$, taking coefficient of discharge 1.2.

Ans. Linear velocity of Disk $= \dfrac{\pi DN}{60} = 3.14$ m/s

$$R_e = \frac{U_0 d\rho}{\mu} = \frac{3.14 \times 0.1 \times 950}{0.08075} = 3694.$$

Drag force $= C_D = \dfrac{1}{2} \times \rho U_0^2 \times A$

$$= 1.2 \times \left(\frac{1}{2} \times 950 \times 3.14^2 \right) \times \left(\frac{1}{4} \times 0.1^2 \right) = 441.16 \text{ N}$$

Torque produced by drag
$$T = 2\, F_0 \times R = 2 \times 441.16 \times 0.5$$
$$= 441.16 \text{ Nm}$$

$$w = \frac{2\pi N}{60} = 2\pi \times \frac{60}{60} = 6.28 \text{ rad/s}$$

Power required for rotation
$$P = TW = 441.16 \times 628 = 2.771 \ w = 2.77 \text{ K}$$

EXERCISE

1. A cylindrical tower has a diameter of 2.5 m and is 50 m high. Estimate the drag force on the tower and the bending moment at its bottom when a wind at 80 km/hr blows across it. (if $C_D = 0.33$, $\rho = 1.2$ kg/m^3 for air) **[Ans.: 12.2 kN 305 kN]**

2. An Airplane wing having a span of 12.5 m and a chord of 2.5 m travels at a velocity of 320 km/hr, at the angle of attack producing maximum lift drag ratio. Find lift and drag of the wing. **[Ans.: 60248 N, 319 SN]**

3. An automobile having a projected area of 1.6 m^2 and drag coefficient $C_D = 0.35$ travels at a uniform speed of 60 km/hr is still air at 20°C and 100 K Pa. Find power required to overcome the air resistance if $\rho_{air} = 1.2$ kg/m^3. **[Ans.: 1.556 kw]**

Previous Years' Solved Papers
from Different Universities

Fluid Mechanics
Fourth Semester B.E. Degree Examination, December 2011 (VTU)

Time: 3 hrs.   Max. Marks: 100

Note: Answer any five full questions, selecting at least two questions from each part.

Part A

1. (a) Define the following terms with their units :
   - (i) Capillarity
   - (ii) Surface tension
   - (iii) Mass density
   - (iv) Pressure intensity
   - (v) Kinematic viscosity.   **(10 Marks)**
   (b) Derive the relation for pressure intensity and the surface tensile force, in case of soap bubble.   **(04 Marks)**
   (c) A steel shaft of 30 mm diameter rotates at 240 rpm, in a bearing of diameter 32 mm. Lubricant oil of viscosity 5 poise is used for lubricant of shaft in the bearing. Determine the torque required at the shaft and power lost in maintaining the lubrication. Length of bearing is 90 mm.   **(06 Marks)**

2. (a) State and prove Pascal's law.   **(04 Marks)**
   (b) Show that, for a submerged plane surface, the centre of pressure, lies below the centre of gravity of the submerged surface.   **(08 Marks)**
   (c) A differential mercury manometer is used for measuring the pressure difference between two pipes A and B. Pipe A is 500 mm above the pipe B and deflection in Hg manometer is 200 mm. Pressure intensity in pipe A is greater than pipe B. Pipes carry oil of specific gravity 0.90. Find the pressure difference between the two pipes. Sp.gr. of mercury = 13.6.   **(08 Marks)**

3. (a) Explain the importance of meta centre with stability of floating bodies.
                                                                               (04 Marks)
   (b) A wooden block (barge) 6 mts in length, 4 mts in width and 3 mts deep, floats in fresh water with depth of immersion 1.5 mts. A concrete block is placed centrally on the surface of the wooden block, so that the depth of immersion with concrete is 2.8 mts. Find the volume of the concrete block placed centrally, if the specific gravity of concrete is 2.75. Find also the volume of water displaced.   **(08 Marks)**
   (c) Differentiate between:
   - (i) Steady flow and uniform flow
   - (ii) Laminar and turbulent flow
   - (ii) Streamline and streakline
   - (iv) Rotational and irrotational flow
                                                                               (08 Marks)

4. (a) Show that streamlines and equipotential lines are orthogonal to each other.
                                                                               (04 Marks)

(b) Torque developed by a disc of diameter D, rotating at a speed N is dependant on fluid viscosity 'μ' and fluid density 'ρ'. Obtain an expression for torque,

$$T = \rho N^2 D^5 \;\phi \left[\frac{\mu}{\rho ND^2}\right].$$ **(08 Marks)**

(c) For a two dimensional fluid flow, velocity potential is $\phi = y + x^2 - y^2$. Find the stream function and velocity at a point P (2, 3). Check irrotationality of flow. **(08 Marks)**

Part B

5. (a) Derive Bernoulli's equation and state the assumptions made. Mention the statement of Bernoulli's equation. **(10 Marks)**

(b) A pipe gradually tapers from a diameter of 0.4 mts to diameter 0.25 mts at the upper end. The pipe carries oil of specific gravity 0.90 and rate of flow is 45 kg/sec. Elevation difference between two sections is 5.0 metres. If the pressure intensities at the bottom and the upper sections are 225 kN/m^2 and 105 kN/m^2 respectively, find the direction of flow and also loss of head between the two sections. **(10 Marks)**

6. (a) Sketch and derive the relation for actual discharge through an orifice meter. **(08 Marks)**

(b) A pitot static probe measures the velocity of water flow through a pipe of diameter 7.5 cm. If the mean velocity of water flow is 6.5 m/sec and coefficient of pilot tube is 0.98, find deflection in mercury manometer connected across the pitot – tube. Determine the mass rate of water flow. **(08 Marks)**

(c) List the types of losses, with a neat sketch and equations for head losses. **(04 Marks)**

7. (a) Derive the relation for the pressure drop in a viscous flow through a circular pipe. **(10 Marks)**

(b) Sketch the total energy line and the hydraulic gradient line for a pipeline connecting two reservoirs. **(04 Marks)**

(c) A pipeline 50 m long, connects two reservoirs, having water level difference of 10m. Diameter of the pipe is 300 mm. Find rate of water flow, considering all the losses. Coefficient of friction for pipe material is 0.01. **(06 Marks)**

8. (a) Explain following terms:
   (i) Lift
   (ii) Drag
   (iii) Boundary layer separation
   (iv) Momentum thickness
   (v) Displacement thickness. **(10 Marks)**

(b) Derive a relation for the velocity of sound in a compressible fluid. **(06 Marks)**

(c) Find the velocity of a bullet fired in the air, if the Mach angle is 30°. Temperature of air is 22°C, density of air is 1.2 kg/m^2. Assume $\gamma = 1.4$ and R = 287 J/kg K. **(04 Marks)**

Fluid Mechanics
Fourth Semester B.E. Degree Examination, December 2010 (VTU)

Time: 3 hrs.   Max. Marks: 100

Note: 1. Answer any FIVE full questions.
2. Missing data, if any. may suitably be assumed.

1. (a) Define the following and write the correct units:
    (i) Weight density.
    (ii) Specific volume.
    (iii) Kinematic viscosity.
    (iv) Surface tension.
    (v) Compressibility.   **(10 Marks)**
   (b) What is capillarity? Derive an expression for capillary rise of a liquid in a tube.
                                                                                  (05 Marks)
   (c) The surface tension of a water droplet in contact with the air at 26°C is 0.072 N/m.
       If the diameter of the droplet is 1.45 mm, calculate the pressure within the droplet.
                                                                                  (05 Marks)

2. (a) Write the two fundamental characteristics of fluid pressure and prove it.
                                                                                  (10 Marks)
   (b) Prove that, for a plane vertical plate, the centre of pressure lies at a depth of two third
       of the height of the immersed surface,   **(05 Marks)**
   (c) The absolute pressure at a certain depth in the ocean is 240 kN/m². What is the gauge
       pressure 15 m below this level? Density of sea water is 1026 kg/m³ and atmospheric
       pressure is 75.8 cm of mercury.   **(05 Marks)**

3. (a) Define the following:
    (i) Buoyancy and centre of buoyancy.
    (ii) Meta centre and meta centric height.
    (iii) Absolute and gauge pressure.
    (iv) Simple and differential manometer.
    (v) Pressure and centre of pressure.   **(10 Marks)**
   (b) With a neat sketch, explain the different states of equilibrium.   **(05 Marks)**
   (c) The left limb of a mercury U-tube manometer is open to the atmosphere and the right
       limb is connected to a pipe carrying water, under pressure. The centre of the pipe is
       at the level of the free surface of mercury. Find the difference in the level of mercury
       limbs of U-tube, if the absolute pressure of water in the pipe is 14.5 m of water.
       Atmospheric pressure is 760 mm of mercury.   **(05 Marks)**

4. (a) Derive the general three dimensional continuity equation and then reduce it to
       continuity equation for steady, two dimensional incompressible flow.   **(10 Marks)**
   (b) Explain the terms:
    (i) Total acceleration
    (ii) Convective acceleration
    (iii) Local acceleration   **(05 Marks)**

(c) A stream function is given by the expression, $\psi = 2x^2 - y^2$. Find the components of velocity and the resultant velocity at a point (4. 2). **(05 Marks)**

5. (a) State the Buckingham's π-theorem Using the π-theorem show that in the expression for power, $P = \dfrac{\rho Q^3}{D^4} \varphi \left[\eta, \dfrac{\Delta PD}{\rho Q} \right]$, the power input to a pump depends on the discharge Q. the pressure rise ΔP. the fluid density ρ, the size D and the efficiency η. **(10 Marks)**

(b) Derive an expression for the velocity distribution for a laminar flow, between two parallel fixed plates. **(10 Marks)**

6. (a) Derive the Euler's equation of motion for steady flow and obtain Bernaulli's equation from it. State the assumptions made. **(10 Marks)**

(b) Derive an expression for the head loss due to sudden contraction. **(05 Marks)**

(c) A reservoir has been built, 4 km away from a college campus, having 4000 inhabitants. Water is supplied from the reservoir to the campus. It is estimated that, each inhabitant will consume 200 ltrs per day and half of the daily supply is pumped within 8 hours. Calculate the size of the supply main if the loss of head due to friction in the pipeline is 20 m. Assume co-efficient of the friction for the pipe-line is 0.008. **(05 Marks)**

7. (a) Explain the construction and working of a venturimeter, with a neat sketch and derive an expression for discharge. Why the venturimeter is better than orifice meter? **(10 Marks)**

(b) Define the following.
  (i) Mach number
  (ii) Subsonic flow
  (iii) Supersonic flow
  (iv) Compressible flow
  (v) Adiabatic flow **(05 Marks)**

(c) Determine the velocity of a bullet fired in air, if the Mach angle is observed to be 30°. Given that the temperature of air is 22°C, density 1.2 kg/m^3. K = 1.4 and R = 287.4 J/kg K. **(05 Marks)**

8. (a) Explain the following:
  (i) Boundary layer thickness.
  (ii) Displacement thickness.
  (iii) Momentum thickness.
  (iv) Lift.
  (v) Drag. **(10 Marks)**

(b) An automobile has a projected area of 2.7 m^2 and moves at 70 kmph still air of 20°C. If $C_D = 0.5$, what is the power required to overcome the resistance of air? (density of air is 1.16 kg/m^3 al 20°C) **(05 Marks**

(c) Find the displacement thickness and the momentum thickness for the velocity distribution in the boundary layer given by, $\dfrac{u}{\upsilon} = \dfrac{y}{\delta}$. **(05 Marks**

Fluid Mechanics
(Jan./Feb. 2009)
Semester - IV (ME/IP/IM/MA/AU)

Time: Three Hours   Maximum Marks: 100

Note: Answer five full questions, selecting at least two questions from each part.

Part A

Q.1 (a) Differentiate between: (i) Newtonian and Non-Newtonian fluids. (ii) Ideal and Real fluids. (iii) Dynamic and Kinematic viscosity of fluids. (iv) Vapour pressure and Cavitation.

(8)

Ans.: (i) **Newtonian and Non-Newtonian fluids.**

Sr. No.	Newtonian fluid	Non-newtonian fluid
1.	A real fluid in which shear stress is directly proportional to velocity gradient.	A real fluid in which shear stress is not directly proportional to velocity gradient.
2.	*e.g.* air, water, kerosene, gasoline.	*e.g.* blood, mud slurries, paints, sludges.

(ii) Ideal and Real fluids.

Sr. No.	Ideal fluid	Real fluid
1.	Does not offer any resistance to deformation under action of shear stress.	Does offer resistance to motion.
2.	No viscosity.	Has viscosity.
3.	Incompressible.	Compressible.
4.	No surface tension.	Has surface tension.

(iii) **Dynamic and Kinematic viscosity of fluids:** Please refer Q. 1 (a) (ii) of July/Aug.-2007.

(iv) Vapour pressure and Cavitation.

Sr. No.	Vapour pressure	Cavitation
1.	Partial pressure exerted by vapour molecules on the liquid surface is called vapour pressure.	When cavity collapses, the surrounding water rushes towards its center thus produces turbulence tubule this is called as cavitation.
2.	Vapour pressure plays important role in cavitation.	Cavitation causes noise and vibration.

(b) Derive an expression for capillary rise in water.   (4)
Ans.: Please refer Q.1 (b) of Jan/Feb.-2005.

(c) A cubical block of sides 1 m and weighing 350 N slides down an inclined plane with a uniform velocity of 1.5 m/s. The inclined plane is laid on a slope of 5 vertical to 12 horizontal and has an oil film of 1.0 mm thickness. Calculate the dynamic viscosity of oil.   (8)

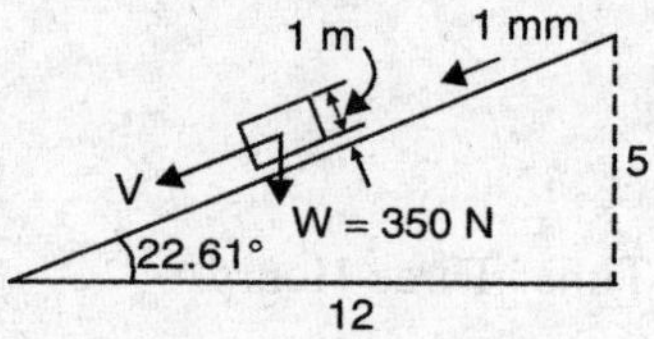

Ans.: Since the block is sliding with a uniform velocity, the force on the block in the direction of motion must be equal to force in opposite direction.

∴ Force in the direction of motion

$$= 350 \times \sin 22.61$$
$$= 134.56 \text{ N.}$$

Fig. 1

∴ Force against the direction of motion,

= Frictional resistance between block and inclined plane

$$= \tau \times \text{Area of contact}$$

$$= \mu \frac{v}{t} \times 1 \times 1 \; (v = \text{Velocity of block})$$

∴

$$134.56 = \mu \times \frac{1.5}{1 \times 10^{-3}} \times 1 \times 1$$

∴

$$\boxed{\mu = 0.0897 \frac{\text{NS}}{\text{m}^2}} \quad \text{Dynamic viscosity of oil.}$$

Q.2 (a) Prove that the centre of pressure lies below the centre of gravity of a vertically immersed plane surface in a static fluid. (8)

Ans.: Consider a vertically immersed plane surface in a static fluid as shown in Fig. 2.

Intensity of pressure at top will be zero and at the base of lamina will be γh. Thus $\Delta \, abc$ represents the pressure diagram for this plane.

Total pressure on plane $P = \gamma A \bar{x}$

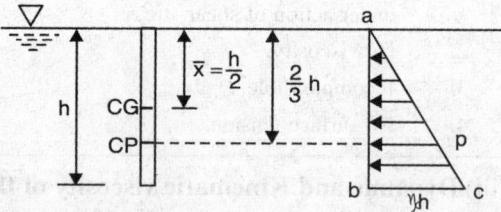

$$= \gamma \times h \times 1 \times \frac{h}{2}$$

∴

$$P = \gamma h$$

∴ Depth of centre of pressure is given by,

∴

$$\bar{h} = \bar{x} \frac{I_G}{A\bar{x}}$$

Fig. 2

$$= \frac{h}{2} + \frac{1 \times h^3}{12} \times \frac{1}{(h \times 1) \times \frac{h}{2}}$$

∴

$$\bar{h} = \frac{h}{2} + \frac{h}{6} = \frac{2}{3}h.$$

Hence it is proved that centre of pressure lies below centre of gravity.

(b) An inverted U-tube manometer is connected to two horizontal pipes *A* and *B* through which water is flowing. The vertical distance between the axes of these pipes is 30 cm. When an oil (*S* = 0.8) is used as a gauge fluid, the vertical height of water columns in the two limbs of the inverted manometer (when measured from the respective centre lines of the pipes) are found to be same and equal to 35 cm. Determine the difference of pressure between the pipes. Pipe *B* is lying below the pipe *A*.

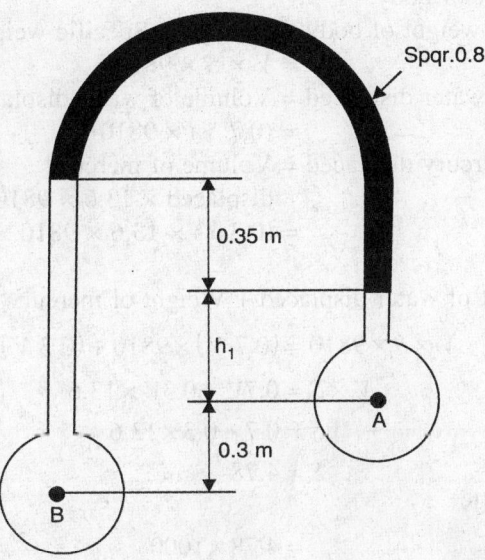

Spqr.0.8

0.35 m

h_1

A

0.3 m

B

Fig. 3

Writing monometric equation

$$\frac{P_A}{\gamma} - h_1 - 0.35 \times 0.8 + 0.35 + h_1 + 0.3 = \frac{P_2}{\gamma}$$

∴ $$\frac{P_A}{\gamma} - \frac{P_2}{\gamma} = -0.37 \text{ m of water.}$$

∴ $$\frac{P_A}{\gamma} - \frac{P_2}{\gamma} = -3629.7 \text{ N/m}^2$$

(c) A metallic body floats at the interface of mercury ($S = 13.6$) and water such that 30% of its volume is submerged in mercury and remaining in water. Estimate the density of the metal.
Ans.:

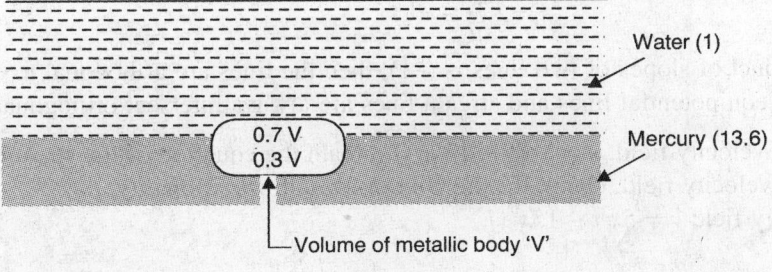

Water (1)

0.7 V
0.3 V

Mercury (13.6)

Volume of metallic body 'V'

Fig. 4

Let volume of metallic body = V m^3
Volume of body submerged in water = 0.7 m^3
Volume of body submerged in mercury = 0.3 m^3
Let 'S' be specific gravity of body,

Now, weight of body = Volume × Specific weight of body
$$= V \times S \times 9810$$

Weight of water displaced = Volume of water displaced × 1 × 9810
$$= (0.7 \ V) \times 9810$$

Weight of mercury displaced = Volume of mercury
displaced × 13.6 × 9810
$$= (0.3 \ V) \times 13.6 \times 9810$$

For equilibrium of body,
Weight of body = Weight of water displaced + Weight of mercury displaced.

∴ $$V \times S \times 9810 = (0.7 \ V) \times 9810 + (0.3 \ V) \times 13.6 \times 9810$$

∴ $$V \times S = 0.7V + 0.3V \times 13.6$$

∴ $$S = 0.7 + 0.3 \times 13.6$$

∴ $$S = 4.78$$

∴ Density of metallic body

$$= 4.78 \times 1000$$

$$= 4780 \ \text{kg/m}^3$$

Q.3 (a) Prove that the equipotential lines and the stream lines are always intersect orthogonally.
Ans.: We known that,
Slope of equipotential line is,

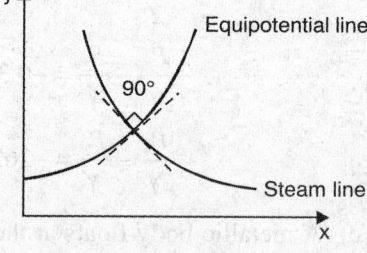

$$\frac{dy}{dx} = \frac{-u}{v}$$

Slope of stream line is.

$$\frac{dy}{dx} = \frac{v}{u}$$

Taking product and slopes of equipotential lines and stream lines,

Fig. 5

$$\frac{-u}{v} \times \frac{v}{u} = -1$$

If the product of slopes of two lines is (−1), then the lines are arthogonal to each other. Therefore equipotential lines and stream lines are always intersect orthogonally.

(b) given the velocity field, $v = 5 \ x^3 i - 15 \ x^2 yj$, obtain the equation of the streamlines. For the above given velocity field, check for the continuity and irrotationality. (8)
Ans.: Velocity field $v = 5x^3 i - 15x^2 yj$,

∴ $$u = 5x^3$$

$$v = -15x^2 y$$

$\therefore$
$$\frac{\partial \psi}{\partial x} = v = -15x^2 y$$

$\therefore$
$$\frac{\partial \psi}{\partial y} = -u = -5x^3$$

Now
$$\frac{\partial \phi}{\partial x} = -15x^2 y$$

Integrating we get,

$$\psi = \frac{-15yx^3}{3} + f(y)$$

Also,
$$\frac{\partial \psi}{\partial y} = -5x^3 + f^1(y)$$

$$\frac{\partial \psi}{y} = -5x^3$$

$\therefore$
$$f'(y) = 0$$

$\therefore$
$$f(y) = 0 + c$$

$$\boxed{\psi = -5x^3 y + c}$$

Equations of streamlines.

Also,
$$u = 5x^3$$

$\therefore$
$$\frac{\partial u}{\partial x} = 15x^2$$

$$v = -15x^2 y$$

$$\frac{\partial v}{\partial y} = -15x^2$$

$\therefore$
$$\frac{\partial u}{\partial x} + \frac{\partial v}{\partial y} = 15x^2 - 15x^2 = 0$$

$\therefore$ Continuity is satisfies.

To check irrotationality.

$$u = 5x^3$$

$\therefore$
$$\frac{\partial u}{\partial y} = 0$$

$$v = -15x^2 y$$

$$\therefore \qquad \frac{\partial v}{\partial x} = -15y \times 2x = -30xy$$

$$\therefore \qquad \frac{\partial v}{\partial x} \neq \frac{\partial u}{\partial y}$$

$\therefore$ Flow is rotational

(c) The velocity potential function is given by the expression, $\phi = \dfrac{xy^3}{3} - x^2 + \dfrac{x^3y}{3} + y^2$.

    (i) Find the velocity components in x and y directions.
    (ii) Show that ϕ represents a possible case of flow.   (6)

Ans.: We have, (i) $\qquad \phi = \dfrac{-xy^3}{3} - x^2 + \dfrac{x^3y}{3} + y^2$

$$\therefore \qquad \frac{\partial \phi}{\partial x} = \frac{-y^3}{3} - 2x + \frac{y}{3}3x^2 + 0$$

$$\therefore \qquad \frac{\partial \phi}{\partial y} = \frac{-y^3}{3} - 2x + x^2y$$

Also, $\qquad \dfrac{\partial \phi}{\partial y} = -\dfrac{x}{3}\cdot 3y^2 - 0 + \dfrac{x^3}{3} + 2y$

$$\therefore \qquad \frac{\partial \phi}{\partial y} = -xy^2 + \frac{x^3}{3} + 2y$$

but $\qquad \dfrac{\partial \phi}{\partial x} = -u$

$$\therefore \qquad u = \frac{y^3}{3} + 2x - x^2y \qquad\qquad \text{Velocity component in } x \text{ direction}$$

Also $\qquad \dfrac{\partial \phi}{\partial y} = -V$

$$\therefore \qquad V = xy^2 - \frac{x^3}{3} - 2y \qquad\qquad \text{Velocity component in } y \text{ direction}$$

(ii) Now, $\qquad \dfrac{\partial u}{\partial x} = 0 + 2 - y(2x) = 2 - 2xy$

$$\frac{\partial v}{\partial y} = x(2y) - 0 - 2 = 2xy - 2$$

$$\therefore \qquad \frac{\partial u}{\partial x} + \frac{\partial v}{\partial y} = 2 - 2xy + 2xy - 2 = 0$$

∴ Flow is satisfying continuity equations and possible.

Q.4 (a) What do you mean by: (i) Geometric similarity (ii) Kinematic similarity (iii) Dynamic similarity (iv) Dimensional homogeneity. (4)

Ans.: (i) **Geometric similarity:** This is the similarity of size and shape of model and prototype.

If the ratios of corresponding lengths in the model and prototype are same, the model is geometrically similar to prototype.

$$\frac{L \text{ model}}{L \text{ prototype}} = \frac{L_m}{L_p} = Lr$$

(ii) **Kinematic similarity:** Please refer Q.4 (a) (ii) of July/Aug.-2008.

(iii) **Dynamic similarity:** This is similarity of forces. If the ratios of all corresponding forces in the model and prototype are same and they are geometrically and Kinematically similar, they are said to be dynamically similar.

For complete dynamic similarity,

$$\frac{F_{Gm}}{F_{Gp}} = \frac{F_{Vm}}{F_{Vp}} = \frac{F_{Em}}{F_{Ep}} = \frac{F_{sm}}{F_{sp}} = \frac{F_{Pm}}{F_{pp}} = \frac{F_{lm}}{F_{lp}} = \text{constant}$$

(iv) **Dimensional Homogeneity:** Please refer Q.4 (a) (i) of July/Aug.-2008.

(b) The thrust (T) of a propeller is assumed to depend on the axial velocity of the fluid (V), the density (ρ) and viscosity (μ) of fluid, the rotational speed (N) r.p.m., and the diameter of the propeller (D). Find the relationship for T by using dimensional analysis. (10)

Ans.: T is a function of V, e, μ, N, D

$$\therefore \qquad T = f(V, e, \mu, N, D)$$

or

$$f_1(T, V, e, \mu, N, D) = 0$$

Total variables $\qquad n = 6$

Writing dimensions of each variables,

$$T = MLT^{-2}$$

$$V = LT^{-1}$$

$$e = ML^{-3}$$

$$\mu = ML^{-1}T^{-1}$$

$$N = T^{-1}$$

$$D = L$$

Fundamental dimensions $m = 3$

$$\therefore \qquad n - m = 6 - 3 = 3$$

$\therefore$ Number of π terms = 3

$\therefore$ $\qquad f_1\left(\pi_1,\ \pi_2,\ \pi_3\right)=0$

Choosing e, N, D as repeating variables.

$$\pi_1 = e^{a_1} n^{b_1} D^{c_1} T$$

$$\pi_2 = e^{a_2} N^{b_2} D^{c_2} V$$

$$\pi_3 = e^{a_3} N^{b_3} D^{c_3} \mu$$

First π term,

$$\pi_1 = e^{a_1} N^{b_1} D^{c_1} T$$

Substituting dimensions on both sides.

$$\left[M^0 L^0 T^0\right] = \left[ML^{-3}\right]^{a_1} \left[T^{-1}\right]^{b_1} \left[L\right]^{c_1} \left[ML^{-2}\right]$$

Equating power of M, L, T,

$$0 = q_1 + 1 \qquad\qquad\qquad \therefore a_1 = -1$$

$$0 = -3a_1 + c_1 + 1 \qquad\qquad \therefore 0 = 4 + c_1 \therefore c_1 = -4$$

$$0 = -b_1 - 2 \qquad\qquad\qquad \therefore b_1 = -2$$

$\therefore$ $\qquad\qquad \pi_1 = e^{-1} N^{-2} D^{-4} T$

$\therefore$ $\qquad\qquad \boxed{\pi_1 = \dfrac{T}{eN^2 D^4}}$

Second π term,

$$\pi_2 = e^{a_2} N^{b_2} D^{c_2} V$$

$\therefore$ $\quad \left[M^0 L^0 T^0\right] = \left[ML^{-3}\right]^{a_2} \left[T^{-1}\right]^{b_2} \left[L\right]^{c_2} \left[LT^{-1}\right]$

$\therefore$ $\qquad\qquad\qquad 0 = a_2$

$$0 = -3a_2 + c_2 + 1 \qquad\qquad \therefore c_2 = -1$$

$\therefore$ $\qquad\qquad 0 = -b_2 - 1 \qquad\qquad \therefore b_2 = -1$

$\therefore$ $\qquad\qquad \pi_2 = e^0 N^{-1} D^{-1} V$

$\therefore$ $\qquad\qquad \boxed{\pi_2 = \dfrac{V}{ND}}$

Third π term,

$$\pi_3 = e^{a_3} N^{b_3} D^{c_3} \mu$$

$\therefore$ $\quad \left[M^0 L^0 T^0\right] = \left[ML^{-3}\right]^{a_3} \left[T^{-1}\right]^{b_3} \left[L\right]^{c_3} \left[ML^{-1} T^{-1}\right]$

$$\therefore \qquad 0 = a_3 + 1 \qquad\qquad \therefore a_3 = -1$$

$$\therefore \qquad 0 = -3a_3 + c_3 - 1 \qquad \therefore c_3 = -1$$

$$\therefore \qquad 0 = -b_3 - 1 \qquad\qquad \therefore b_3 = -1$$

$$\therefore \qquad \pi_3 = e^{-1}N^{-1}D^{-1}\mu$$

$$\therefore \qquad \boxed{\pi_3 = \frac{\mu}{eND^2}}$$

$$\therefore \qquad f_1\left(\pi_1, \pi_2, \pi_3\right) = 0$$

$$\therefore \qquad f_1\left(\frac{T}{eN^2D^4}, \frac{V}{ND}, \frac{\mu}{eND^2}\right) = 0$$

$$\therefore \qquad \frac{T}{eN^2D^4} = \frac{V}{ND} \times \frac{\mu}{eND^2}$$

$$\therefore \qquad T = \frac{V}{ND} \times \frac{\mu}{e\,ND^2} \times eN^2D^4$$

$$\therefore \qquad \boxed{T = V\mu D}.$$

(c) A model of an air duct operating with water produces a pressure drop of 10 kN/m^2 over 10 m length. If the scale ratio is 1/50, $\rho\omega = 1000$ kg/m^3 and $\mu\omega = 0.01$ P_{a-s}, $\mu_a = 0.00002$ P_{a-s}, estimate the corresponding pressure drop in a 20 m long air duct. (6)

Ans.: Euler's law,

$$\frac{V_m}{\sqrt{P_m/e_m}} = \frac{V_P}{\sqrt{P_P/e_P}}$$

$$\therefore \qquad R_e = \frac{eVL}{\mu}$$

$$\therefore \qquad V = \frac{R_e \cdot \mu}{eL}$$

$$\therefore \qquad \frac{\dfrac{\mu_m}{e_m L_m}}{\sqrt{P_m/e_m}} = \frac{\dfrac{\mu_P}{e_P L_P}}{\sqrt{P_P/e_P}}$$

$$\therefore \qquad = \frac{\dfrac{0.001}{1000 \times 10}}{\sqrt{10 \times 10^3/1000}} = \frac{\dfrac{0.00002}{1.2 \times 20}}{\sqrt{P_P/1.2}}$$

$$\boxed{P_P = 833.33 \text{ N/m}^2} \text{ Corresponding pressure drop.}$$

Part B

Q.5 (a) Derive Euler's equation of motion along stream line and hence obtain the Bernoulli's equation for incompressible fluids. (6)

Ans. Please refer the Q.4 (c) of July/Aug.-2004 and Q.5 (b) of July/Aug.-2005.

(b) Using the Euler's equation of motion, derive the Bernoulli's equation for a compressible fluid under going (i) Isothermal process and (ii) Adiabatic process. (6)

Ans.: Euler's equation,

$$gd_2 + \frac{dp}{e} + Vdv = 0$$

(i) **Isothermal process:**

Integrating Euler's equation,

$$gz + \frac{P}{e} + \frac{V^2}{2} = \text{constant}$$

∴ Dividing by 'g'

$$Z + \frac{P}{\gamma} + \frac{V^2}{2g} = \text{constant}$$

(ii) **Adiabatic process:**

Integrating Euler's equation

$$gz + \frac{P}{e} + \frac{V^2}{z} = \text{constant}$$

∴ Dividing by 'g',

$$Z + \frac{P}{\gamma} + \frac{V^2}{2g} = \text{constant}$$

∴
$$\frac{P_1}{\gamma} + \frac{V_1^2}{2g} + Z_1 = \frac{P_2}{\gamma} + \frac{V_2^2}{2g} + Z_2 + h_1$$

where h_1 is the energy loss.

(c) A conical tube is fixed vertically with its small end upwards. Velocity of flow down the tube is 4.5 m/s at the upper end and 1.5 m/s at the lower end. Tube is 1.5 m long and the pressure at the upper end is 24.3 *kPa* (*ab*). Loss in the tube expressed as head is $0.3(V_1 - V_2^2)^2/2g$, where V_1 and V_2 are the velocities of fluid ($S = 0.8$) flow at the upper and lower ends respectively. What is the pressure head at the lower end? (8)

Ans.: Applying Bernoulli's equation,

$$\frac{P_1}{\gamma} + \frac{V_1^2}{2g} + Z_1 = \frac{P_2}{\gamma} + \frac{V_2^2}{2g} + Z_2 + hf$$

∴
$$\frac{24.3 \times 10^3}{0.8 \times 9810} + \frac{4.5^2}{2 \times 9.81} + 1.5 = \frac{P_2}{0.8 \times 9810} + \frac{1.5^2}{2 \times 9.81} + 0 + hf$$

$$\therefore \quad 5.628 = \frac{P_2}{7848} + 0.11467 + hf$$

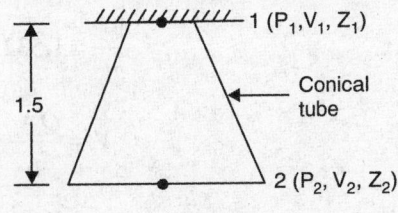

$$hf = \frac{0.3(V_1 - V_2)^2}{2g}$$

Fig. 6

$$\therefore \quad = \frac{0.3[4.5 - 1.5]^2}{2 \times 9.81}$$

$$\therefore \quad = 0.1376$$

$$5.628 = \frac{P_2}{7848} + 0.11467 + 0.1376$$

$$\therefore \quad P_2 = 42188.72 \text{ N/m}^2$$

Pressure at lower end

$\therefore$ Pressure head at lower end

$$\frac{P_2}{\gamma} = \frac{42188.72}{0.8 \times 9810} = 5.375 \text{ m}$$

Q.6 (a) Derive an expression for the actual discharge through orifice meter. (8)

Ans.: Please refer Q.5 (c) of July/Aug.-2007.

(b) Water is to be supplied to a town of 4 lakhs inhabitants. The reservoir is 6.4 km away from the town and the loss of head due to friction is measured as 15 m. Calculate the size of the supply main if each inhabitant consumes 180 liters of water per day and half of the daily supply is pumped is 8 hour. Take the coefficient of friction for the pipe, $f = 0.0075$.

Ans.: $Q_{total} = $ (4 lakh population) × (180 litres per head/day)

$$Q_{total} = (4 \text{ lakh population}) \times (180 \text{ litres per head/day})$$

$$= \frac{4 \times 10^5 \times 180 \times 10^{-3}}{8 \times 3600}$$

$$\therefore \quad Q_{total} = 2.500 \text{ m}^3/\text{s}$$

$$A = \frac{\pi}{4} d^2 = 0.7853 \, d^2$$

loss of head due to friction,

$$hf = \frac{4fLV^2}{2gd}$$

$$\therefore \quad 15 = \frac{4 \times 0.0075 \times 6.4 \times 10^3 \times V^2}{2 \times 9.81 \times d}$$

$$\therefore \qquad \frac{V^2}{d} = 1.532$$

$$\therefore \qquad V = \frac{Q_{total}}{A}$$

$$\therefore \qquad V = \frac{2.5}{0.7853\,d^2}$$

$$\therefore \qquad \frac{\left[\dfrac{2.5}{0.7853\,d^2}\right]^2}{d} = 1.532$$

$$\therefore \qquad \frac{10.134}{d^5} = 1.532$$

$$\therefore \qquad \boxed{d = 1.459 \text{ m}}$$

This is the size of supply main.

(c) A venturimeter is to be installed in a 180 mm pipeline horizontally at a section where the pressure is 110 kPa (gauge). If the maximum flow rate of water in the pipe is 0.15 m³/s find the least diameter of the throat so that the pressure at the throat does not fall below 80 kPa (vacuum). Assume that 4% of the differential head is lost between inlet and the throat. (6)

Ans.:
$$d_1 = 180 \text{ mm} = 0.18 \text{ m}$$

$$\therefore \qquad a_1 = \frac{\pi}{4}(d_1)^2 = \frac{\pi}{4}(0.18)^2 = 0.0254 \text{ m}^2$$

Differential head,

$$h = \frac{P_1}{\gamma} - \frac{P_2}{\gamma}$$

$$\therefore \qquad h = \frac{110 \times 10^3}{9810} - \left(-\frac{80 \times 10^3}{9810}\right)$$

$$\therefore \qquad h = 19.36 \text{ m of water}$$

$$\therefore \qquad \text{Head lost, } hf = 4 \text{ \%}h$$

$$\therefore \qquad = \frac{4}{100} \times 19.36$$

$$hf = 0.7744 \text{ m}$$

$$cd = \sqrt{\frac{h - hf}{h}}$$

$$= \sqrt{\frac{19.36 - 0.7744}{19.36}}$$

$$cd = 0.9797 \qquad \text{coefficient of discharge}$$

$$\therefore \qquad Q = cd \frac{a_1 a_2}{\sqrt{a_1^2 - a_2^2}} \sqrt{2gh}$$

$$\therefore \qquad 0.15 = \frac{0.9797 \times 0.0254 \times a^2}{\sqrt{0.0254^2 \times a_2^2}} \times \sqrt{2 \times 9.81 \times 19.36}$$

$$\therefore \qquad 0.3092 = \frac{a_2}{\sqrt{0.0254^2 - a_2^2}}$$

$$0.09565 = \frac{a_2^2}{0.0254^2 - a_2^2} \times 10^{-5} - 0.09565\, a_2^2 = a_2^2$$

$$\therefore \qquad a_2 = 7.504 \times 10^{-3} \text{ m}$$

$$\therefore \qquad a_2 = \frac{\pi}{4} d_2^2$$

$$\therefore \qquad 7.504 \times 10^{-3} = \frac{\pi}{4} \times d_2^2$$

$$\therefore \qquad \boxed{d^2 = 0.0977 \text{ m}} \quad \text{This is the least diameter of throat.}$$

Q.7 (a) Derive Hagen Poiseuille equation for a laminar flow in a circular tube. (10)
Ans.: Please refer Q.6 (c) of Jan./Feb.-2006.

(b) Water at 15°C flows between two large parallel plates at a distance of 1.6 mm apart. Determine (i) The maximum velocity (ii) Pressure drop per unit length and (iii) Shear stress at the walls of the plates of the average velocity is 0.2 m/s. The viscosity of water at 15°C is given as 0.01 paise. (10)
Ans.: (i) Maximum velocity,

$$u_{max} = \frac{3}{2} \times u_{av}$$

$$= \frac{3}{2} \times 0.2$$

$$\therefore \qquad u_{max} = 0.3 \text{ m/s}$$

(ii) Pressure drop per unit length,

$$\text{i.e.} \qquad \frac{\Delta P}{L} = \frac{12 \mu\, u_{av}}{B^2} = \frac{12 \times 0.01 \times 10^{-1} \times 0.2}{1.6^2 \times (10^{-3})^2}$$

$\therefore$ $$\frac{\Delta P}{L} = 937.5 \, \text{N/m}^3$$

(iii) Shear stress at walls of plates,

$$\tau_0 = \left(\frac{-\partial P}{\partial x}\right)\frac{B}{2}$$

But,

$$u_{av} = \frac{B^2}{12\mu} \times \left(\frac{-\partial P}{\partial x}\right)$$

$\therefore$ $$0.2 = \frac{1.6 \times (10^{-3})^2}{12 \times 0.01 \times 10^{-1}} \times \left(\frac{-\partial P}{\partial x}\right)$$

$\therefore$ $$\left(\frac{-\partial P}{\partial x}\right) = \frac{\partial P}{L} = 937.5 \, \text{N/m}^3$$

$\therefore$ $$\tau_0 = \frac{937.5 \times 1.6 \times 10^{-3}}{2}$$

$\therefore$ $$\boxed{\tau_0 = 0.75 \, \text{N/m}^3}$$

Q. 8 (a) We know that the velocity of sound wave is the square root of the ratio of change pressure to the change of density of a fluid. Using this definition, derive the expressions for a velocity of sound in a compressible fluid when it undergoes process (i) Isothermal and (ii) Reversible adiabatic. (10)

Ans.: Please refer Q. 6 (b) of July/Aug 2006.

(b) define the following and write their equations:
  (i) Drag   (ii) Lift
  (iii) Displacement thickness   (iv) Momentum thickness

Ans.: (i) **Drag:** Please refer Q.8 (a) of July/Aug. 2006.
(ii) **Lift:** Please refer Q. 8 (a) of July/Aug. 2006.

(iii) **Displacement thickness:**
Definition: Please refer Q.8 (a) (i) of July/Aug. 2005.
Equation: Please refer Q.8 (b) of July/Aug. 2006.

(iv) **Momentum thickness:**
Definition: Please refer Q.8 (a) (ii) of July/Aug. 2005.
Equation: Please refer Q.8 (b) of July/Aug. 2006.

(c) A man descends to the ground from an aeroplane with the help of a parachute which is hemispherical having a diameter of 4 m against the resistance of air with a uniform velocity of 25 m/s. Find the weight of the man if the weight of parachute is 9.81 N. Take $C_D = 0.6$ and density of air = 1.25 kg/m³.

Ans.: Drag force: $F_D = C_D \times A \times \dfrac{ev^2}{2}$

$\therefore \qquad (9.81 + w) = 0.6 \times \dfrac{\dfrac{\pi}{4} \times 4^2}{2} \times \dfrac{1.25 \times 25^2}{2}$

$\therefore \qquad w + 9.81 = \dfrac{2945.24}{2}$

$\therefore \qquad \boxed{w = 1462.81 \text{ N}}$ This is weight of man.

Fluid Mechanics
June-July 2009
Fourth Semester B.E. Degree Examination (VTU)

Time: 3 hrs.   Max. Marks: 100

Note: Answer any Five full questions choosing at least two questions from each unit.

Part A

1. (a) Give reasons:
   (i) Viscosity of liquids varies with temperature.
   (ii) Thin objects float on free surface of static liquid.
   (iii) Metacentric height determines stability of floating body.
   (iv) Rise of water in a Capillary tube.
   (v) Mercury is used as Manometric liquid.   **(05 Marks)**
   (b) Define following terms with their units.
   (i) Specific weight;   (ii) Kinematic viscosity; (iii) Surface Tension
   (iv) Specific gravity;   (v) Capillarity   **(05 Marks)**
   (c) The space between two square flat parallel plates is filled with oil. Each side of the plates is 800 mm. Thickness of the oil film is 20 mm. The upper plate moves at a uniform velocity of 3.2 m/sec when a force of 50 N applied to upper plate. Determine:
   (i) Shear stress
   (ii) Dynamic viscosity of oil in poise
   (iii) Power absorbed in moving the plate
   (iv) Kinematic viscosity of oil if specific gravity of oil is 0.90.   **(10 Marks)**
2. (a) State and prove Hydrostatic law.   **(05 Marks)**
   (b) With neat sketch, explain working of differential u-Tube Manometer and derive relation for measuring pressure difference between two pipes.   **(05 Marks)**
   (c) A wooden block of size 6m × 3m height floats in freshwater. Find the depth of immersion and determine the metacentric height. Specify gravity of wood is 0.70. Find the volume of concrete block placed on the wooden block, so as to completely submerge the wooden block in water. Take specific gravity of concrete as 3.0.
                                                                   (10 Marks)
3. (a) Explain experimental procedure to determine the metacentric height of a floating vessel.   **(04 Marks)**
   (b) Derive continuity equation for a three dimensional fluid flow in Cartesian co-ordinates.   **(08 Marks)**
   (c) Velocity potential function for a two dimensional fluid flow is given by $\phi = x\,(2y - 1)$. Check the existence of flow. Determine the velocity of flow at a $P\,(2, 3)$ and the stream function.   **(08 Marks)**
4. (a) Show that streamlines and equipotential lines are orthogonal to each other.
                                                                   (05 Marks)
   (b) Explain Model Similitude and Non-dimensional numbers.   **(05 Marks)**
   (c) The pressure difference Δp for a viscous flow in a pipe depends upon the diameter of

the pipe 'D', length of pipe 'L', velocity of flow 'V', viscosity of fluid μ and the density of fluid 'ρ'. Using Buckingham's theorem, show that the relation for pressure difference Δp is given by $\Delta p = \rho V^2 f\left(\dfrac{1}{Re}, \dfrac{L}{D}\right)$ **(10 Marks)**

Part B

5. (a) State and prove Bernoulli's equation for a fluid flow. Mention assumptions made in derivation. **(10 Marks)**

   (b) Water is flowing through a taper pipe of length 150 m, having diameter 500 mm at the upper end and 250 mm at the lower end. Rate of flow is 70 liters per sec. The pipeline has a slope of 1 in 30. Find the pressure at the lower end if the pressure at higher level is 2.5 bar. **(10 Marks)**

6. (a) Explain with neat sketch, working of pitot-static tube. **(05 Marks)**

   (b) Differentiate between Orificemeter and venturimeter with neat sketches.

   (05 Marks)

   (c) A horizontal venturimeter with 50 cm diameter at inlet and 20 cm throat diameter is used for measuring rate of water flow, if the pressure at inlet is 1.8 Bar and vacuum pressure at the throat is 30 cm of mercury, find the rate of flow. Assume 10% differential pressure head is lost between the inlet and throat section. Assume coefficient of discharge is 0.96. **(10 Marks)**

7. (a) Derive Hagen-Poiseulle's equation for viscous flow through a circular pipe.

   (10 Marks)

   (b) Rate of water flow through a horizontal pipe is 0.030 m^3/sec. Length of pipe is 1000 meters. Diameter of pipe for first half of length is 200 mm and suddenly changes to 400 mm for remaining length. Find the elevation difference between the two reservoirs connected by the horizontal pipeline. Take $f = 0.01$ for material of pipeline.

   (10 Marks)

8. (a) Explain terms:

   (i) Lift (ii) Drag

   (iii) Displacement thickness (iv) Momentum thickness **(08 Marks)**

   (b) Explain Mach angle and Mach cone. **(04 Marks)**

   (c) A projectile travels in air of pressure 15 N/cm^2 at 10°C, at a speed of 1500 km/hr. Find the Mach number and Mach angle. Assume $\gamma = 1.4$ and R = 287 J/kg°K

   (08 Marks)

Fluid Mechanics
Jan./Feb. 2008
Semester - IV (Mechanical Engineering) (UV)

Time: 3 Hours   Maximum Marks: 100

Note: 1. Answer any five full questions.
2. Missing data if any can be suitably assumed.

Q.1 (a) Define the following and mention their S.I. units:
    (i) Density   (ii) Dynamic viscosity
    (iii) Surface tension   (iv) Vapour pressure
    (v) Bulk modulus.   (10)
Ans.: (i) **Density:** Please refer Q.1 (a) (i) of July/Aug.-2007.
    (ii) **Dynamic viscosity:** Please refer Q.1 (a) (ii) of July/Aug.-2007.
    (iii) **Surface tension:** Please refer Q.1 (a) (i) of July/Aug.-2005.
    (iv) **Vapour pressure:** Please refer Q.1 (a) (iii) July/Aug.-2005.
    (v) **Bulk modulus:** Please refer Q.1 (a) (iv) July/Aug.-2005.

(b) Derive an expression for capillary rise of liquid in a tube.   (5)
Ans.: Please refer Q.1 (b) of Jan./Feb.-2005.

(c) The surface tension of water droplet in contact with air at 20°C is 0.071 N/m. If the diameter of droplet is 1.45 mm, calculate the pressure within the droplet.   (5)
Ans.: Data: σ = surface tension = 0.071 N/m.
           d = diameter of droplet = 1.45 mm.
           P = pressure within droplet = ?
Formula:

$$P = \frac{4\sigma}{d} = \frac{4 \times 0.071}{1.45 \times 10^{-3}}$$

$\therefore$ $\boxed{P = 195.862\,\text{N}/\text{m}^2}$

Q.2 (a) Derive an expression for hydrostatic force on an inclined submerged plane surface and depth of centre of pressure.   (10)
Ans.: Please refer Q.3 (a) of Jan./Feb.-2005 and
    Please refer Q. 2(b) of July/Aug.-2005.

(b) A circular plate of 2 m diameter is immersed in an oil of specific gravity of 0.8, such that its surface is 30° to the free surface. Its top edge is 2.5 m below the free surface. Find the force and centre of pressure.
Ans.: Data:

$$A = \text{Area of plate} = \frac{\pi}{4} \times d^2 = \frac{\pi}{4} \times 2^2 = 3.142 \text{ m}^2.$$

$$\frac{\gamma_{oil}}{\gamma_{water}} = \text{S.G. of oil.}$$

$$\therefore \qquad \frac{\gamma_{oil}}{9810} = 0.8$$

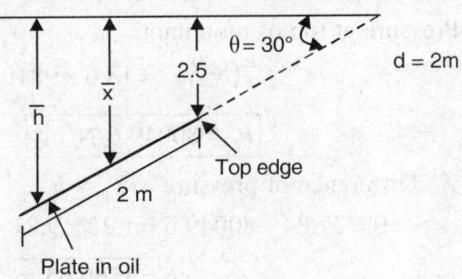

$$\therefore \qquad \gamma_{oil} = 0.8 \times 9810 = 7848 \ N/m^3$$

Referring above Fig. 1

$$\sin 30 = \frac{c_d}{1}$$

$$\therefore \qquad \boxed{c_d = 0.5 \ m}$$

$$\therefore \qquad \overline{x} = 2.5 + c_d$$

$$= 2.5 + 0.5 = 3 \ m$$

Fig. 1

& $\therefore$ Force on the surface

$$P = \gamma_{i,j} A \overline{x}$$

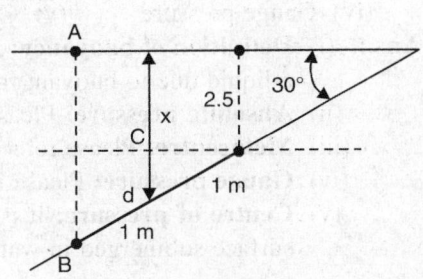

$$\therefore \qquad = 7848 \times 3.142 \times 3$$

$$\therefore \qquad P = 73975.248 \ N$$

$$\boxed{P = 73.975 \ kN}$$

Centre of pressure,

Fig. 2

$$\overline{h} = \overline{x} + \frac{I_G \sin^2 \theta}{A\overline{x}}$$

$$I_G = \frac{\pi \times d^4}{\sigma 4} = \frac{\pi}{\sigma 4} \times 2^4 = \frac{\pi}{4} m^4$$

$$\therefore \qquad \boxed{I_G = 0.7853 \ m^4}$$

$$\therefore \qquad \overline{h} = 3 + \frac{0.7853 \times \sin^2 30}{3.142 \times 3} = 3 + 0.0208$$

$$\therefore \qquad \boxed{\overline{h} = 3.0208 \ m}$$

(c) Measurements of pressure at the base and top of a mountain are 74 cm and 60 cm of mercury respectively. Calculate the height of the mountain if air has a specific weight of 1.22 kg/m³. (5)

Ans.:

Pressure at base of mountain

$$P_1 = \gamma h_1 = 13.6 \times 9810 \times 0.74$$

$$\therefore \qquad \boxed{P_1 = 98727.84 \ N/m^2}$$

Pressure at top of mountain,

$$P_1 = \gamma h_2 = 13.6 \times 9810 \times 0.60$$

$$\boxed{P_2 = 80049.6 \text{ N/m}^2}$$

$\therefore$ Difference of pressure $= \gamma_{air} \times h$

$\therefore \quad 98727.84 - 80049.6 = 1.22 \times 9.81 \times h$

$\therefore \qquad \boxed{h = 1560.65 \text{ m}}$

$\therefore \quad$ Height of mountain $= 1560.65$ m

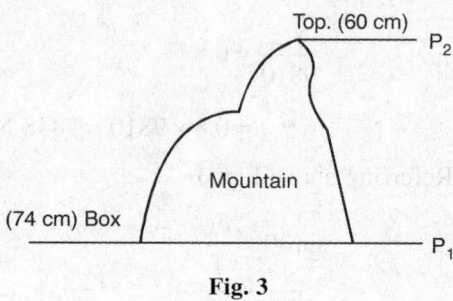

Fig. 3

Q.3 (a) Define the following:

    (i) Buoyancy   (ii) Absolute pressure   (iii) Metacentre

    (iv) Gauge pressure   (v) Centre of pressure.   (10)

Ans.: (i) **Definition of buoyancy:** It is defined as the tendency of submerged body to rise in a liquid due to buoyant force.

    (ii) **Absolute pressure:** Please refer Q.2 (c) of Jan./Feb.-2005.

    (iii) **Metacentre:** Please refer Q.2 (c) of July/Aug.-2005.

    (iv) **Gauge pressure:** Please refer Q. 2(c) Jan./Feb.-2005.

    (v) **Centre of pressure:** It is defined as the point of application of total pressure on surface submerged in water.

It is given as, $\bar{h} = \bar{x} + \dfrac{I_G \sin^2 \theta}{A \bar{x}}$

(b) A block of wood of specific gravity 0.8 floats in water. determine the metacentric height of block if its size is 3 m long, 2 m wide and 1 m height. State whether equilibrium is stable or unstable.   (5)

Ans.: See Fig. 4 on next page.

    Weight causing displacement,

$$W = (3 \times 2 \times 1) \times 0.8 \times 9810$$

$\therefore \qquad W = 47088$ N

Volume of liquid displaced,

$$\therefore \qquad V = \frac{47088}{9810} = 4.3 \text{ m}^3$$

Depth of immersion,

$$d = \frac{4.8}{(3 \times 2)} = 0.8 \text{ m.}$$

So

$\therefore \qquad OB = 0.4$ m $\qquad OG = 0.5$ m

$\therefore \qquad BG = OG - OB = 0.5 - 0.4 = 0.1$ m

$\therefore \qquad BM = \dfrac{I}{V} = \dfrac{3 \times (2)^3}{12} \times \dfrac{1}{4.8} = 0.4167 \text{ m}$

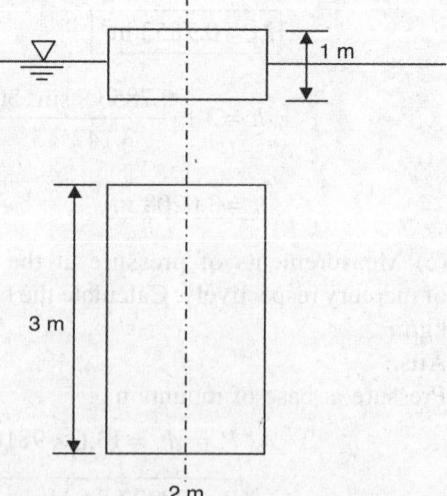

Fig. 4

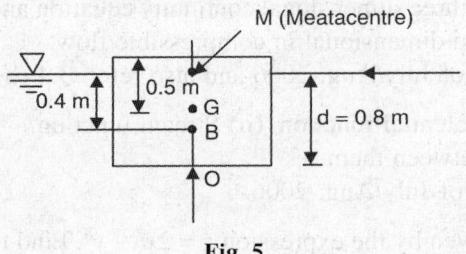

Fig. 5

$\therefore$ $GM = BM - BG = 0.4167 - 0.1$

              $GM = 0.3167$ m

GM is positive.

So block is in stable equilibrium.

(c) The left limb of a mercury U-tube manometer is open to atmospheric and the right limb is connected to a pipe carrying water under pressure. The centre of the pipe is at the level of the free surface of mercury. Find the difference in level of mercury limbs of U-tube if the absolute pressure of water in the pipe is 14.5 m of water, atmospheric pressure is 760 mm of mercury.

(5)

Ans.: Pressure of water = 14.5 m of water = hm.

$\therefore$ $hm = s_2 h_2 - s_1 h_1$

$\therefore$ $14.5 = 13.6 \times h_2 - 1 \times h_1$

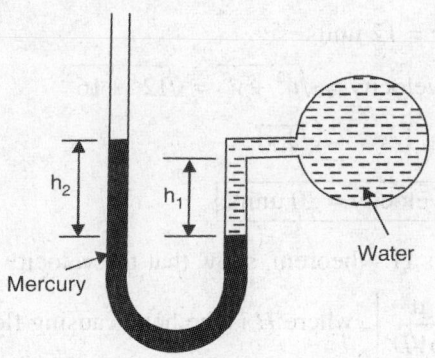

Fig. 6

But $h_1 = h_2$

$\therefore$ $14.5 = 13.6 h_2 - h_2$

$\therefore$ $14.5 = 12.6 h_2$

$\therefore$ $\boxed{h_2 = 1.1507 \text{ m.}}$

$\therefore$ $\boxed{h_1 = 1.1507 \text{ m.}}$

$\therefore$ Difference of level of mercury limbs of U-tube is 1.1507 m.

Q.4 (a) Derive the general three-dimensional continuity equation and then reduce it to continuity equation for steady, two dimensional in compressible flow. (10)

Ans.: Please refer Q.4 (a) of July/Aug.-2006 and also refer Q.4 (b) of July/Aug.-2005.

(b) Explain: (i) Velocity potential function. (ii) Stream function.
Write down the relation between them. (5)

Ans.: Please refer Q.4 (b) of July/Aug.-2006.

(c) A stream function is given by the expression $z = 2x^2 - y^3$. Find the components of velocity and the resultant velocity at a point (4, 2). (5)

Ans.: Stream function $z = 2x^2 - y^3$.

$$\therefore \qquad \frac{\partial z}{\partial x} = V$$

$$\therefore \qquad V = 2 \times 2x - 0$$

$$\therefore \qquad \boxed{V = 4x}$$

at (4, 2), $V = 4 \times 4 = 16$ units

Also, $\qquad \dfrac{\partial z}{\partial y} = -u$

$$\therefore \qquad u = -\left[0 - 3y^2\right]$$

$$\therefore \qquad \boxed{u = 3y^2}$$

$\therefore$ at (4, 2) $\qquad u = 3 \times 2^2 = 12$ units

$$\therefore \qquad \text{Resultant velocity} = \sqrt{u^2 + v^2} = \sqrt{12^2 + 16^2}$$

$$= \sqrt{400}$$

$$\therefore \qquad \boxed{\text{Resultant velocity} = 20 \text{ units.}}$$

Q.5 (a) Using Buckingham's Π - theorem, show that the velocity through a circular orifice is given by $V = \sqrt{2gh}\ \phi\left[\dfrac{D}{H}, \dfrac{\mu}{\rho VD}\right]$, where H is the head causing flow, D is the diameter of the orifice, μ is the coefficient of viscosity, ρ is the mass density and g is the acceleration due to gravity. (10)

Ans.: Please refer Q.5 (a) of July/Aug.-2005.

(b) Derive the Euler's equation of motion for steady flow and obtain Bernoulli's equation from it. State the assumptions made in the derivation of Bernoulli's equation. (10)

Ans.: Please refer Q.4 (c) of July/Aug.-2004, Q.5 (a) of July/Aug.-2006 and also refer Q.5 (b) of July/Aug.-2005.

Q.6 (a) Explain a venturimeter. Drive an expression for discharge. Why venturimeter is better than orifice meter? (10)

Ans.: Venturimeter is a device used to measure discharge through a pipe line. It consists of

1. Converging Section:

It is a conical tube used to accelerate the flow. Angle is almost 19° to 20°.

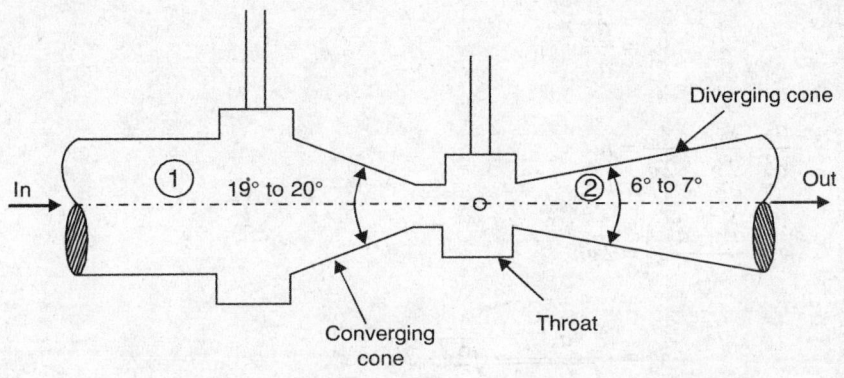

Fig. 7

2. Throat:

It is having uniform cross-section. Length of throat is equal to its diameter. Diameter of throat is $\frac{1}{3}$ to $\frac{3}{4}$th of diameter at inlet.

3. Diverging Section:

It is also conical portion, which is diverging in the direction of fluid flow.

    Used to convert K.E. into P.E.

    Angle is 6° to 7°.

Expression for Discharge:

Applying Bernoulli's equation at (1) and (2).

$$z_1 + \frac{p_1}{\gamma} + \frac{v_1^2}{2g} = z_2 + \frac{p_2}{\gamma} + \frac{v_2^2}{2g}$$

Consider $\qquad h_f = 0$

$$\therefore \qquad \frac{v_2^2 - v_1^2}{2g} = \left(z_1 + \frac{p_1}{\gamma}\right) - \left(z_2 + \frac{p_2}{\gamma}\right)$$

For horizontal venturimeter,

$$z_1 = z_2$$

$$\therefore \qquad \frac{v_2^2 - v_1^2}{2g} = \frac{p_1}{\gamma} - \frac{p_2}{\gamma} = h$$

But by continuity formula,

$$Q = a_1 v_1 = a_2 v_2$$

$$\therefore \qquad v_2 = \frac{a_1}{a_2} v_1$$

$$\therefore \qquad \frac{a_1^2}{a_2^2} \cdot \frac{v_1^2}{2g} - \frac{v_1^2}{2g} = h$$

$$\therefore \qquad v_1^2 \left(\frac{a_1^2 - a_2^2}{a_2^2} \right) = 2gh$$

$$\therefore \qquad v_1 = \frac{a_2}{\sqrt{a_1^2 - a_2^2}} \cdot \sqrt{2g} \cdot \sqrt{h}$$

$$\therefore \qquad Q = a_1 v_1$$

$$\therefore \qquad \boxed{Q = \frac{a_1 a_2 \sqrt{2g} \cdot \sqrt{h}}{\sqrt{a_1^2 - a_2^2}}}$$

This is the expression for discharge in a venturimeter.

Venturimeter is better than orifice meter:

1. Losses are low and hence coefficient of discharge is high.
2. It is used for measurement of higher rates of flow.
3. Used for measurement of flow in large pipes.

(b) Derive Darcy-Weisbach formula to calculate the frictional head loss in pipe in terms of friction factor.

Ans.: Please refer Q.6 (b) of Jan./Feb.-2005.

Q.7 (a) Explain:

(i) Mach number	(ii) Subsonic flow	(iii) Supersonic flow
(iv) Laminar flow	(v) Turbulent flow	(10)

Ans.:

   (i) **Mach number:** Please refer Q.7 (c) of July./Aug.-2005.

   (ii) **Subsonic flow:** In subsonic flow, mach number is less than one.

$$\therefore \qquad M < 1$$

   i.e., velocity of fluid is less than velocity of sound.

   (iii) **Supersonic flow:** in supersonic flow, mach number is more than one.

(iv) **Laminar flow:** Please refer Q.7 (a) of July/Aug.-2005 and Q.8 (c) (ii) of July/Aug.-2004.

(v) **Turbulent flow:** Please refer Q.7 (a) July/Aug.-2005 and refer Q.8 (c) (ii) of July/Aug.-2004.

(b) Water at 15°C flows between two large parallel plates at a distance of 1.6 mm apart. Determine
  (i) The maximum velocity
 (ii) The pressure drop per unit length and
(iii) The shear stress at the walls of the plates if the average velocity is 0.25 m/s. The viscosity of water at 15°C is given as 0.01 poise. (5)

Ans.: (i) Maximum velocity,

$$u_{max} = \frac{3}{2} \times u_{av}$$

$$u_{max} = \frac{3}{2} \times 0.2 = 0.3 \ m/s.$$

(ii) Pressure drop per unit length,

i.e.
$$\frac{\Delta p}{L} = \frac{12 \mu u_{av}}{B^2} = \frac{12 \times 0.01 \times 10^{-1} \times 0.2}{1.6^2 \times (10^{-3})^2}$$

∴
$$\boxed{\frac{\Delta p}{L} = 9.375 \times 10^2 \ N/m^3}$$

(iii) Shear stress at walls of plates,

$$\tau_0 = \left(-\frac{\partial p}{\partial x} \right) \cdot \frac{B}{2}$$

But,
$$u_{av} = \frac{B^2}{12 \mu} \times \left(\frac{-\partial p}{\partial x} \right)$$

∴
$$0.2 = \frac{1.6^2 \times (10^{-3})^2}{12 \times 0.01 \times 10^{-1}} \times \left(\frac{-\partial p}{\partial x} \right)$$

∴
$$\left(\frac{-\partial p}{\partial x} \right) = 9.375 \times 10^2 \ N/m^3$$

∴
$$\tau_0 = 9.375 \times 10^2 \times \left(\frac{1.6 \times 10^{-3}}{2} \right)$$

∴
$$\boxed{\tau_0 = 0.75 \ N/m^2}$$

(c) Find the velocity of bullet fired in standard air if the Mach angle is 30°. Take $R = 287.14$ J/kg K and $K = 1.4$ for air. Assume temperature at 15°C. (5)

Ans.: Please refer Q.7 (d) of July/Aug.-2005.

Q.8 (a) Define
    (i) Drag   (ii) Lift
    (iii) Boundary layer thickness   (iv) Displacement thickness
    (v) Momentum thickness. (10)

Ans.:
    (i) **Drag:** Please refer Q.3 (b)(ii) July/Aug.-2004.
    (ii) **Lift:** Please refer Q.3 (b)(ii) July/Aug.-2004.
    (iii) **Boundary layer thickness:** Please refer Q.8 (a)(i) Jan./Feb.-2007.
    (iv) **Displacement thickness:** Please refer Q.8 (a)(ii) July/Aug.-2005.
    (v) **Momentum thickness:** Please refer Q.8 (a)(ii) July/Aug.-2005.

(b) A circular disc 3 m in diameter is held normal to a 26.4 m/s wind of density 0.0012 gm/cc. What force is required to hold it at rest? Assume co-efficient of drag of disc = 1.1. (5)

Ans.: Data: $C_d = 1.1$

F_D = Force required to hold it at rest.

$$\rho = 0.0012 \ \text{gm/C}_c = \frac{0.0012 \times 10^{-3}}{\left(10^{-2}\right)^3}$$

$$\rho = 1.2 \ \text{kg/m}^3$$

$$V = 26.4 \ \text{m/s}$$

$\therefore$
$$C_d = \frac{F_D}{\dfrac{1}{2}\rho A v^2}$$

$\therefore$
$$F_D = C_d \times \frac{1}{2}\rho A v^2 = 1.1 \times \frac{1}{2} \times 1.2 \times \left(\frac{\pi}{4} \times 3^2\right) \times 26.4^2$$

$\therefore$
$$\boxed{F_D = 3251.5 \ \text{N}}$$

This is the force required to hold it at rest.

(c) Find the displacement thickness and the momentum thickness for the velocity distribution is the boundary layer given by $\dfrac{u}{v} = 2\left(\dfrac{y}{\delta}\right) - \left(\dfrac{y}{\delta}\right)^2$ where u is the velocity at a distance y from the plate and $u = U$ at $y = \delta$, where δ is the boundary layer thickness. (5)

Ans.: Please refer Q.7 (c) of Jan./Feb.-2006.

Fluid Mechanics
June-July 2008
Fourth Semester B.E. Degree Examination (VTU)

Time: 3 hrs.   Max. Marks: 100

Note: Answer any five questions selecting at least Two questions from each part.

Part A

1. (a) Define the following fluid properties and state their units.   **(06 Marks)**
       (i) Specific weight   (ii) Kinematic viscosity
       (iii) Surface tension   (iv) Vapour pressure.
   (b) Define capillarity, obtain an expression for capillary rise of a liquid.
                                                                        (06 Marks)
   (c) An oil film of thickness 1.5 mm is used for lubrication between a square plate of size
       0.9 m × 0.9 m and an inclined plane having an angle of inclination 20°. The weight
       of the square plate is 392.4 N and it slides down the plane with a uniform velocity of
       0.2 m/s. Find the dynamic viscosity of the oil.   **(08 Marks)**

2. (a) State and prove the Pascal's law.   **(07 Marks)**
   (b) A Caission for closing the entrance to a dry dock is of trapezoidal form 16 m wide
       at the top and 10 m wide at the bottom and 6 m deep. Find the total pressure and
       centre of pressure on the caisson if the water on the outside is just with the top and
       dock is empty.   **(08 Marks)**
   (c) Explain the conditions of equilibrium for a floating body with neat sketches.
                                                                        (05 Marks)

3. (a) Differentiate between:
       (i) Steady and unsteady flow   (ii) Uniform and non-uniform flow
       (iii) Laminar and turbulent flow.   **(06 Marks)**
   (b) Obtain an expression for continuity equation for a three-dimensional flow.
                                                                        (08 Marks)
   (c) A stream function is given by $\psi = 5x - 6y$. Calculate the velocity components and
       also magnitude and direction of the resultant velocity at any point.   **(06 Marks)**

4. (a) Explain the following terms in brief:
       (i) Dimensionally homogeneous equation
       (ii) Kinematic similarity.
   (b) State Buckingham's Π - theorem. The efficiency η of a fan depends on density ρ,
       dynamic viscosity μ of the fluid, angular velocity ω, diameter D of the rotor and the
       discharge Q. Express η in terms of dimensionless parameters.   **(10 Marks)**
   (c) Define the following dimensionless numbers and state their significance:
       (i) Froude's number (ii) Euler's number (iii) Mach's number.   **(06 Marks)**

Part B

5. (a) Name the various forces present in a fluid flow.   **(02 Marks)**

(b) Derive the Euler's equation of motion of ideal fluids and hence deduce Bernoulli's equation of motion. Mention the assumptions made. **(10 Marks)**

(c) The water is flowing through a taper pipe of length 100 m having diameters 600 mm at the upper end and 300 mm at the lower end, at the rate of 50 litres/s. The pipe has a slope of 1 in 30. Find the pressure at the lower end if the pressure at the higher level is 19.62 N/cm^2. **(08 Marks)**

6. (a) Derive the Darey—Weisbach equation for the loss of head due to friction in a pipe. **(10 Marks)**

(b) Define the terms: Hydraulic gradient and Total energy line. **(04 Marks)**

(c) A horizontal venturimeter with inlet diameter 20 cm and throat diameter 10 cm is used to measure the flow of water. The pressure at inlet is 17.658 N/cm^2 and the vacuum pressure at the throat is 30 cm of mercury. Find the discharge of water through venturimeter. Take $C_d = 0.98$. **(06 Marks)**

7. (a) Define Reynold's number. What is its significance? **(04 Marks)**

(b) Prove that the velocity distribution for a viscous flow between two parallel plates, when both plates are fixed across a section is parabolic in nature. **(10 Marks)**

(c) An oil of viscosity 0.1 Ns/m^2 and relative density 0.9 is flowing through a circular pipe of diameter 50 mm and of length 300 m. The rate of flow of fluid through the pipe is 3.5 litres/s. Find the pressure drop in a length of 300 m pipe. **(06 Marks)**

8. (a) Derive an expression for displacement thickness of a flow over a plate. **(08 Marks)**

(b) Define Mach number, Mach angle and Mach cone. **(06 Marks)**

(c) A flat plate 1.5 m × 1.5 m moves at 50 km/hr in stationary air of density 1.15 kg/m^3. If the coefficient of drag and lift are 0.15 and 0.75 respectively, determine.
(i) Lift force   (ii) Drag force   (iii) Resultant force. **(06 Marks)**

1. An orifice meter with orifice dia 10 cm is inserted in a pipe of 20 cm dia. The p_r gauges fitted upstream and downstream of orifice meter given readings of 19.62 N/cm^2 and 9.81 N/cm^2 respectively, $c_d = 0.6$, find discharge of water through pipe.

dia of orifice, $d_0 = 10$ cm

area, $a_0 = \dfrac{\pi}{4}(10)^2 = 78.54$ cm^2

dia of pipe, $d_1 = 20$ cm

area, $a_1 = \dfrac{\pi}{4}(20)^2 = 314.16$ cm^2

$$P_1 = 19.62 \text{ N/cm}^2 = 19.62 \times 10^4 \text{ N/m}^2$$

$$\frac{P_1}{W} = \frac{19.62 \times 10^4}{9810} = 20 \text{ m of water,}$$

$$\frac{P_2}{W} = \frac{9.81 \times 10^4}{9810} = 10 \text{ m of water.}$$

$$h = \frac{P_1}{W} - \frac{P_2}{W} = (20.0 - 10.0) = 10 \text{ m of water} = 1000 \text{ cm.}$$

$$= 1000 \text{ cm of water.}$$

$$c_d = 0.6$$

$$Q = C_d \frac{a_0 a_1}{\sqrt{a_1^2 - a_0^2}} \times \sqrt{2gh}$$

$$= 68213.28 \text{ cm}^3/\text{s.}$$

Water is flowing through a pipe of 5 cm dia under a p_r of 29.43 N/cm² (gauge) and with mean velocity of 2.0 m/s. Find the total head or total energy per unit weight of the water at a cross section, which is 5 m above the datum line.

$$\text{dia of pipe } = 5 \text{ cm} = 0.5 \text{ m}$$

$$p_r, p = 29.43 \text{ N/cm}^2 = 29.43 \times 10^4 \text{ N/m}^2$$

$$V = 2 \text{ m/s}, Z = 5 \text{ m.}$$

$$pr \text{ lead} = \frac{P}{\rho g} = \frac{29.43 \times 10^4}{9810} = 30 \text{ m}$$

$$\text{Kinetic head} = \frac{V^2}{2g} = \frac{2 \times 2}{2 \times 9.81} = 0.204 \text{ m}$$

$$\therefore \quad \text{Total head} = \frac{P}{\rho g} + \frac{V^2}{2g} + z = 30 + 0.204 + 5$$

$$= 35.204 \text{ m.}$$

B. Equation for Real Fluid

$$\frac{P_1}{\rho g} + \frac{V_1^2}{2g} + Z_1 = \frac{P_2}{\rho g} + \frac{V_2^2}{2g} + Z_2 + h_L$$

h_L = cross of energy from (1) to (2)

A pipe of dia 40 cm caries water at a velocity of 25 m/sec. The p_r at points A and B are given as 3 kgf/cm² and 2.3 kgf/cm², respectively while the datum lead at A and B are 28 m and 30 m. Find the less of lead between A and B.

Given

$$\text{dia of pipe,} \qquad D = 40 \text{ cm} = 0.4 \text{ m.}$$

$$\overset{*}{\forall} = 25 \text{ m/sec}$$

$$\text{at point } A, \qquad P_A = 3 \times 10^4 \text{ kgf/m}^2$$

$$Z_A = 28 \text{ m}$$

$$V_A = V = 25 \text{ m/sec}$$

$$\therefore \quad \text{Total energy } E_A = \frac{PA}{\rho g} + \frac{V_A^2}{2g} + Z_A$$

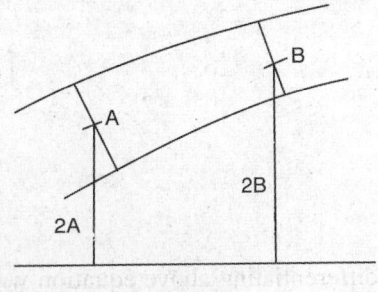

Fig. 1

$$= \frac{3 \times 10^4}{1000} + \frac{25^2}{2 \times 9.81} + 28$$

$$= 30 + 31.85 + 28 = 89.85 \text{ m}$$

At point B, $P_B = 2.3$ Kgf/cm^2 = 2.3×10^4 kgf/cm^2

$$Z_B = 30 \text{ m.}$$

$$V_B = V = V_A = 25 \text{ m/sec.}$$

$\therefore$ Total Energy at B,

$$E_B = \frac{P_B}{W} + \frac{V_B^2}{2g} + 2B$$

$$= \frac{2.3 \times 10^4}{1000} + \frac{25^2}{2 \times 9.81} + 30$$

$$= 23 + 31.85 + 30 = 84.85 \text{ m}$$

$\therefore$ Loss of Energy $= E_A - E_B$

$$= 89.85 - 84.84 = 5 \text{ m.}$$

If for a 2D potential flow, velocity potential is given by $\phi = x(2y - 1)$ determine the velocity at point (4, 5), Determine also the value of stream function ψ at point p.

$$\phi = x(2y - 1).$$

$$u = \frac{\partial \phi}{\partial x} = 2y - 1$$

$$v = -\frac{\partial \phi}{\partial x} = 2x$$

$$u = -1 + 2 \times 5 = +9$$

$$v = 2 \times 4 = 8$$

$$V = \sqrt{9^2 + 8^2} = 12.00 \text{ unit/sec}$$

$$\frac{\partial \psi}{\partial y} = -4 = 1 - 2y$$

$$\frac{\partial \psi}{\partial x} = V = -2x,$$

$$\int d\psi = \int (2y - 1) \, dy$$

$$\psi = \frac{2y^2}{2} - y + \text{constant of integer}$$

$$\psi = y^2 - y + k$$

differentiating above equation w.r.f. to n.

$$\frac{\partial \psi}{\partial x} = \frac{\partial k}{\partial x}$$

But

$$\frac{\partial \psi}{\partial x} = -2x$$

$$\frac{\partial k}{\partial x} = -2x$$

$$K = -x^2$$

$$\psi = y^2 - y - x^2$$

$$\psi\,(4,\ 5)\ \psi = 4\ \text{m}$$

In a 100 mm dia horizontal pipe a venturi meter of 0.5 contraction ratio has been fixed. The head of water on the metre when there is no flow in 3 m (gauge). Find the rate of flow for which the throat p_r will be 2 m of water absolute. The co-efficient of meter is 0.97. Take atmospheric p_r lead = 10.5 m of H_2O.

Ans.

$$\text{dia of pipe, } d_1 = 100\ \text{mm} = 10\ \text{cm.}$$

$$\text{Area } a_1 = \frac{\pi}{4} d_1^2 = \frac{\pi}{4}(10)^2 = 78.54\ \text{cm}^2$$

$$\text{dia at throat, } d_2 = 0.5\ d_1 = 0.5 \times 10 = 5\ \text{cm}$$

$$\text{Area } a_1 = \frac{\pi}{4}(5)^2 = 19.635\ \text{cm}^2$$

$$\text{Head of water for no flow,} = \frac{P_1}{W} = 3\ \text{m(gauge)} = 3 + 10.3 = 13.3\ \text{m (abs)}$$

$$\text{Throat } pr, = \frac{P_2}{W} = 2\ \text{m of water abs, difference of } p_r \text{ lead, } h = \frac{P_1}{W} - \frac{P_2}{W}$$

$$= 13.3 - 2.0 = 11.3\ \text{m} = 1130\ \text{cm}$$

$$Q = C_d \frac{a_1 a_2}{\sqrt{a_1^2 - a_2^2}} \times \sqrt{2gh}$$

$$= 29.306\ \text{litres/sec.}$$

Solution.

Dia at inlet $d_1 = 20$ cm

$$a_1 = \frac{\pi}{4} \times (20)^2 = 314.16\ \text{cm}^2$$

Dia at throat,

$$d_2 = 10\ \text{cm, } a_2 = \frac{\pi}{4}(10)^2 = 78.74\ \text{cm}^2$$

$$P_l = 17.658 \ \text{N/cm}^2 = 17.658 \times 10^4 \ \text{N/m}^2$$

w for water, $9.81 \times 1000 \dfrac{\text{N}}{\text{m}^2}$.

$$\frac{P_1}{w} = \frac{17.658 \times 10^4}{9.81 \times 1000} = 18 \ \text{m of water}$$

$$\frac{P_2}{w} = -30 \ \text{cm of mercury}$$

$$= -0.30 \ \text{m of mercury}$$

$$= -0.30 \times 13.6$$

$$= -4.08 \ \text{m of water}$$

Differential head $= h = \dfrac{P_1}{w} - \dfrac{P_2}{w}$

$$= 18 - (-4.08)$$

$$= 22.08 \ \text{m of water}$$

$$= 2208 \ \text{cm of water}$$

discharge $Q = C_d \dfrac{a_1 a_2}{\sqrt{a_1^2 - a_2^2}} \times \sqrt{2gh}$

$$= 0.98 \times \frac{314.16 \times 7874}{\sqrt{(314.16)^2 - (78.74)^2}} \times \sqrt{2 \times 981 \times 2208} \ \text{cm}^2/\text{s}$$

$$= 165.5 \ \text{Lit/s}$$

Fluid Mechanics
July/August 2008
Semester - IV (ME/IP/IM/MA/AU)

Time: Three Hours   Maximum Marks: 100

Note: Answer any five questions selecting at least two questions from each part.

Part A

Q.1 (a) Define the following fluid properties and state their units. (6)

    (i) Specific weight   (ii) Kinematic viscosity

    (iii) Surface tension   (iv) Vapour pressure.

Ans.: (i) Specific weight:

Definition: It is defined as the ratio of weight per unit volume of fluid.

$$\textbf{Unit:} \quad \text{Specific weight} = \frac{\text{Weight}}{\text{Volume}}$$

It's unit is N/m^3.

(ii) Kinematic viscosity:

Definition: It is defined as the ratio of dynamic viscosity to the mass density.

$$\textbf{Unit:} \quad \text{Kinematic viscosity} = \frac{\text{Dynamic viscosity}}{\text{Mass density}}$$

It's unit is m^2/s.

(iii) Surface tension: Please refer Q.1 (a) (i) of July/Aug.-2005.

(iv) Vapour pressure: Please refer Q.1 (a) (iii) of July/Aug.-2005.
Unit of vapour pressure is N/m^2.

(b) Define capillarity, obtain an expression for capillary rise of a liquid. (6)

Ans.: Please refer Q.1 (b) of Jan./Feb.-2005.

(c) An oil film of thickness 1.5 mm is used for lubrication between a square plate of size 0.9×0.9 m and an inclined plane having an angle of inclination 20°. The weight of the square plate is 392.4 N and it slides down the plate with a uniform velocity of 0.2 m/s. Find the dynamic viscosity of the oil. (8)

Ans.: Data:

$u = 0.2$ m/s

$y = 1.5$ mm

$W = 392.4$ N

$A = 0.9 \times 0.9 = 0.81$ m^2

$\therefore \quad \tau = \dfrac{W\sin 20}{A} = \dfrac{392.4\sin 20}{0.81}$

$\therefore \quad \tau = 165.68 \ \text{N}/\text{m}^2$

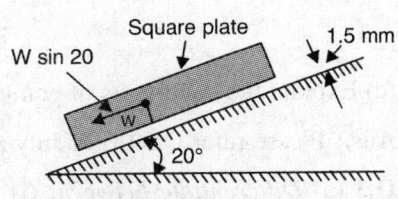

Fig. 1

$$\tau = \frac{\mu \times u}{y}$$

$$\therefore \qquad 165.68 = \frac{\mu \times 0.2}{1.5 \times 10^{-3}}$$

$$\therefore \qquad \mu = 1.242 \frac{Ns}{m^2} \text{ Dynamic viscosity of oil.}$$

Q.2 (a) State and prove the Pascal's law. (7)

Ans.: Please refer Q.2 (a) of July/Aug.-2005.

(b) A caisson for closing the entrance to a dry dock is of trapezoidal form 16 m wide at the top and 10 m wide at the bottom and 6 m deep. Find the total pressure and centre of pressure on the caisson if the water on the outside is just with the top and dock is empty. (8)

Ans.:

$\therefore$ Total pressure

$$= rA\bar{x}$$

$$= 9810 \times \left[\frac{1}{2} \times (10 + 16) \times 6 \right] \times 2.25$$

Fig. 2

$$= 1721655 \text{ N}$$

$$P = 1721.655 \text{ kN}$$

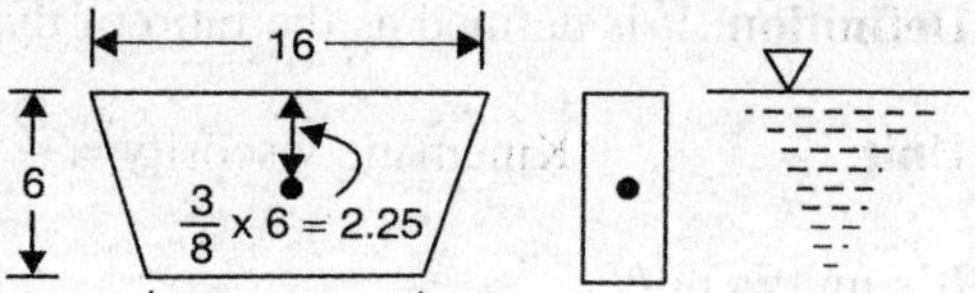

Depth of centre of pressure,

$$\bar{h} = \bar{x} + \frac{I_G}{A\bar{x}}$$

$$= 2.25 + \frac{\frac{13 \times 6^3}{12}}{78 \times 2.25}$$

$$\therefore \qquad \bar{h} = 3.583 \text{ m}$$

(c) Explain the conditions of equilibrium for a floating body with neat sketches. (5)

Ans.: Please refer Q.5 (a) of July/Aug.-2004.

Q.3 (a) *Differentiate between:* (i) Steady and un-steady flow (ii) Uniform and non-uniform flow (iii) Laminar and turbulent flow. (6)

Ans.: Steady and un-steady flow: (i) Please refer Q.4 (c) (i) of Jan./Feb.-2005.

(ii) Uniform and non-uniform flow:

Sr. No.	Uniform flow	Non-uniform flow
1.	If velocity of flow at a given instant is same in magnitude and direction at different points in the flowing fluid, then it is called as uniform flow.	If the velocity of flow, at a given instant, changes from point to point, the flow is called as non-uniform flow.
2.	Mathematically $\frac{\partial v}{\partial s} = 0$	Mathematically $\frac{\partial v}{\partial s} \neq 0$
	Example: Flow through a long straight pipe of uniform diameter.	Example: Flow through a long pipe with varying cross-section.

(iii) Laminar and turbulent flow: Please refer Q.7 (a) of July/Aug. 2005.

(b) Obtain an expression for continuity equation for a three-dimensional flow. (8)

Ans.: Please refer Q.4 (b) of July/Aug.-2005 OR Q. 4(a) of July/Aug. 2006.

(c) A stream function is given by $\psi = 5x - 6y$. **Calculate the velocity components and also magnitude and direction of the resultant velocity at any point.** (6)

Ans.: Stream function, $\psi = 5x - 6y$

∴ Velocity components,

$$u = -\frac{\partial x}{\partial y}$$

∴

$$= -\left[\frac{\partial}{\partial y}(5x - 6y)\right]$$

$$= -(-6)$$

∴

$$u = 6 \text{ m/s}.$$

$$V = \frac{r\psi}{rx}$$

$$= \frac{r}{\partial x}[5x - 6y]$$

∴

$$V = 5 \text{ m/s}$$

∴ Magnitude of resultant velocity at any point,

$$R = \sqrt{u^2 + v^2}$$

$$= \sqrt{6^2 + 5^2}$$

∴

$$R = 7.81 \text{ m/s}$$

Direction of resultant velocity at any point,

$$\therefore \qquad \theta = \tan^{-1}\left(\frac{v}{u}\right)$$

$$\therefore \qquad \theta = \tan^{-1}\left(\frac{5}{6}\right)$$

$$\therefore \qquad \theta = 39.8°$$

Q.4 (a) Explain the following terms in brief: (i) Dimensionally homogeneous equation (ii) Kinematic similarity. (4)

Ans. (i) **Dimensionally homogeneous equation:** An equation is said to be dimensionally homogeneous when the dimensions of left hand side (LHS) are same as dimensions of right hand side (RHS).

Let us check up dimensional homogeneity of following equation,

$$F = ma$$

Dimension of LHS $= N = kg.m/s^2 = [MLT^{-2}]$

$\therefore$ Dimension of RHS $= kg.m/s^2 = [MLT^{-2}]$

$\therefore$ Dimensions of LHS = Dimensions of RHS.

(ii) **Kinematic similarity:** If the ratios of velocities of accelerations at all the corresponding points in the model and prototype are same, the model is said to be kinematically similar to prototype.

$$\therefore \qquad \text{Velocity ratio} = V_r = \frac{V_m}{V_p} = \frac{L_m/T_m}{L_p/T_p}$$

$$\text{Acceleration ratio} = a_r = \frac{a_m}{a_p} = \frac{\dfrac{V_m}{T_m}}{\dfrac{V_p}{T_p}}$$

(b) State Buckingham's π theorem. The efficiency η of a fan depends on density ρ, dynamic viscosity μ of the fluid, angular velocity ω, diameter D of the rotor and the discharge Q. Express η in terms of dimensionless parameters. (10)

Ans.: Buckingham's π theorem: Please refer Q.5 (a) of July/Aug.-2005. η is a function of ρ, μ, ω, D, Q.

$$\therefore \qquad \eta = f(\rho, \mu, \omega, D, Q)$$

or $$f_1(\eta, \rho, \mu, \omega, D, Q) = 0$$

Total variables $n = 6$.

Writing dimensions of each variables.

$$\eta = M^0 L^0 T^0$$

$$\rho = ML^{-3}$$

$$\mu = ML^{-1}T^{-1}$$
$$\omega = T^{-1}$$
$$D = L$$
$$Q = L^3T^{-1}$$

Fundamental dimensions $m = 3$

$\therefore$ Number of 'π' terms $= n - m = 6 - 3 = 3$

$\therefore$ $f_1(\pi_1, \pi_2, \pi_3) = 0$

Each π term contains $m + 1$ variables where $m = 3$ and is also equal to repeating variables. Choosing ρ, ω, Q as repeating variables.

$\therefore$
$$\pi_1 = \rho^{a_1}\omega^{b_1}Q^{c_1} \cdot \eta$$
$$\pi_2 = \rho^{a_2}\omega^{b_2}Q^{c_2} \cdot \mu$$
$$\pi_3 = \rho^{a_3}\omega^{b_3}Q^{c_3} \cdot D$$

First π term:
$$\pi_1 = \rho^{a_1}\omega^{b_1}Q^{c_1} \cdot \eta$$

Substituting dimensions on both sides.

$$M^0L^0T^0 = (ML^{-3})^{a_1}(T^{-1})^{b_1}(L^3T^{-1})^{c_1} M^0L^0T^0$$

$\therefore$ Equating the powers of M, L, T on both sides,

Power of M,
$$0 = a_1 + 0 \qquad\qquad \therefore a_1 = 0$$

Power of L,
$$0 = -3a_1 + 3c_1 \qquad\qquad \therefore c_1 = 0$$
$$0 = 3c_1 \qquad\qquad \therefore b_1 = 0$$

Power of T,
$$0 = -b_1 - c_1$$

$\therefore$
$$\pi_1 = \eta$$

Second π term:
$$\pi_2 = \rho^{a_2}\omega^{b_2}Q^{c_2} \cdot \mu$$

$\therefore$
$$M^0L^0T^0 = (ML^{-3})^{a_2}(T^{-1})^{b_2}(L^3T^{-1})^{c_2} ML^{-1}T^{-1}$$

$\therefore$
$$0 = a_2 + 1 \therefore a_2 = -1$$

$\therefore$
$$0 = -3a_2 + 3c_2 - 1$$

$\therefore$
$$0 = -3 + 3c_2 - 1$$

$\therefore$
$$3c_2 = -2 \qquad\qquad \therefore c_2 = -2/3$$

$\therefore$
$$0 = -b_2 - c_2 - 1$$

$$\therefore \qquad 0 = -b_2 + \frac{2}{3} - 1 \qquad\qquad \therefore b_2 = -\frac{1}{3}$$

$$\therefore \qquad \pi_2 = \rho^{-1}\omega^{-1/3}Q^{-2/3}\mu = \frac{\mu}{\rho\omega^{1/3}Q^{2/3}}$$

Third π term:

$$\therefore \qquad \pi_3 = \rho^{a_3}\omega^{b_3}Q^{b_3}D$$

$$\therefore \qquad M^0L^0T^0 = \left(ML^{-3}\right)^{a_3}\left(T^{-1}\right)^{b_3}\left(L^3T^{-1}\right)^{c_3}(L)$$

$$\therefore \qquad 0 = a_3$$

$$\therefore \qquad 0 = -3a_3 - 3c_3 + 1$$

$$\therefore \qquad c_3 = \frac{-1}{3}$$

$$\therefore \qquad 0 = -b_3 - c_3$$

$$\therefore \qquad b_3 = 1/3$$

$$\therefore \qquad \pi_3 = \rho^0\omega^{1/3}Q^{-1/3}D$$

$$\therefore \qquad f_1\left(\pi_1, \pi_2, \pi_3\right) = 0$$

$$\therefore \qquad f_1\left(\eta_1\frac{\mu}{\rho\omega^{1/3}Q^{2/3}}, \frac{\omega^{1/3}D}{Q^{1/3}}\right) = 0$$

$$\therefore \qquad \eta = \frac{\mu D}{\rho Q}$$

(c) Define the following dimensionless numbers and state their significance:

    (i) Froude's number   (ii) Euler's number   (iii) Mach's number.   (6)

Ans.: Please refer Q.6 (a) of July/Aug.-2006.

Significance:

  (i) **Froude's number:** In the study of free surface flows. *e.g.* open channel flow, which are dominated by gravity forces.

  (ii) **Euler's number:** is important where pressure forces are predominate. *e.g.* in cavitation studies.

 (iii) **Mach's number:** It is important in aerodynamic and in phenomenon like water hammer which are concerned with compressible fluids.

Part B

Q.5 (a) Name the various forces present in a fluid flow.   (2)

Ans.: Various forces present in fluid flow:

    1. Gravity force   2. Pressure force

3. Viscous force   4. Turbulent force

5. Surface tension force   6. Elastic force.

(b) Derive the Euler's equation of motion for ideal fluids and hence deduce Bernoulli's equation of motion. Mention the assumptions made.   (10)

Ans.: Please refer Q.4 (b) of Jan./Feb.-2007.

(c) The water is flowing through a taper pipe of length 100 m having diameters 600 mm at the upper end and 300 mm at the lower end, at the rate of 50 litre/s. The pipe has a slope of 1 in 30. Find the pressure at the lower end if the pressure at the higher level is 19.62 N/cm².

Ans.:

Applying Bernoulli's equation at 1 and 2.

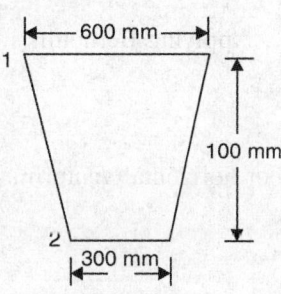

Fig. 3

$$\frac{P_1}{r} + \frac{V_1^2}{2g} + z_1 = \frac{P_2}{r} + \frac{V_2^2}{2g} + z_2$$

$$\therefore \quad \frac{19.62 \times 10^4}{9810} + \frac{\left(\dfrac{50 \times 10^{-3}}{\dfrac{\pi}{4} \times 0.6^2}\right)^2}{2 \times 9.81} + 100$$

$$= \frac{P_2}{9810} + \frac{\left(\dfrac{50 \times 10^{-3}}{\dfrac{\pi}{4} \times 0.3^2}\right)^2}{2 \times 9.81} + 0$$

$$\therefore \quad 20 + 1.593 \times 10^{-3} + 100 = \frac{P_2}{9810} + 0.0255$$

$$\therefore \quad P_2 = 117.7 \times 10^4 \, \text{N/m}^2$$

This is the pressure at the lower end.

Q.6 (a) Derive the Darey-Weisbach equation for the loss of head due to friction in a pipe.   (10)

Ans.: Please refer Q.6 (b) of Jan./Feb.-2005.

(b) Define the terms: Hydraulic gradient and total energy line.   (4)

Ans.: Please refer Q.6 (c) of Jan./Feb.-2005 and Q.7(a) of Jan./Feb.-2006.

(c) A horizontal venturimeter with Inlet diameter 20 cm and throat diameter 10 cm is used to measure the flow of water. The pressure at inlet is 17.658 N/cm² and the vacuum pressure at the throat is 30 cm of mercury. Find the discharge of water through venturimeter. Take $C_d = 0.98$.   (6)

Ans.:

Given

$$d_1 = 20 \text{ cm} = 0.2 \text{ m}$$

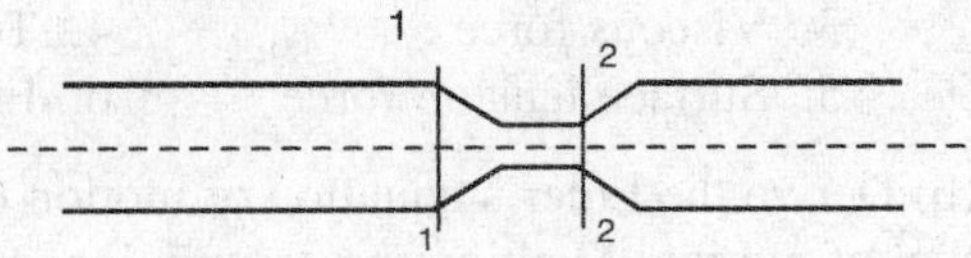

$$d_2 = 10 \text{ cm} = 0.1 \text{ m}$$

$$P_1 = 17.658 \times 10^4 \text{ N/m}^2$$

$P_2 = 30$ cm of mercury

$= 30 \times 13.6 = 408$ cm of water

Fig. 4

$$= \frac{408}{100} \times 9810$$

$$= 40024.8 \text{ N/m}^2.$$

∴ Applying Bernoulli's theorem,

$$\frac{P_1}{r} + \frac{V_1^2}{2g} + z_1 = \frac{P_2}{r} + \frac{V_2^2}{2g} + z_2$$

For horizontal venturimeter,

$$z_1 = z_2$$

∴ $$\frac{17.658 \times 10^4}{9810} + \frac{\left(\dfrac{Q}{\dfrac{\pi}{4} \times 0.2^2}\right)}{2 \times 9.81} = \frac{40024.8}{9810} + \frac{\left(\dfrac{Q}{\dfrac{\pi}{4} \times 0.1^2}\right)^2}{2 \times 9.81}$$

∴ $$13.92 + 51.64Q^2 = 826.26Q^2$$

∴ $$13.92 = 774.62Q^2$$

∴ $$Q = 0.134 \text{ m}^3/\text{s}$$

Actual discharge $= C_d \cdot Q = 0.98 \times 0.134 = 0.13132 \text{ m}^3/\text{s}$

Q.7 (a) Define Reynold's number. What is its significance?   (4)
Ans.: Please refer Q.7 (b) of Jan./Feb.-2005.

(b) Prove that the velocity distribution for a viscous flow between two parallel plates, when both plates are fixed across a section is parabolic in nature.   (10)
Ans.: Please refer Q.7 (b) of July/Aug.-2006.

(c) An oil of viscosity 0.1 Ns/m² and relative density 0.9 is flowing through a circular pipe of diameter 50 mm and of length 300 m. The rate of flow of fluid through the pipe is 3.5 litres/s. Find the pressure drop in a length of 300 m pipe.   (6)
Ans.:

Discharge, $$Q = \frac{\pi}{4} D^2 u_{av}$$

∴ $$3.5 \times 10^{-3} = \frac{\pi}{4} \times (50 \times 10^{-3}) \times u_{av}$$

$$\therefore \qquad u_{av} = 1.782 \text{ m/s}$$

$\therefore$ Average velocity,

$$u_{av} = \frac{R^2}{8\mu}\left(-\frac{\partial P}{\partial x}\right)$$

$$1.782 = \frac{\left(25\times10^{-3}\right)^2}{8\times0.1}\left(-\frac{\partial P}{\partial x}\right)$$

$$\therefore \qquad \left(-\frac{\partial P}{\partial x}\right) = 2280.96$$

$$\therefore \qquad (-\partial P) = 2280.96 \times 300$$

$$\therefore \qquad (-\partial P) = 684288 \text{ N/m}^2$$

$\therefore$ Pressure drop in a length of pipe 300 m pipe is 684288 N/m².

Q.8 (a) Derive an expression for displacement thickness of a flow over a plate (8)
Ans.: Please refer Q.8 (b) of July/Aug.-2006.

(b) Define Mach number, Mach angle and Mach cone. (6)
Ans.: Mach number: Please refer Q.7 (c) of July/Aug.-2005.
Mach angle: Please refer Q.8 (b) of Jan./Feb.-2006.
Mach cone: Please refer Q.8 (b) of Jan./Feb.-2006.

(c) A flat plate 1.5 *m* × 1.5 m moves at 50 km/hr in stationary air of density 1.15 kg/m³. If the coefficient of drag and lift and 0.15 and 0.75 respectively, determine.
    (i) Lift force   (ii) Drag force   (iii) Resultant force. (6)
Ans.: Area of plate = 1.5 × 1.5 = 2.25 *m*²
Velocity of plate, $V = 50$ km/hr

$$V = 50 \times \frac{5}{18}$$

$$\therefore \qquad V = 13.88 \text{ m/s}$$

$$\rho = 1.15 \text{ kg/m}^3$$

$$C_D = 0.15$$

$$C_L = 0.75$$

(i) Lift force, $$F_L = C_L A \frac{\rho v^2}{2}$$

$$= \frac{0.75 \times 2.25 \times 1.15 \times 13.88^2}{2}$$

$$\therefore \qquad F_L = 186.93 \text{ N}$$

(ii) Drag force,

$$F_D = C_D A \frac{\rho v^2}{2}$$

$$= \frac{0.15 \times 2.25 \times 1.15 \times 13.88^2}{2}$$

$\therefore \qquad F_D = 37.38 \text{ N}$

(iii) Resultant force,

$$F_R = \sqrt{F_L^2 + F_D^2}$$

$$= \sqrt{186.93^2 + 37.38^2}$$

$\therefore \qquad F_R = 190.63 \text{ N}.$

Fluid Mechanics
(Jan./Feb. 2007)
Semester - IV (ME/IP/AU/IM MA)

Time: 3 Hours   Maximum Marks: 100

Note: (1) Answer any five full questions.
(2) Draw neat sketches wherever necessary.

Q.1. (a) Define compressibility and derive an expression bulk modulus of elasticity for a perfect gas undergoing isentropic process.

Ans.: Compressibility: It is a property of a fluid. Consider a small element of fluid of volume v. The pressure exerted on the element by the neighbouring fluid is P. If the pressure is now increased by an amount dP, the volume of the element will correspondingly be reduced by the amount dv. The compressibility of fluid 'K', i.e. thus defined as

$$K = -\frac{1}{v}\frac{dv}{dP}$$

Liquids have very low value of compressibility, while gases have very high compressibility. Compressibility in terms of density is

$$K = \frac{1}{\rho}\frac{d\rho}{dP}$$

We know that for compressible fluid velocity of sound.

$$C = \sqrt{\frac{dP}{d\rho}} \qquad\qquad\qquad\qquad \text{...(i)}$$

If V_s = Original specific volume

$$\text{Bulk modules} = E = \frac{\text{Increase of pressure}}{\text{Volumetric stress}}$$

$$\therefore \qquad E = \frac{dP}{-\left(dv/V_s\right)} = -V_s\frac{dP}{dv}$$

$$\therefore \qquad \frac{dP}{dv} = -\frac{E}{V_s} \qquad\qquad\qquad\qquad \text{...(ii)}$$

But $V_s = \frac{1}{\omega} = \frac{1}{\rho g}$

$$\therefore \qquad \frac{dP}{dv} = -E\rho \cdot g$$

Since $\rho V_s = \text{constant}$

$$V_s \cdot d\rho + \rho dv = 0$$

$$\therefore \qquad \frac{V_s}{dv} = -\frac{\rho}{d\rho} \qquad \qquad \dots \text{(iii)}$$

From equations (ii) and (iii)

$$E = \rho \cdot \frac{dP}{d\rho} \qquad \qquad \dots \text{(iv)}$$

And from equations (i) and (iv)

$$C = \sqrt{\frac{dP}{d\rho}} = \sqrt{\frac{E}{\rho}}$$

For a perfect gas undergoing isentropic process

$$PV_s^K = \text{constant}$$

Differentiating

$$\therefore \qquad PKV_s^{K-1}dv + V_s^K dP = 0$$

$$K \cdot PV_s^{K-1}dv = -V_s^{K-1}dP$$

$$\therefore \qquad \frac{dP}{dv} = -\frac{KP}{V_s}$$

But

$$\frac{dP}{dv} = -E\rho \cdot g = -\frac{E}{V_s}$$

$$\therefore \qquad \frac{dP}{dv} = -\frac{E}{V_s} = -\frac{KP}{V_s}$$

$$\therefore \qquad E = KP \text{ is required expression.}$$

(b) Define surface tension and show that the gauge pressure within a liquid droplet varies inversely with the diameter of the droplet. (6)

Ans.: Please refer Q.1 (a) of July/Aug.-2005 and Q.1 (c) of Jan./Feb.-2006.

(c) A shaft of 0.1 m diameter rotates at 60 rpm in a 0.2 m long bearing. Taking that the two surfaces are uniformly separated by a distance of 0.5 mm and taking linear velocity distribution in a lubricating oil having dynamic viscosity of 4 CP, find the power absorbed in the bearing. (8)

Ans.: Given $\quad d = 0.1$ m. $\qquad N = 60$ rpm.

$$L = 0.2 \text{ m.}$$

Distance between shaft and bearing = 00.5 mm.

$$= 0.0005 \text{ m.}$$

$$\therefore \qquad D = 0.1 + 2(0.0005) = 0.101 \text{ m.}$$

Viscosity of oil $\mu = \Delta CP = 4 \times 10^{-2}$ N.s/m^2

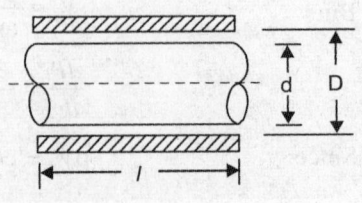

Fig. 1

Tangential velocity $V = \dfrac{\pi d N}{60} = \dfrac{\pi \times 0.1 \times 60}{60}$

$$= 0.314 \text{ m/s}.$$

Shear stress $\tau = \mu \cdot \dfrac{du}{dy}$

$$du = V = 0.314 \text{ m/s}$$

$$dy = \left(\dfrac{D-d}{2}\right) = 0.0005 \text{ m}$$

$$\tau = \dfrac{0.314}{0.0005} \times 4 \times 10^{-2}$$

$$= 25.12 \text{ N/m}^2$$

Shear force $F = \tau \times A = t \times \pi d \times L$

$$= 25.12 \times \pi \times 0.1 \times 0.2$$

$$= 1.5733 \text{ N}$$

Torque $T = F \times \dfrac{d}{2} = 1.5783 \times \dfrac{0.1}{2}$

$$= 0.0739 \text{ N.m.}$$

$$P = \dfrac{2\pi N T}{60} = \dfrac{2\pi \times 60 \times 0.0789}{60}$$

$$= 0.4957 \text{ watts}.$$

Power obsorbed in bearing is 0.4957 watts.

Q.2 (a) Sketch and explain hydrostatic paradox. (4)

Ans.: Hydrostatic paradox

Three containers are filled with same liquid upto the same height H and have the same area A at the bottom.

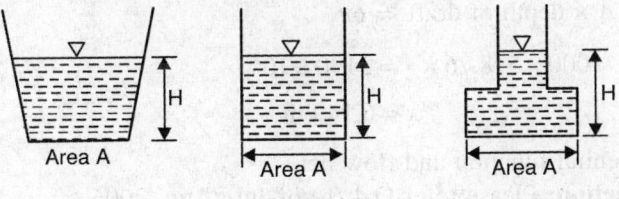

Fig. 2

From hydrostatic equation $P = \gamma h$.

The pressure density depends only upon the height of column and not upon the size of column. Accordingly, in these containers of different shapes and sizes the same unit pressure would be exerted against the bottom of the containers since each container has same area A at the bottom. Pressure force $F = P$. A on the bottom of each container would be same. Even though the weight of the fluid in each container is different. This situation is referred as Hydrostatic paradox.

(b) Define metacentre and derive an expression for a floating body for its metacentric height. (8)

Ans.: Please refer Q. 2(a) (ii) of Jan./Feb. 2006.

(c) A cargo ship weighing 4000 tonnes has a draft of 7 m in seawater (Sp. gr. 1.035). After discharging cargo of 510 tonnes its draft reduces by 0.5 m. What will be its draft in a fresh water harbour after further discharging a cargo of 300 tonnes? Assume no change in cross sectional area for depth under consideration. (8)

Ans.: Weight of cargo ship = 4000 tonnes = 4000 × 1000 kg

$$\text{Sp. gr.} = 1.035 \qquad \rho = 1035 \text{ kg/m}^3$$

$$\text{Upward thrust} = \text{Total weight of cargo}$$

$$\rho \times A \times \text{depth of draft} = \text{wt}$$

$$1035 \times A \times 7 = 4000 \times 10^3$$

∴ Area of cargo immersed

$$A = \frac{4000 \times 10^3}{1035 \times 7} = 552.104 \text{ m}^2$$

After discharging 510 tonnes, New draft is 6.5 m

∴ Area of cargo immersed

$$A = \frac{(4000 - 510) \times 10^3}{1035 \times 6.5}$$

Let

$$\text{Draft in fresh water} = x \text{ m}$$

$$\text{Weight of cargo} = 4000 - 510 - 300$$

$$= 3190 \text{ Tonnes}$$

$$= 3190 \times 10^3 \text{ kg}$$

$$\text{Density of fresh water} = 1000 \text{ kg/m}^3$$

∴

$$\text{Upward thrust} = \text{total weight of cargo}$$

$$\rho \times A \times \text{depth of draft} = \omega t$$

$$1000 \times 518.76 \times x = 3190 \times 10^3$$

$$x = 6.149 \text{ m}$$

Q.3 (a) Explain potential function and flow net. (6)

Ans.: Potential function: Please refer Q.4 (b) of July/Aug.-2006.

Flow net: A flow net is a family of equipotential lines and streamlines with the constants varying in arithmetic progression.

Interval between the adjacent equipotential lines and streamlines is same, ΔC.

Let the distance between two adjacent streamlines be Δn and that between two adjacent equipotential lines be ΔS.

Velocity of flow $V_C = \dfrac{\Delta C}{\Delta S}$ & $\dfrac{\Delta C}{\Delta n}$

The expression for the velocity in terms of ΔC, ΔS and Δn are approximate when the value of ΔC is finite. But when ΔC tends to zero the expression becomes exact as the velocity of flow is same.

$$\frac{\Delta C}{\Delta S} = \frac{\Delta C}{\Delta n} \text{ gives } \Delta S = \Delta n$$

∴ Flow net consists of orthogonal grid and it becomes perfect square as grid size approaches zero.

Flow net is only possible when

(i) The flow is possible

(ii) The flow is irrotational

(iii) The flow is two dimensional.

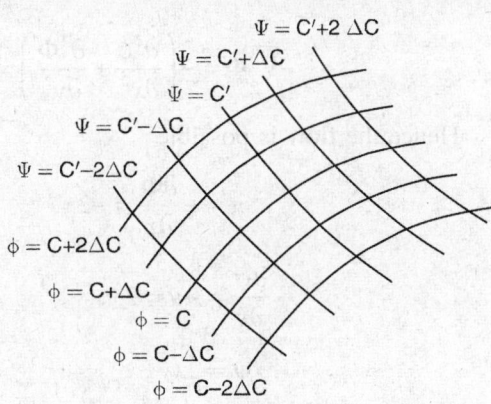

Fig. 3

(b) The velocity in a flow field is given by $u = 3$ m/s, $v = 6$ m/s. Determine the equation of the streamline passing through the origin and the one passing through a point (2 m, 3 m). ...(6)

Ans.: Given $u = 3$ m/s $v = 6$ m/s.

$$\frac{du}{dx} = \frac{v}{u} = \frac{6}{3} = 2$$

∴ Equation of streamline

$$u\,dy - v\,dx = 0$$

$$3\,dy - 6\,dx = 0$$

$$dy - 2\,dx = 0$$

(c) A velocity potential in 2-D flow is $\Phi = y + x^2 - y^2$. Find the stream function for this flow. ...(8)

Ans.: $\Phi = y + x^2 - y^2$

We will have to check for the equation of continuity which is Laplace equation in the case of potential function.

$$-\nabla^2 \Phi = 0 \text{ or } -\left(\frac{\partial^2 \Phi}{\partial x^2} + \frac{\partial^2 \Phi}{\partial y^2} \right) = 0$$

$$\Phi = x^2 - y^2 + y$$

∴

$$\frac{\partial \Phi}{\partial x} = 2x \qquad\qquad \frac{\partial \Phi}{\partial y} = -2y + 1$$

$$\frac{\partial^2 \Phi}{\partial x^2} = 2 \qquad\qquad \frac{\partial^2 \Phi}{\partial y^2} = -2$$

$\therefore$
$$-\nabla^2 \Phi = -\left(\frac{\partial^2 \phi}{\partial x^2} + \frac{\partial^2 \Phi}{\partial y^2}\right) = -(2 - 2) = 0$$

Hence the flow is possible

$$u = -\frac{\partial \Phi}{\partial x} = -2x \qquad\qquad v = -\frac{\partial \Phi}{\partial y} = 2y - 1$$

$$\frac{\partial \psi}{\partial y} = -u = 2x \qquad\qquad \frac{\partial \psi}{\partial x} = v = 2y - 1$$

$$\psi = 2xy \qquad\qquad \psi = 2xy - x$$

Hence the stream function is

$$\psi = 2xy - x + C$$

Where C is numerical constant.

Q.4 (a) The losses $\dfrac{\Delta h}{l}$ per unit length of pipe in a turbulent flow through a smooth pipe depend upon velocity V, diameter D, gravity g, dynamic viscosity μ and density ρ. With dimensional analysis determine general form of the equation for the losses. ...(6)

Ans.: The fundamental dimensions of the variables are

$$\Delta h = L \qquad\qquad\qquad D = L$$

$$l = L \qquad\qquad\qquad V = LT^{-1}$$

$$g = LT^{-2} \qquad\qquad\qquad \mu = ML^{-1}T^{-1}$$

$$\rho = ML^{-3}$$

$$\text{Number of variables } n = 7$$
$$\text{Number of dimensions } = 3$$
$$\text{Number of parameters } n - m = 7 - 3 = 4$$

Choosing ρ, V and l as repeating variables the four parameters are

$$\pi_1 = \frac{\Delta h}{l}$$

$$\pi_2 = \frac{l}{d}$$

$$\pi_3 = \rho^{x_3} V^{y_3} l^{z_3} g$$

$$\pi_4 = \rho^{x_4} V^{y_4} l^{z_4} \mu$$

Substituting the fundamental dimensions.

$$M^0 L^0 T^0 = \left(ML^{-3}\right)^{x_3} \left(LT^{-1}\right)^{y_3} \left(L\right)^{z_3} \left(LT^{-2}\right)$$

Collecting the indices dimensions

For	$Mx_3 = 0$	...(1)
For	$L-3x_3 + y_3 + z_3 + 1 = 0$	...(2)
For	$T-y_3 - 2 = 0$	...(3)

From equation (1) $x_3 = 0$

From equation (3) $y_3 = -2$

From equation (2) $z_3 = 1$

The third parameter is $\pi_3 = \rho^0 V^{-2} l^1 g$

$$= \frac{gl}{V^2}.$$

The equations for the fourth parameter are

For	$Mx_4 = 0$	...(1)
For	$L-3x_4 + y_4 + z_4 - 1 = 0$	...(2)
For	$T-y_4 - 1 = 0$	...(3)

From equation (1) $x_4 = -1$

From equation (3) $y_4 = -1$

From equation (2) $z_4 = -1$

The fourth parameter is $\pi_4 = \rho^{-1} V^{-1} l^{-1} \mu = \dfrac{\mu}{\rho V l}$

The relation between the parameter is

$$\pi_1 = f\left(\pi_2, \pi_3, \pi_4\right)$$

$$\frac{\Delta h}{l} = f\left(\frac{l}{d}, \frac{gl}{V^2}, \frac{\mu}{\rho V l}\right)$$

(b) Derive the Euler's equation of motion for real fluids and hence deduce Bernoulli's equation of motion. Mention the assumptions made.   ...(8)

Ans.: Please refer Q.5 (b) of July/Aug.-2004 and Q.3 (b) of Jan./Feb.-2006.

(c) A pipe gradually tapers from a diameter of 0.3 m to 0.1 m over the length as shown in Fig. 4. It conveys kerosene (Sp. gr. 0.80) at 50 l/s. The pressure at bottom end is 200 kN/m^2. If the pressure at upper end is not to fall below 100 kN/m^2, find the value of Z. (Neglect losses).

(6)

Ans.: Given $D_1 = 0.3$ m $D_2 = 0.1$ m

Sp. gravity of kerosene= 0.80

$$\rho = 0.80 \times 10^3 \text{ kg/m}^3$$
$$Q = 50 \text{ l/s} = 0.05 \text{ m}^3/\text{s}$$
$$P_1 = 200 \text{ kN/m}^2 = 200 \times 10^3 \text{ N/m}^2$$
$$P_2 = 100 \text{ kN/m}^2 = 100 \times 10^3 \text{ N/m}^2$$
$$Z_1 = 0 \qquad Z_2 = Z$$

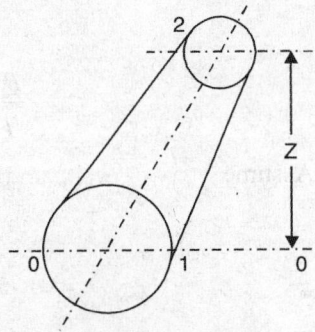

Fig. 4

$$A_1 = \frac{\pi}{4}(0.3)^2 = 0.07968 \text{ m}^2$$

$$A_2 = \frac{\pi}{4}(0.1)^2 = 0.00785 \text{ m}^2$$

$$Q = A_1 V_1 = A_2 V_2$$

$$\therefore \qquad V_1 = \frac{Q}{A_1} = \frac{0.05}{0.07068} = 0.7074 \text{ m/s}$$

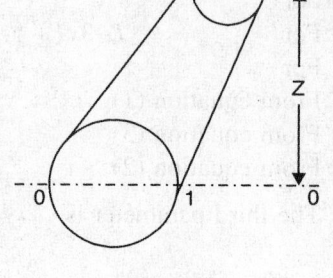

Fig. 4(a)

$$V_2 = \frac{Q}{A_2} = \frac{0.05}{0.00785} = 6.366 \text{ m/s}$$

By Bernoulli's equation between section 1–1 and 2–2

$$\frac{P_1}{\rho \cdot g} + \frac{V_1^2}{2g} + Z_1 = \frac{P_2}{\rho \cdot g} + \frac{V_2^2}{2g} + Z_2$$

$$\frac{200 \times 10^3}{0.8 \times 10^3 \times 9.81} + \frac{(0.7074)^2}{2 \times 9.81} + 0 = \frac{100 \times 10^3}{0.8 \times 10^3 \times 9.81} + \frac{(6.366)^2}{2 \times 9.81} + Z_2$$

$$\therefore \qquad Z_2 = 25.484 + 0.0255 - 12.742 - 2.0655$$

$$Z_2 = 10.702 \text{ m}$$

Q.5 (a) In a 100 mm diameter horizontal pipe a venturimeter of 0.5 contraction ratio has been fitted. The head of water on the meter when there is no flow is 3 m of water. Find the rate of flow for which the throat pressure will be 2 m of water. The co-efficient of the meter is 0.97.

(8)

Ans.: Given d_1 = 100 mm = 0.1 m

$$\frac{d_2}{d_1} = 0.5$$

$$\therefore \qquad d_2 = 0.5 d_1$$

$$= 0.5 \times 0.1 = 0.05 \text{ m}$$

$$\frac{P_1}{\gamma} = 3 \text{ m} \quad \text{Throat pressure } \frac{P_2}{\gamma} = 2 \text{ m}$$

Assume $\qquad P_{atm} = 10.3 \text{ m } C_d = 0.97$

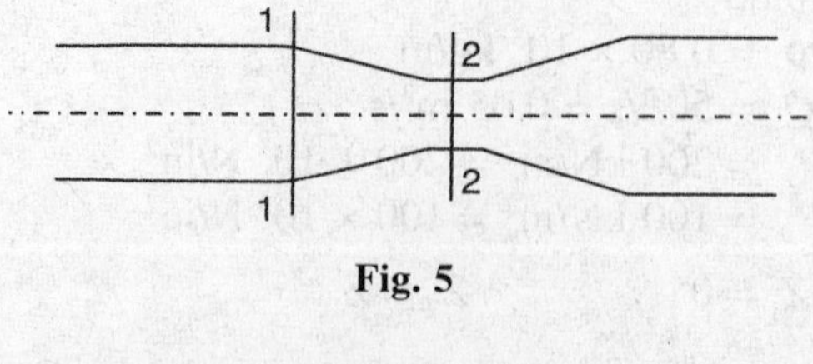

Fig. 5

Applying Bernoulli's theorem

$$P_{atm} + P_1/2 = \frac{P_2}{\gamma} + \frac{V_2^2}{2g}$$

$$10.3 + 3 = 2 + \frac{V_2^2}{2g}$$

$$\therefore \qquad \frac{V_2^2}{2g} = 11.3$$

$$\therefore \qquad V_2 = \sqrt{2 \times 9.81 \times 11.3} = 14.89 \, m/s$$

Theoretical discharge $\qquad Q = a_2 V_2$

$$= \frac{\pi}{4}(0.05)^2 \times 14.89 = 0.02924 \, m^3/s$$

$\therefore$ Actual discharge $\qquad q = C_d \cdot Q$

$$= 0.97 \times 0.02924 = 0.02836 \, m^3/s$$

$$= 28.36 \, lit/sec$$

(b) Derive an expression for a flow through a triangular notch in terms of head over notch.

(6)

Ans.: Please refer Q.6(b) of July/Aug.-2004.

(c) A pitot static probe is used to measure the flow of water in a 5 cm diameter pipe. If the mean velocity is 5 m/s, and the pitot static tube is connected across a mercury filled differential manometer, what should be the level of difference in the mercury column?

Ans.: $d = 5 \, cm = 0.05 \, m \qquad A = \frac{\pi}{4}(0.05)^2 = 0.00136 \, m^2$

Mean velocity $V_0 = 5 \, m/s$

$$\therefore \qquad V_0 = C\sqrt{2g \, \Delta h}$$

Let $\qquad C = 0.98$

$$5 = 0.98\sqrt{2 \times 9.81 \times \Delta h}$$

$$\therefore \qquad \Delta h = \frac{5^2}{(0.98)^2 \times 2 \times 9.81} = 1.3267 \, m$$

Q.6 (a) Derive Darcy-Weisbach equation and deduce it to Chezy's equation. (8)

Ans.: Please refer Q.6 (b) of Jan.- Feb.-2005.

(b) Show that the energy transmitted by a long pipe is maximum when one third of the energy put into the pipe is lost in friction. One hundred kW is to be transmitted through a pipe, the pressure at the inlet of the pipe being 70 bar. If the pressure drop per kilometer is to be 0.44 bar and if $f = 0.02$, find the diameter of the pipe and the efficiency of transmission for 16 km.

(12)

Ans.: Suppose energy is transmitted by means of water under pressure, through a certain distance. The energy transmitted will be proportional to

(i) The quantity of water passing through the pipe/second.

(ii) The total head of water available at the end of the pipe line.

Consider a pipe of length l and diameter D conveying water.

Let H be the total energy head at inlet to the pipe. Let hf be the loss of head due to friction. Let V be the velocity of flow of water.

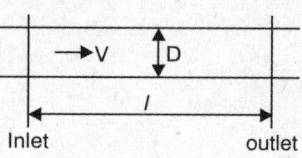

Fig. 6

Total energy head available at outlet of pipe

$$= H - h_f = H - \frac{4flV^2}{2gD}$$

The ratio of the energy head at outlet to the energy head at inlet is called the efficiency of transmission.

$$\text{Efficiency of transmission} = \frac{H - h_f}{H}$$

$$\text{Discharge through the pipe/sec} = \frac{\pi D^2}{4} V$$

∴ H.P. available at the outlet of the pipe

$$= P = \frac{\pi D^2}{4} V \left[H - \frac{4fl \cdot V^2}{2gD} \right] \frac{1}{75}$$

∴

$$P = \frac{\omega \pi D^2}{4 \times 75} \left[HV - \frac{4flV^3}{2gD} \right]$$

Condition of maximum transmission of power

$$\frac{dP}{dV} = 0$$

∴

$$\frac{dP}{dV} = \frac{\omega \pi D^2}{4 \times 75} \left[H - 3 \left(\frac{4flV^2}{2gD} \right) \right] = 0$$

∴

$$H - 3 \left(\frac{4flV^2}{2gD} \right) = 0$$

∴

$$H - 3h_f = 0$$

∴

$$h_f = \frac{H}{3}$$

Thus power transmitted is the maximum when the loss of head due to friction is one third of the total head supplied.

H.P. = 100 kW $P_1 = 70$ bar $= 70 \times 10^5$ N/m^2

$$\frac{\Delta P}{\text{km}} = 0.44 \text{ bar} \quad f = 0.02$$

$$l = 16 \text{ km}$$

$\therefore$ Loss of pressure $= 0.44 \times 16 = 7.04$ bar

$$= 7.04 \times 10^5 \text{ N/m}^2$$

$\therefore$ $P_2 = 70 - 7.04 = 62.92 \times 10^5 \text{ N/m}^2$

Energy head at inlet $= \dfrac{70 \times 10^5}{100 \times 9.81} = 713.56$ m

Energy head at end $= \dfrac{62.92 \times 10^5}{1000 \times 9.81} = 641.39$ m

Loss of energy $= h_f = H_1 - H_2 = 713.56 - 641.39$

$$= 72.17 \text{ m}$$

Efficiency of power transmission $= \dfrac{H - h_f}{H}$

$$= \frac{713.56 - 72.17}{713.56} = 0.8988 \text{ or } 89.88\%$$

Power to be transmitted $= 100$ kW

$$p = \frac{\rho \cdot g \cdot Q \cdot h}{1000}$$

$$100 = \frac{1000 \times 9.81 \times Q \times 713.56}{1000}$$

$\therefore$ $Q = \dfrac{100}{9.81 \times 713.56} = 0.0143 \text{ m}^3/\text{s}$

$$h_f = \frac{flQ^2}{3.025 D^5}$$

$$D^5 = \frac{flQ^2}{3.0257 \times hf} = \frac{0.02 \times 16 \times 1000 \times (0.01428)^2}{3.0257 \times 72.17}$$

$$D^5 = 0.000299$$

$$D = 0.1973 \text{ m}$$

Q.7 (a) Derive an expression for velocity and average velocity for viscous flow between two stationery parallel plates. (6)

Ans.: Please refer Q. 7(b) of July/Aug.-2006

(b) Explain D' Alembert paradox. (4)

Ans.: Please refer Q.2(a) of same paper.

(c) A television transmitter antenna consists of a vertical pipe 20 cm diameter and 30 m high on top of tall structure. Determine the total drag on the antenna and the bending moment about the base in a 30 m/s wind at NTP. Take density of air as 1.22 kg/m³ and viscosity as 1.79×10^{-5} Ns/m², $C_D = 0.2$. (10)

Ans.:

Given

$$V = 30 \text{ m/s}$$

$$\rho_{air} = 1.22 \text{ kg/m}^3$$

$$\mu = 1.79 \times 10^{-5} \text{ Ns/m}^2$$

$$C_D = 0.2$$

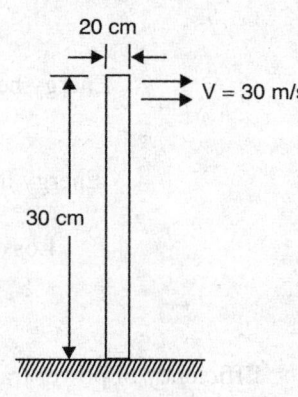

Fig. 7

Drag force on the antenna $= C_d \cdot \rho \cdot A \cdot \dfrac{V^2}{2}$

$$d = 20 \text{ cm} = 0.2 \text{ m}$$

$$l = 30 \text{ m}$$

Cross-sectional area of pole $= l \times d$

$$= 0.2 \times 30 = 6 \ m^2$$

$\therefore$ Drag force on the vertical pipe

$$F_D = C_d \cdot \rho \cdot A \cdot \frac{V^2}{2}$$

$$= 0.2 \times 1.22 \times 6 \times \frac{(30)^2}{2} = 658.8 \text{ N}$$

Bending moment at the base of the pole

$$= F_D l = 658.8 \times 30$$

$$= 19764 \text{ Nm}$$

Q.8 (a) Explain

   (i)   Boundary layer thickness.
  (ii)   Displacement thickness.
 (iii)   Momentum thickness.

   The velocity profile in a laminar boundary layer is approximated by a parabolic profile

$\dfrac{u}{U} = 2\left(\dfrac{y}{\delta}\right) - \left(\dfrac{y}{\delta}\right)^2$ where u is the velocity at y and $u \rightarrow U$ as $y \rightarrow \delta$. Calculate displacement

thickness and momentum thickness. (10)

Ans.: Please refer Q.8(a) of July/Aug.-2005 and Q.8(b) of July/Aug.-2006 for (i), (ii) and (iii).

For Calculation of displacement thickness and momentum thickness: Please refer Q. 7(c) of Jan./Feb.-2006.

(b) Determine the velocity of a bullet fired in the air if the mach angle is observed to be 30°. Given that the temperature of air is 22 °C, density 1.2 kg/m³. Take $\gamma = 1.4$ and $R = 287$ J/kgK. Derive the equation used. ...(10)

Ans.: For derivation of equation $C = \sqrt{\gamma RT}$: Please refer Q.6(b) of July/Aug.2006

$$\alpha = 30°, \ R = 287 \ \text{J/kg/K}, \ \gamma = 1.4$$

$$t = 22°C = 22 + 273 = 295 \ \text{K}$$

Velocity of sound $C = \sqrt{\gamma RT}$

$$C = \sqrt{1.4 \times 287 \times 295} = 344.28 \ \text{m/s}$$

$$\sin \alpha = \frac{C}{V} \quad \therefore V = \frac{C}{\sin \alpha}$$

$$\therefore \qquad V = \frac{344.28}{\sin 30°} = 688.56 \ \text{m/s}$$

∴ Velocity of bullet in still air = 688.56 m/s.

Fluid Mechanics
(July/August 2007)
Semester-IV Mechanical Engineering (VTU)

Time: 3 Hours   Maximum Marks: 100

Note: 1. Answer any five full questions.
2. Any missing data may be assumed suitably.

Q.1 (a) Define and differentiate between the following:
    (i) Weight density and mass density   (ii) Kinematic viscosity and dynamic viscosity
  (iii) Compressibility and bulk modulus   (iv) Surface tension and capillarity   (12)
Ans.: (i) **Weight density:** It is defined as the ratio of weight of fluid to volume of fluid.
$\therefore$ It is symbolised as 'γ'

$$\therefore \qquad \gamma = \text{Weight density} = \frac{\text{Weight}}{\text{Volume}} = \frac{W}{V}$$

and its unit is $\dfrac{\text{N}}{\text{m}^3}$

Mass density: It is defined as the ratio of mass of fluid to volume of fluid. It is symbolised as 'ρ'

$$\therefore \qquad \rho = \text{mass density} = \frac{\text{Mass}}{\text{Volume}} = \frac{m}{V}$$

and its unit is $\dfrac{\text{kg}}{\text{m}^3}$

Note $\qquad \boxed{\gamma = \rho \times g}$

i.e. Weight density = Mass density $\times$ Acceleration due to gravity.

(ii) **Kinematic viscosity:** It is defined as the ratio of dynamic viscosity to the mass density.

$$\therefore \qquad \text{Kinematic viscosity} = \frac{\text{Dynamic viscosity}}{\text{Mass density}}$$

$$\therefore \qquad \nu = \frac{\mu}{e}$$

$$\boxed{\text{It's unit is} = \frac{\text{m}^2}{\text{s}}} \qquad \therefore \left(\frac{\text{N.s}/\text{m}^2}{\text{kg}/\text{m}^3} \right) = \frac{\text{m}^2}{\text{s}}$$

Dynamic viscosity: We know that,

$$\tau = \mu \frac{du}{dy}$$

$$\therefore \qquad \mu = \text{dynamic viscosity}$$

$$\therefore \qquad \boxed{\mu = \frac{\tau}{(du/dy)}}$$

$$\therefore \qquad \text{unit of } \mu = \frac{\text{N/m}^2}{(\text{m/s})/\text{m}} = \frac{\text{Ns}}{\text{m}^2}$$

(iii) **Compressibility:** Please refer Q.1(a) of Jan./Feb.-2007.
Bulk modulus: Please refer Q.1(a) (iv) of July/August-2005.
(iv) **Surface Tension:** Please refer Q.1(a) (i) of July/August-2005.
Capillarity: Please refer Q.1(b) of Jan./Feb.-2005.

(b) The dynamic viscosity of an oil used for lubrication between a shaft and sleeve is 6 poise. The diameter of the shaft is 0.4 m and rotates at 190 rpm. Calculate the power lost in the bearing for a sleeve length of 90 mm. The thickness of the oil film is 1.5 mm. (8)

Ans.: Given data:

$$N = 190 \text{ rpm}, \qquad D = 0.4 \text{ m}$$

$$\therefore \qquad r = 0.2 \text{ m}$$

$$l = 90 \text{ mm} = 0.09 \text{ m}$$

Film thickness = 1.5 mm = 0.0015 m

Outer diameter = 0.4 + 2 × 0.0015

$$D_0 = 0.403 \text{ m}$$

$\therefore$ Angular velocity of shaft

$$\omega = \frac{2\pi N}{60} = \frac{2 \times \pi \times 190}{60} = 19.89 \text{ rad/s}$$

Linear velocity of shaft

$$v = r\omega = 0.2 \times 19.89$$

$$\therefore \qquad \boxed{v = 3.978 \text{ m/s}}$$

$$dy = 1.5 \text{ mm} = 0.0015 \text{ m}$$

$$\therefore \qquad \text{Velocity gradient} = \frac{dv}{dy} = \frac{3.978}{0.0015} = 2652\,(1/\text{s})$$

$$\text{Viscosity } (\mu) = 6 \text{ poise} = 0.6 \frac{\text{Ns}}{\text{m}^2}$$

$\therefore$ Shear stress developed (τ)

$$\therefore \qquad \tau = \frac{\mu\, dv}{dy} = 0.6 \times 2652$$

$$\therefore \qquad \boxed{\tau = 1591.2 \text{ N/m}^2}$$

Force acting on the shaft is,

$$F = \text{Shear stress} \times \text{Area} = 1591.2 \times \pi \times D_0 \times l$$
$$= 1591.2 \times \pi \times 0.403 \times 0.09$$

∴
$$\boxed{F = 181.31 \text{ N}}$$

Torque acting on the shaft,

$$T = \text{Force} \times \text{radius of shaft} = 181.31 \times 0.2$$

∴
$$\boxed{T = 36.062 \text{ Nm}}$$

∴ Power required to overcome torque.
i.e. power lost in bearing

$$= T \times \omega = 36.062 \times 19.89$$

∴
$$\boxed{P = 717.27 \text{ Watt}}$$

Q.2(a) State and prove hydrostatic law.

Ans.: Hydrostatic law: The intensity of pressure at any point in a liquid varies directly with the depth of the point below the free liquid surface (or the height of free liquid surface above the point).

It is given as $P = \gamma h$

Proof: We know the equation of fluid statics.

$$\frac{\partial p}{\partial z} = -\gamma$$

In case of incompressible fluid 'r' is constant and therefore integrating above equation,

$$\int_{P_1}^{P_2} dP = \int_{Z_1}^{Z_2} -\gamma dz$$

∴
$$P_2 - P_1 = -\gamma \left(z_2 - z_1 \right)$$

∴
$$\frac{P_2}{\gamma} + z_2 = \frac{P_1}{\gamma} + z_1$$

Further, it can be integrated to give,

$$P = -\gamma z + C$$

But at
$$z = H,$$

$$P = P_a$$

∴
$$P_a = -\gamma H + C$$

∴
$$C = P_a + \gamma H$$

∴
$$P = P_a + \gamma (H - Z)$$

If distances are measured from liquid surface, positive vertically downwards and sag it is 'h' then $(H - Z) = h$

∴
$$P = P_a + \gamma h$$

If we consider the pressure because of liquid only,

Then $\boxed{P = \gamma h}$ Hence proved.

(b) Write a note on differential manometers. (6)

Ans.: Differential manometers. See Fig. 1.

It is measuring pressure difference between two points (A and B)

The two limbs are connected to points A and B, the difference of pressure between which is to be measured.

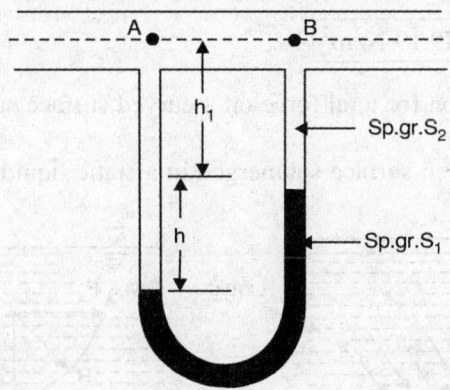

Fig. 1

The two points may be on same pipe or two different pipes.

For large pressure differences, mercury is used.

For small pressure differences, carbon tetrachloride can be used.

(c) The right limb of a simple U-tube manometer containing mercury is open to the atmosphere while the left limb is connected to a pipe in which a fluid of *SG* 0.9 is flowing. The center of the pipe is 12 cm below the level of mercury in the right limb. Find the pressure of fluid in the pipe if the difference of mercury level in two limbs is 20 cms. (8)

Ans.:

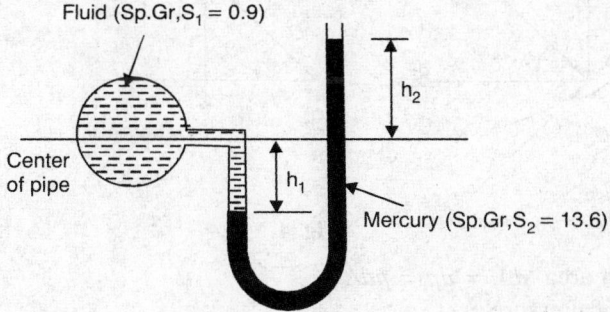

Fig. 2

$\therefore$ $\qquad$ $S_1 = 0.9$ $\quad h_2 = 12$ cm $= 0.12$ m

$\qquad\qquad h_1 + h_2 = 20$ cm

$\therefore$ $\qquad h_1 = 8$ cm

$\therefore$ $\qquad hm = S_2\left(h_2 + h_1\right) - S_1 h_1$ m of water $= 13.6 \times 0.2 - 0.9 \times 0.08$

$\qquad\qquad hm = 2.648$ m of water

$\therefore$ Pressure of fluid in pipe

$\qquad\qquad P = \gamma hm = \rho g\, hm = 1000 \times 0.9 \times 9.81 \times 2.648$

$$\boxed{P = 23379.19 \text{ N/m}^2}$$

Q.3 (a) Derive an expression for total force on a curved surface submerged in a static fluid.

(5)

Ans.: Total force on a curved surface submerged in a static liquid:

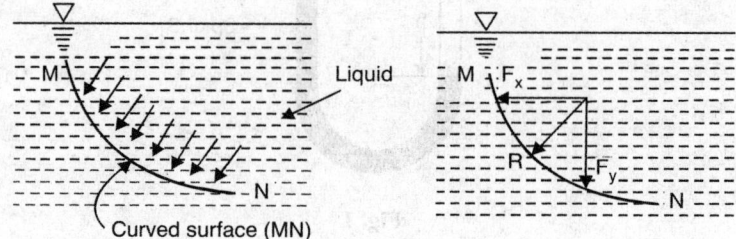

Fig. 3

Let F_x – Horizontal component of total force

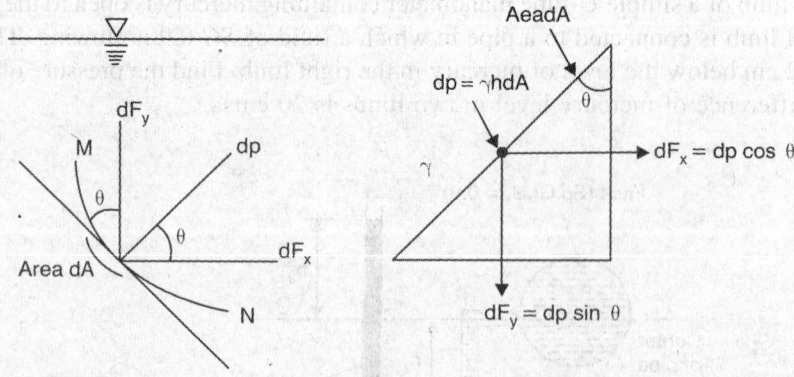

Fig. 4

Total force acting on area 'dA' $= dp = pdA$
where 'p' intensity of pressure

$\therefore$ $\qquad\qquad dF_x = dp\cos\theta = pdA\cos\theta$

$$dF_y = dp\sin\theta = p\,dA\sin\theta$$

$$\therefore \qquad dF_x = \gamma h\,dA\cos\theta$$

$$\therefore \qquad F_x = \int \gamma h\,dA\cos\theta$$

But '$dA\cos\theta$' is the projection of curved surface dA on vertical plane.

$\therefore \gamma h dA\cos\theta$ = Total force on projection of 'dA' on vertical plane.

$$\therefore \qquad F_x = \int \gamma h\,dA\cos\theta$$

$$= \text{Total pressure on the projection of curved}$$
$$\text{surface } MN \text{ on vertical plane}$$

$$\therefore \qquad \boxed{F_x = \gamma A\bar{x}}$$

where $\bar{x}$ = depth of C.G. of projected area vertically below free liquid surface.

Similarly,

$$F_y = \int \gamma h\,dA\sin\theta$$

$dA\sin\theta$ is the projection of curved surface dA on horizontal plane.

$(dA\sin\theta\ h)$ – volume of liquid supported vertically by curved surface dA.

$\gamma dA\sin\theta h$ – weight of this supported liquid.

$$\therefore \qquad F_y = \text{weight of liquid supported vertically by curved surface } MN$$

$\therefore$ Resultant force $R = \sqrt{F_x^2 + F_y^2}$

and $\qquad \theta = \tan^{-1}\left(F_y/F_x\right)$

(b) A tank contains water upto a height of 0.5 m above the base. An immiscible liquid of *SG* 0.8 is filled on the top of water upto 1 m height. Calculate

  (i) Total pressure on one side of the tank.

  (ii) The position of center of pressure for one side of the tank, which is 2 m wide. (10)

Ans.:

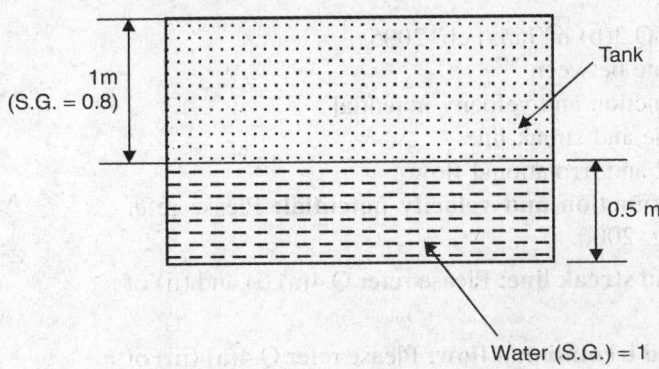

Fig. 5

Pressure by water on left side of tank

$$F_1 = \gamma_1 A_1 \bar{x}_1 = 9810 \times (2 \times 0.5) \times \left(\frac{0.5}{2}\right)$$

$$F_1 = 2452.5 \text{ N}$$

F_1 will be acting at a distance of $y_1 = \dfrac{0.5}{3} = 0.167$ m from base of tank.

Pressure by immiscible liquid on left side of tank

$$F_2 = \gamma_2 A_2 \bar{x}_2 = (1000 \times 0.8) \times 9.81 \times (1 \times 2) \times \frac{1}{2}$$

$$F_2 = 7848 \text{ N}$$

F_2 will be acting at a distance of,

$$y_2 = \frac{1}{3} = 0.333 \text{ m from bottom level of immiscible liquid.}$$

∴ Resultant total pressure on one side of

Tank $= F_1 + F_2 = 2452.5 + 7848$

∴ $\boxed{F = 10300.5 \text{ N}}$

∴ Taking moment @ 'O'

$$F_1 \times 0.167 + F_2 \times 0.333 = F \times y$$

∴ $2452.5 \times 0.167 + 7848 \times (0.333 \times 0.167)$

$= 10300.5 \times y$

∴ $\boxed{y = 0.4207 \text{ m}}$ ...Ans.

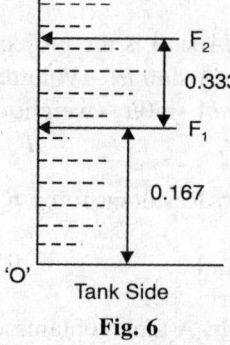

Fig. 6

∴ Position of centre of pressure for one side of tank is **0.4207 m** from bottom of tank.

(c) How will you determine the metacentric height of a floating body experimentally with a neat sketch? (5)

Ans.: Please refer Q.3(b) of Jan./Feb.-2005.

Q.4 (a) Differentiate between (6)
   (i) Stream function and velocity potential
   (ii) Stream line and streak line
   (iii) Rotational and irrotational flow

Ans.: (i) **Stream function and velocity potential:** Please refer Q.4(b) of July/Aug.-2006.

(ii) **Stream line and streak line:** Please refer Q.4(a) (i) and (ii) of Jan./Feb.-2005.

(iii) **Rotational and irrotational flow:** Please refer Q.4(a) (iii) of July/Aug.-2005.

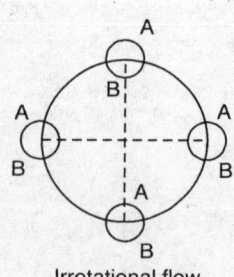

Irrotational flow

Fig. 7

Irrotational flow: The flow is called as an irrotational flow, if the fluid particles do not rotate about their mass centres while moving in the direction of motion.

e.g., Flow of liquid in an emptying wash basin.

(b) Obtain an expression for continuity equation for a 3 dimensional flow in cartesian coordinates. (6)

Ans.: Please refer Q.4(a) of July/Aug.-2006.

(c) The velocity components in a two dimensional flow field for an incompressible fluid are as follows

$$u = \frac{y^3}{3} + 2x - x^2y \text{ and } v = xy^2 - 2y - x^3/3$$

Obtain an expression for the stream function ψ. (8)

Ans.: $\dfrac{\partial \psi}{\partial x} = v = xy^2 - 2y - \dfrac{x^3}{3}$

$$\frac{\partial \psi}{\partial y} = -u = \frac{-y^3}{3} - 2x + x^2y$$

$$\frac{\partial \psi}{\partial x} = xy^2 - 2y - \frac{x^3}{3}$$

$\therefore$
$$\psi = \int xy^2 - 2y - \frac{x^3}{3}$$

$\therefore$
$$\psi = \frac{y^2 x^2}{2} - 2yx - \frac{1x^4}{3 \times 4} + f(y)$$

$\therefore$
$$\frac{\partial \psi}{\partial y} = x^2y - 2x + 0 + f'(y)$$

But
$$\frac{\partial \psi}{\partial y} = -u = \frac{-y^3}{3} - 2x + x^2y$$

$\therefore$
$$x^2y - 2x + f'(y) = \frac{-y^3}{3} - 2x + x^2y$$

$\therefore$
$$\boxed{f'(y) = \frac{-y^3}{3}}$$

$\therefore$
$$\boxed{f(y) = \frac{-y^4}{12} + C}$$

$\therefore$
$$\boxed{\psi = \frac{x^2 y^2}{2} - 2xy - \frac{x^4}{12} - \frac{y^4}{12} + C} \text{ Stream function}$$

Q.5 (a) State Buckingham's π theorem. (2)

Ans.: Please refer Q.5(a) of July/August-2005.

(b) Find the expression for the power developed by a pump (P) when it depends upon the head (H), discharge (Q) and specific weight (W) of the fluid. (8)

Ans.: $f(P, Q, H, \gamma) = 0$ or constant

Units:
$$P\left[M L^{-2} T^{-3}\right] \quad Q\left[L^{3}T^{-1}\right]$$
$$H[L] \qquad \omega\left[M L^{-2} T^{-2}\right]$$

No. of variables $n = 4$

No. of primary dimensions $m = 3$

No. of π terms $= n - m = 4 - 3 = 1$

$\therefore f_1(\pi_1) = 0$ or constant

Taking Q, H, ω as repeating variables,

$$\pi_1 = Q^{x_1}, H^{y_1}, \omega^{z_1}, P$$

$\therefore$
$$\left[M^0 L^0 T^0\right] = \left[L^3 T^{-1}\right]^{x_1} [L]^{y_1} \left[ML^{-2}T^{-2}\right]^{z_1} \left[ML^2 T^{-3}\right]$$

$\therefore$ For M,
$$0 = z_1 + 1$$
$$\boxed{\therefore \ z_1 = -1}$$

For L,
$$0 = 3x_1 + y_1 - 2z_1 + 2$$
$\therefore$
$$\boxed{0 = 3x_1 + y_1 + 4}$$

For T,
$$0 = -x_1 - 2z_1 - 3 = -x_1 - 2(-1) - 3$$
$\therefore$
$$0 = -x_1 + 2 - 3$$
$\therefore$
$$0 = -x_1 - 1$$
$$\boxed{\therefore \ x_1 = -1}$$

Also,
$$0 = 3(-1) + y_1 + 4$$
$\therefore$
$$0 = y_1 + 1$$
$\therefore$
$$\boxed{y_1 = -1}$$

$\therefore$
$$\pi_1 = Q^{-1} H^{-1} \omega^{-1} P$$

$\therefore$
$$\pi_1 = \frac{P}{QH\omega}$$

$\therefore$
$$f_1\left(\frac{P}{QH\omega}\right) = 0 \text{ or constant}$$

$\therefore$
$$\boxed{P = QH\omega} \text{ Expression for power developed by pump}$$

(c) Derive an expression for discharge through as orifice. (7)

Ans.: Using Bernoulli's theorem in between points (1) and (2),

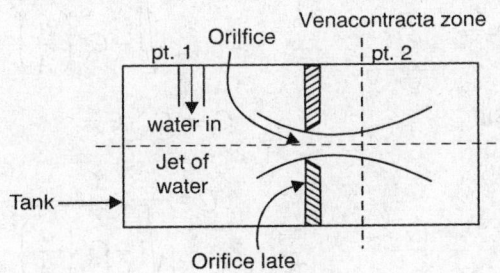

Fig. 8

$$Z_1 + \frac{P_1}{\gamma} + \frac{V_1^2}{2g} = Z_2 + \frac{P_2}{\gamma} + \frac{V_2^2}{2g}$$

$$\because \qquad Z_1 = Z_2$$

$$\therefore \qquad \frac{V_2^2 - V_1^2}{2g} = \frac{p_1}{\gamma} - \frac{p_2}{\gamma} = h$$

where 'h' is the pressure head difference between points (1) and (2)

$$\therefore \qquad V_2^2 - V_1^2 = 2gh$$

By continuity equation,

$$AV_1 = a_c V_2$$

where,

A = area of pipe

a_C = area of jet at vena-contracta

Also

$$C_C = \frac{a_c}{a}$$

where

a = area of orifice

C_C = Coefficient of contraction for orifice

$$\therefore \qquad a_c = C_C \cdot a$$

$$\therefore \qquad AV_1 = C_C a \cdot V_2$$

$$V_1 = C_C \cdot \frac{a}{A} \cdot V_2$$

$$\therefore \qquad V_2^2 - C_C^2 \cdot \frac{a^2}{A^2} V_2^2 = 2gh$$

$$V_2 = \sqrt{\frac{2gh}{1 - C_C^2 \left(\dfrac{a}{A}\right)^2}}$$

But

$$C_V = \frac{\text{Actual velocity}}{\text{Theoretical velocity}}$$

$$\therefore \qquad V_{2\,\text{actual}} = C_V \cdot V_2 \text{ theoretical}$$

$$\therefore \qquad V_{2\,\text{actual}} = C_V \cdot \sqrt{\frac{2gh}{1 - C_C^2 (a/A)^2}}$$

Q = Actual discharge

$$\therefore \qquad Q = C_C \cdot a \cdot V_2$$

$$\therefore \quad Q = \frac{C_C \cdot C_V}{\sqrt{1 - C_C^2 \left(\dfrac{a}{A}\right)^2}} \cdot a \cdot \sqrt{2gh}$$

But $\quad C_d = C_C \cdot C_V$

$$\therefore \quad Q = \frac{C_d}{\sqrt{1 - C_C^2 \left(\dfrac{a}{A}\right)^2}} \cdot a \cdot \sqrt{2gh}$$

$$\left(\frac{a}{A}\right)^2 = \left(\frac{d}{D}\right)^4$$

$$\therefore \quad \boxed{Q = \frac{C_d \cdot a \cdot \sqrt{2gh}}{\sqrt{1 - C_C^2 \left(d/D\right)^4}}}$$

This is the expression for discharge through an orifice.

(d) Why coefficient of discharge (C_d) of venturimeter is higher than that of an orificemeter? (3)

Ans.: Coefficient of discharge is the ratio of actual discharge to the theoretical discharge.

$$C_d = \frac{Q_{actual}}{Q_{theoretical}}$$

In, orificemeter,

Q_{actual} is less than $Q_{theoretical}$ in venturimeter.

Because, losses are more in orificemeter.

$\therefore$ Q_{actual} is less.

$\therefore$ C_d is less in orificemeter.

As losses are very low and hence coefficient of discharge of venturimeter (C_d) is very high and is about 0.98. (Higher than orificemeter).

Q.6 (a) State Bernoulli's theorem for steady flow of an incompressible fluid and derive an expression for the same. State the assumptions for such a derivation. (10)
Ans: Please refer R.5 (a) of July/Aug.-2006 and Q.4 (c) of July/Aug.-2004.

(b) Find the diameter of a pipe of length 2000 m when the rate of flow of water through pipe is 200 lts/sec and the head lost due to friction is 4 m. Take the value of C = 50 in Chezy's formulae.
Ans.: Chezy's equation,

$$V = C\sqrt{ms}$$

Head lost due to friction,

$$h_f = \frac{4V^2 L}{dC^2}$$

Where, $h_t = 4$ m, $L = 2000$ m, $C = 50$

Also, $Q = 200$ lit/sec $= 200 \times 10^{-3}$ m³/s

$$Q = 0.2 \text{ m}^3/\text{s}$$

$$Q = \frac{\pi}{4} d^2 \cdot V$$

∴ $0.2 = \frac{\pi}{4} d^2 \times V$

∴ $= 0.2546 = d^2 \times V$

∴ $\boxed{V = \frac{0.2546}{d^2}}$

∴ $$4 = \frac{4 \times \left(\dfrac{0.2546}{d^2}\right)^2 \times 2000}{d \times 50^2}$$

∴ $10000 \times d = 4 \times \left(\dfrac{0.2546}{d^2}\right)^2 \times 2000$

∴ $19.283 = \dfrac{1}{d^5}$

∴ $d^5 = 0.0518$

∴ $\boxed{d = 0.553 \text{ m}}$

∴   Diameter of pipe $= 0.553$ m

∴ $\boxed{d = 533 \text{ mm}}$

Q.7 (a) What is Hagen Poiseuille's formula? Derive an expression for the same.   (6)
Ans.: Please refer Q.6 (c) of Jan./Feb.-2006.

(b) Obtain an expression for velocity of the sound wave in a compressible fluid in terms of change in pressure and change of density.   (8)
Ans.: Please refer Q.6 (b) of July/Aug.-2006.

(c) Define Mach number, Mach angle and Mach cone.   (6)
Ans. Mach number: Please refer Q.7 (c) of July/Aug.-2005.
    Mach angle: Please refer Q.8 (b) of Jan./Feb.-2006.

Q.8 (a) Differentiate between

(i) Streamline body and bluff body (ii) Friction drag and pressure drag   (8)

Ans.: (i) **Streamline body:**

– In streamline body, wake is small.

– So pressure drag also is reduced.

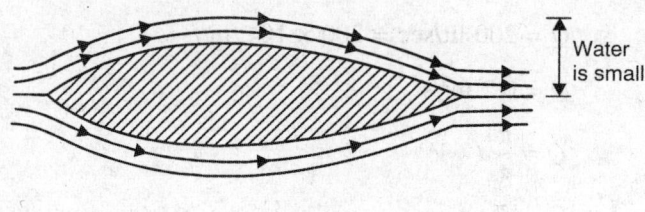

Fig. 9

– In streamline body, body shape will move.

The point of separation as much downstream as possible. e.g. aero foils.

Bluff body:

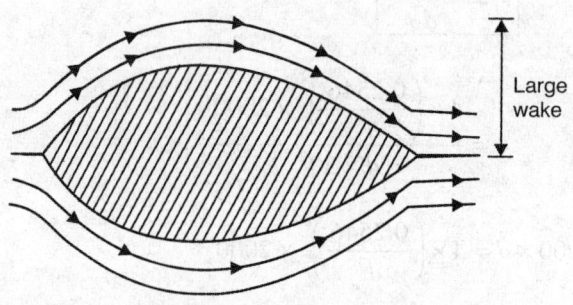

Fig. 10

- In bluff body, wake is large.
- So pressure drag is also increased.
- In body like bluff, body shape will move the point of separation as up as possible e.g. Padmini car.

(ii) **Friction drag and pressure drag:** Please refer Q.8 (b) of July/Aug. 2005.

(b) A man weighing 981 N descends to the ground from an aeroplane with the help of a parachute against the resistance of air. The shape of the parachute is hemispherical of 2 m diameter. Find the velocity of the parachute with which it comes down. Assume C_d = 0.5 and ρ for air 0.00125 gm/cc and γ = 0.015 stoke.   (8)

Ans. Average drag coefficient is,

$$C_d = \frac{1.328}{\sqrt{R_e}}$$

$$\therefore \qquad \sqrt{R_e} = \frac{1328}{C_d} = \frac{1.328}{0.5}$$

$\therefore$ $\qquad \sqrt{R_e} = 2.656$

$\therefore$ $\qquad \boxed{R_e = 7.054}$

$\therefore$ $\qquad R_e = \dfrac{VD}{\gamma}$

$$7.054 = \dfrac{V \times 2}{0.015 \times 10^{-4}} \qquad \left(\because \quad 1 \text{ stoke} = 10^{-4}\, \text{m}^2/\text{s} \right)$$

$$\boxed{V = 5.2905 \times 10^{-6}\, \text{m/s}}$$

(c) Define displacement thickness and momentum thickness. (4)

Ans. Please refer Q. 8 (a) (i) and (ii) of July/Aug. - 2005.

Fluid Mechanics
(Jan./Feb. 2006)
Semester - IV (ME/IP/AU/IM/MA)

Time: 3 Hours   Maximum Marks: 100

Note: Answer any five full questions.

Q.1 (a) Give reasons for the following:
  (i) The miniscus of water is concave upwards while the miniscus of mercury is convex upwards.
  (ii) Viscosity of liquids decrease on heating whereas viscosity of gases increase on heating.
  (iii) Rain drops and tiny dew drops are spherical in shape.   (6)
Solution. (i) Water has greater adhesion than cohesion, so it will wet a solid surface with which it is in contact and will tend to rise at the point of contact, with the result that the liquid (water) surface is concave upward and angle of contact θ is less than 90°.

On the other hand in mercury cohesion predominates, and it does not wet the solid surface and it will depress at the point of contact with the result that the liquid surface is convex upwards and angle of contact θ is greater than 90°.

(ii) Viscosity of liquids is governed by the cohesive forces between the molecules of the liquid. On increasing the temperature cohesion force decreases. Therefore viscosity decreases with increase in temperature. In case of gases molecular activity decides the viscosity of fluid. On increasing the temperature molecular activity in gases increases thus viscosity increases.

(iii) When a rain drop and tiny dew drops is separated initially from the surface of the main body of liquid, then due to surface tension there is net inward force exerted over the entire surface of the droplet which causes the surface of droplet to contract from all sides and result in increasing the internal pressure within the droplet. The contraction of the droplet continues till the inward force due to surface tension is in balance with the internal pressure and the droplets forms into sphere which is the shape for minimum surface area.

(b) Two litres of petrol weighs 14 N. Calculate specific weight, mass density, specific volume and specific gravity of petrol.   (2)
Solution. Vol. = 2 lit W = 14 N

$$V = 2 \times 10^{-3} \text{ m}^3$$

$$\text{Specific weight } (w) = \frac{W}{V} = \frac{14}{2 \times 10^{-3}} = 7 \times 10^3 \text{ N}/\text{m}^3$$

$$\text{Density } (\rho) = \frac{w}{g} = \frac{7000}{9.81} = 713.5 \text{ kg}/\text{m}^3$$

$$\text{Specific gravity} = \frac{\text{Density of liquid}}{\text{Density of water}} = \frac{713.5}{1000}$$

$$= 0.7135$$

$$\text{Specific volume} = \frac{1}{\rho} = \frac{1}{713.5} = 1.40 \times 10^{-3} \text{ m}^3/\text{kg}$$

(c) Derive the expressions for surface tension on a liquid droplet and soap bubble. (4)

Solution. Surface tension on a liquid droplet.

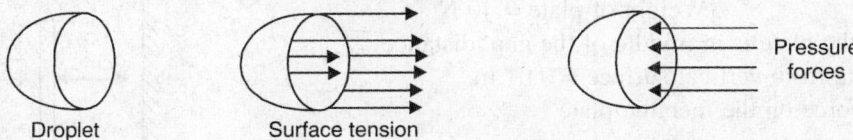

Droplet   Surface tension   Pressure forces

Fig. 1 Forces on droplet

Consider a small spherical droplet of radius r.

σ = Surface tension of the liquid

p = Pressure intensity inside the droplet

The forces acting on one half will be

(i) Tensile force due to surface tension acting around the circumference of the cut portion.

$$= \sigma \times \text{circumference} = \sigma \times \pi d$$

(ii) Pressure force on the area $\dfrac{\pi}{4} d^2 = p \times \dfrac{\pi}{4} d^2$

These forces will be equal and opposite under equilibrium conditions.

$$P \times \frac{\pi}{4} d^2 = \sigma \pi d \quad \therefore P = \frac{6 \times \pi d}{\dfrac{\pi}{4} d^2}$$

$$P = \frac{4\sigma}{d} \qquad \text{...Expression for surface tension}$$

Surface tension on a Hollow soap bubble.

It has two surfaces is contact with air, the inside and other outside. Thus two surfaces are subjected to surface tension.

$$\therefore \qquad \boxed{P = 2 \times \frac{4\sigma}{d} = \frac{8\sigma}{d}} \qquad \text{...Required expression}$$

(d) A vertical gap 22 mm wide of infinite extension contains a fluid of viscosity 20 N.s/m^2 and specific gravity 0.9. A metallic plate 1.2 m × 1.2 m × 0.2 cm is to be lifted up with a constant velocity of 0.15 m/s, through the gap. If the plate is assumed to be at the middle of the gap, determine the force required. The weight of the plate is 40 N. (8)

Solution.

$$\text{Width of gap} = 22 \text{ mm}$$
$$\mu = 2.0 \text{ Ns/m}^2$$
$$\text{Specific gravity of fluid} = 0.9$$
$$\text{Weight density of fluid} = 0.9 \times 1000 = 900 \text{ kg f/cm}^3$$
$$= 980 \times 9.81 \text{ N/m}^3$$
$$\text{Volume of plate} = 1.2 \times 1.2 \times 0.002$$
$$= 0.00288 \text{ m}^2$$

Thickness of plate = 0.2 cm

Velocity of plate = 0.15 m/sec

Weight of plate = 40 N

When the plate is in middle of the gap, distance of the plate from vertical surface = 0.01 m.

Shear force on the metallic plate

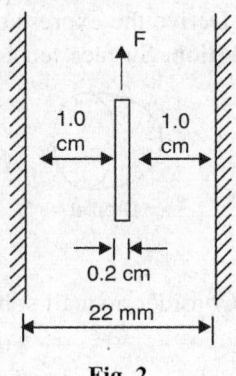

Fig. 2

$$= 2 \times \mu \times \left(\frac{du}{dy}\right) \times \text{Area}$$

$$= 2 \times 2.0 \times \left(\frac{0.15}{0.01}\right) \times 1.2 \times 1.2 \text{ N}$$

$$= 86.4 \text{ N}$$

upward thrust = weight of fluid displaced

= (weight density of fluid) × Volume of fluid displaced

= 9.81 × 980 × 0.00288

= 25.43 N

Net force acting in the downward direction

= weight of plate − upward thrust

= 40 − 25.43 = 14.57 N

∴   Total force required to lift the plate up

= Total shear force + Net downward force

= 86.4 + 14.57 = 100.97 N

Q.2 (a) (i) State stability criterion for submerged and floating bodies.

(ii) Derive an expression for metacentric height of floating body using analytical method.

(10)

Solution. (i) **Stability criterion for submerged body:** Please refer Q. 3 (c) of July/Aug. 2005.

(ii) **Metacentric height of floating body using analytical method:** When a floating body (a) in equilibrium position is given a small angular displacement in the clockwise direction, the new centre of buoyancy is at B_1.

The vertical line through B_1 cuts the normal axis at m. Hence m is the metacentre and Gm is metacentric height.

Consider towards the right of the axis a small strip of thickness dx at a distance x from O. The height of strip × $\angle BOB' = x \times \theta$

∴ Area of strip = Height × Thickness = $x \cdot \theta \cdot dx$

If L is the length of the floating body.

Volume of strip = Area × L = $x \cdot \theta \cdot L \cdot dx$

Weight of strip = $\rho \cdot g \times$ volume

$= \rho \cdot g \cdot x \cdot \theta \cdot L \cdot dx$

Similarly, if a small trip of thickness dx at a distance x from 0 towards the left of the axis is considered the weight of strip will be $\rho \cdot g \cdot x \cdot \theta \cdot L \cdot dx$. The two weights are acting in the opposite direction and hence constitute a couple.

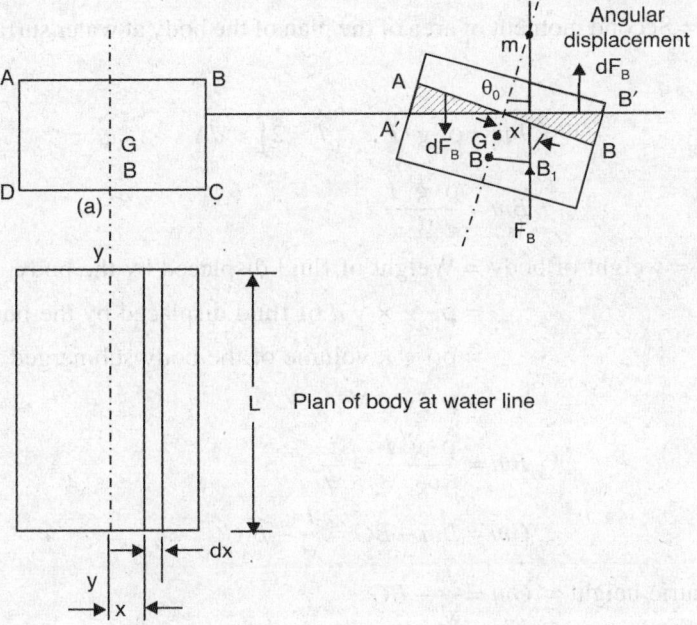

Plan of body at water line

Fig. 3

Moment of this couple = Weight of each strip × distance between two weights

$$= \rho \cdot g \cdot x \cdot \theta \cdot L \cdot dx \, [x + x]$$

$$= 2 \cdot \rho \cdot g \cdot x^2 \cdot \theta \cdot L \, dx$$

∴ Moment of couple for the whole wedge

$$= \int 2\rho \cdot g \cdot x^2 \theta \cdot L \, dx \qquad \qquad ...(i)$$

Moment of couple due to shifting of centre of buoyancy from B to B_1

$$= F_B \times B B_1 = F_B \times Bm \times \theta$$

$$= W \times Bm \times \theta \qquad \qquad ...(ii)$$

$$B B_1 = Bm \times \theta$$

∵ θ is very small and $F_B = W$.

But these two couples are the same

Hence equating equation (i) and (ii)

$$W \times Bm \times \theta = \int 2\rho \cdot g \cdot x^2 \cdot \theta \cdot L \, dx$$

$$W \times Bm = 2 \cdot \rho \cdot g \int x^2 L \, dx$$

$$L \, dx = \text{Elemental area on the water line}$$

$$W \times Bm = 2 \cdot \rho \cdot g \int x^2 \, dA$$

But $2 \int x^2 dA$ = Second moment of area of the plan of the body at water surface about the axis $y - y$.

$\therefore$
$$W \times Bm = \rho \cdot g \cdot I \qquad I = 2 \int x^2 dA$$

$$Bm = \frac{\rho \cdot g \cdot I}{W}$$

W = weight of body = Weight of fluid displaced by the body

$$= \rho \cdot g \times v \, d \text{ of fluid displaced by the body}$$

$$= \rho \cdot g \times \text{volume of the body submerged in water}$$

$$= \rho \cdot g \times \forall$$

$\therefore$
$$Bm = \frac{\rho \cdot g \cdot I}{\rho \cdot g \cdot \forall} = \frac{i}{\forall}$$

$$Gm = Bm - BG = \frac{I}{\forall} - B.G.$$

$\therefore$ metacentric height = $Gm = \dfrac{I}{\forall} - BG$

(b) A cone of specific gravity S floating in water with its apex downwards, has radius R and vertical height H. Show that for stable equilibrium of the cone.

$$H < \left[\frac{R^2 S^{1/3}}{1 - S^{1/3}} \right]^{1/2} \qquad\qquad (10)$$

Solution. Consider a cone of Bax radius R, vertical height H and specific gravity S.

Let G = Centre of gravity of cone

         B = Centre of buoyancy

         2θ = Apex angle

         A = Apex of cone

         h = height of immersion

         d = diameter of cone at water surface

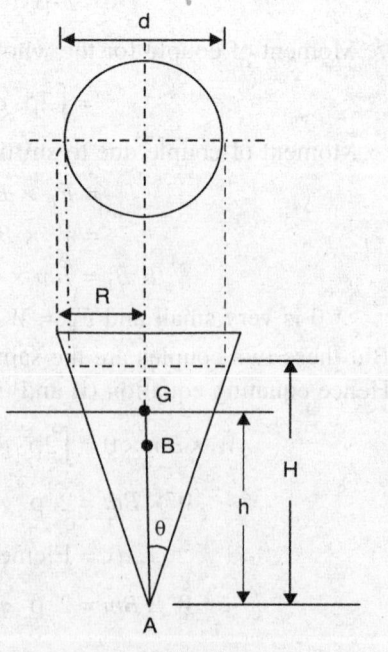

Then $AG = \dfrac{3}{4} H$

$$AB = \frac{3}{4} h$$

Weight of cone = Weight of water displaced

$$1000 \times s \times g \times \frac{I}{3} \pi R^2 \times H = 1000 \times g \times \frac{1}{3} \pi r^2 \times h$$

i.e. $S \, R^2 H = r^2 h$

$$h = \frac{SR^2 H}{r^2}$$

Fig. 4

But $\qquad \tan \theta = \dfrac{R}{H} = \dfrac{r}{h}$

$\therefore \qquad R = H \tan \theta, \; r = h \tan\theta$

$\therefore \qquad h = \dfrac{S\left(H \tan \theta\right)^2}{\left(h \tan \theta\right)^2} = \dfrac{SH^3}{h^2}$

or $\qquad h^3 = S\,H^3$

or $\qquad h = S^{1/3}H$

Distance $\qquad BG = AG - AB$

$$= \frac{3}{4}H - \frac{3}{4}h = \frac{3}{4}(H - h)$$

$$= \frac{3}{4}\left(H - S^{1/3}H\right)$$

$$= \frac{3}{4}H\left(1 - S^{1/3}\right)$$

Also I = Moment of inertia of the plan of body at the water surface $= \dfrac{\pi}{64}d^4$

$\forall$ = Volume of cone in water

$$= \frac{1}{3} \times \frac{\pi}{4} \times d^2 \times h = \frac{1}{3} \times \frac{\pi}{4} \times d^2 \left(HS^{1/3}\right)$$

$$\frac{I}{\forall} = \frac{\dfrac{\pi}{64}d^4}{\dfrac{1}{3} \times \dfrac{\pi}{4} \times d^2 HS^{1/3}} = \frac{3d^2}{16\,HS^{1/3}}$$

Now metacentric height $GM = \dfrac{I}{\forall} - BG$

$\therefore \qquad Gm = \dfrac{3d^2}{16HS^{1/3}} - \dfrac{3H}{4}\left(1 - S^{1/3}\right)$

Gm should be positive for stable equilibrium or Gm > 0

i.e. $\qquad \dfrac{3d^2}{16HS^{1/3}} - \dfrac{3H}{4}\left(1 - S^{1/3}\right) > 0$

$$\frac{3d^2}{16HS^{1/3}} > \frac{3H}{4}\left(1 - S^{1/3}\right)$$

Also $$R = H \tan\theta \text{ and } r = h \tan\theta$$

$$\frac{R}{r} = \frac{H}{h} = \frac{D}{d}$$

$$d = \frac{Dh}{H} = \frac{D}{H} HS^{1/3} = DS^{1/3}$$

Substituting the value of d in equation

$$\frac{3\left(DS^{1/3}\right)^2}{16HS^{1/3}} > \frac{3H}{4}\left(1 - S^{1/3}\right)$$

or

$$\frac{D^2 S^{1/3}}{4H} > H\left(1 - S^{1/3}\right)$$

$$\frac{b^2 S^{1/3}}{4\left(1 - S^{1/3}\right)} > H^2$$

$\therefore$

$$H^2 < \frac{\dfrac{D^2}{4} S^{1/3}}{\left(1 - S^{1/3}\right)} \quad \text{But} \quad \frac{D^2}{4} = R^2$$

$\therefore$

$$H^2 < \frac{R^2 S^{1/3}}{1 - S^{1/3}}$$

$\therefore$

$$H < \left(\frac{R^2 S^{1/3}}{1 - S^{1/3}}\right)^{1/2} \qquad \text{...Hence proved.}$$

Q.3 (a) Derive the expression for depth of centre of pressure from the surface of liquid of the inclined plane surface immersed in fluid and hence prove that centre of pressure is always below the centre of gravity for an immersed surface. (6)

Solution. Please refer Q. 2(b) of July/Aug.-2005.

(b) A U-tube manometer is used to measure the pressure of water in a pipe line which is in excess of atmospheric pressure. The right limb of the manometer contains mercury and is open to atmosphere. The contact between water and mercury is in the left limit Determine the pressure of water in the main line, if the difference in level of mercury in the limbs of U-tube is 10 cm and the free surface of mercury is in level with the centre of the pipe. If the pressure of water in pipe line is reduced to 9810 N/m^2 calculate the new difference in level of mercury, Sketch the arrangements in both the cases. (8)

Solution. Difference in mercury = 10 cm = 0.1 m

Let P_A = Pressure of water in pipe line.

$$\text{Pressure at } B = \text{pressure at } C$$

$\because$ Both are at same level.

$\therefore$ P_B = Pressure of A + Pressure due to 10 cm of water

$= P_A + \rho \cdot g \cdot h$

$P_B = P_A + 1000 \times 9.81 \times 0.1 = (P_A + 981) \, \text{N/m}^2$ ...(i)

Pressure at C = Pressure at D + Pressure due to 10 cm of mercury

$= 0 + \rho + g \times h_0$

$= 13.6 \times 10^3 \times 9.81 \times 0.1 = 13341.6 \, \text{N}$ ...(ii)

Equating equation (i) and (ii)

$P_A + 981 = 13341.6$

$\therefore$ $P_A = 12360.6 \, \text{N/m}^2$ **Ans.**

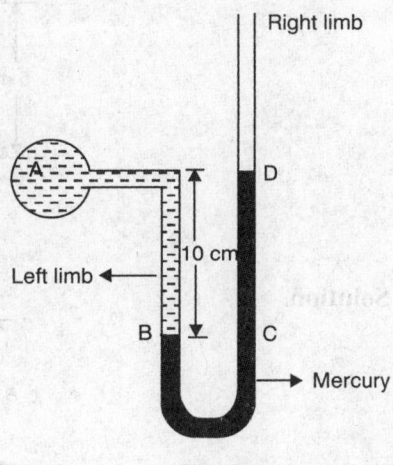

Fig. 5

If pressure of water in pipe line is reduced to 9810 N/m³

i.e. $P_A = 9810 \, \text{N/m}^2$

Let x = rise of mercury in left limb in cm

= fall of mercury in right limb in cm

Pressure at B' = Pressure at C'

Pressure at A + Pressure due to $(10 - x)$ cm of water
= Pressure at D' + pressure due to $(10 - 2x)$ cm of mercury.

or $P_A - \rho_1 g h_1 = P_D' + \rho_2 g \cdot h_2$

$9810 + 1000 \times 9.81 \times \dfrac{(10 - x)}{100}$

$= 0 + (13.6 \times 10^3) \times 9.81 \times \dfrac{(10 - 2x)}{100}$

$272x - 10x = 1360 - 1100$

$262x = 260$

$x = \dfrac{260}{262} = 0.992 \, \text{cm}$

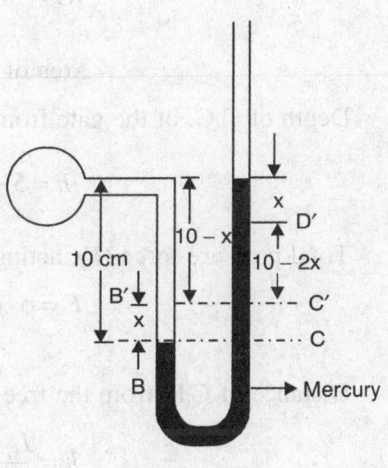

Fig. 6

Now difference of mercury = $10 - 2x$

$= 10 - 2 \times 0.992 = 8.016 \, \text{cm}$ **Ans.**

(c) An inclined rectangular sluice gate AB hinged at A of size 1.2 m × 1 m, as shown in the figure is installed to control the discharge of water. Determine the normal force to be applied at B to open the gate. (6)

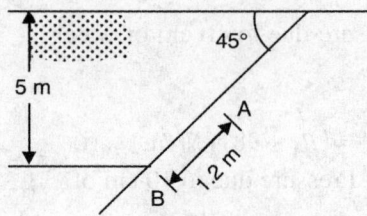

Fig. 7

Solution.

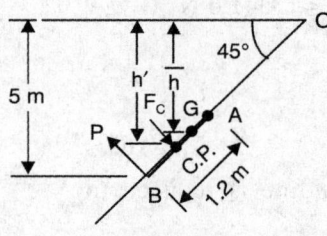

Fig. 8

$$BG = \frac{1}{2}AB = 0.6 \text{ m}$$

$$IG = MOI \text{ of gate} = \frac{1 \times AB}{12} = \frac{1 \times (1.2)^3}{12} = 0.144 \text{ m}$$

Area of gate = 1.2 × 1 = 1.2 m²

Depth of C.G. of the gate from the free surface

$$h = 5 - BG \sin 45° = 5 - 0.6 \times \frac{1}{\sqrt{2}} = 4.576 \text{ m}$$

Total pressure force (*F*), acting on the gate of C.P.

$$F = \rho \cdot g \cdot A \cdot \overline{h} = 1000 \times 9.81 \times 1.2 \times 4.576$$

$$= 53868.67 \text{ N}$$

Distance of C.P. from the free surface

$$h = \frac{I_G \sin^2 \theta}{A\overline{h}} + \overline{h} = \frac{0.144 \times \sin^2 45°}{1.2 \times 4.576} + 4.576$$

$$= 4.589 \text{ m}$$

Distance of CP from O $CPO = \dfrac{h'}{\sin 45°}$

$$= \frac{4.589}{\sin 45°} = 6.4899 \text{ m}$$

Distance of B from O $\quad B = \dfrac{5}{\sin 45°} = 7.071$ m

$\therefore \qquad$ Distance $BCP = BO - CPO$

$\qquad\qquad\qquad = 7.071 - 6.4890 = 0.5811$ M

and distance $\qquad\quad ACP = 1.2 - BCP$

$\qquad\qquad\qquad = 1.2 - 0.5811 = 0.6188$ m

Taking moment about the hinge A

$$P \times AB = F \times ACP$$

$\therefore \qquad\qquad\qquad P = $ Force applied at B to open the gate

$$= \frac{F \times ACP}{AB} = \frac{53868.67 \times 0.6188}{1.2}$$

$$= 27779.72 \text{ N} \qquad\qquad\qquad\qquad \textbf{Ans.}$$

Q.4 (a) Explain Lagrangian and Eulerian approaches of fluid flow study. (6)

Solution. Two methods are adopted to describe fluid motion namely Lagrangian method and Eulerian method.

The Lagrangian method describes the motion of a single fluid particles. The Eulerian method describes the velocity, pressure and all such characteristics at a given point in space.

In Lagrangian method study of single fluid particle with the passage of time, its path traced, its velocities and accelerations is included.

The Eulerian method is commonly adopted. Suppose x, y and z denotes the space coordinates.

Let v be the resultant velocity at any point in space in a fluid body. Let μ, v and w be the x components, the y components and z components of the resultant velocity v.

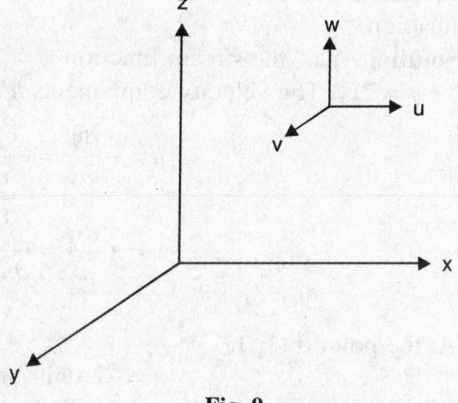

Fig. 9

In Eulerian method the velocity at a point (x, y, z) at time t can be expressed as

$$\mu = f_1(x, y, z, t) \qquad v = f_2(x, y, z, t)$$
$$w = f_3(x, y, z, t) \qquad v = f_4(x, y, z, t)$$

The velocity components μ is some function of the space co-ordinates x, y, z and the time t. Hence for certain values of x, y, z and t there is a corresponding value of μ.

(b) Define:

(i) Path line (ii) Stream line (iii) Streak line (iv) Streak tube. (4)

Solution. (i) **Path line:** Please refer Q. 4(a) (ii) of July/Aug.-2005.

(ii) **Stream line:** Please refer Q. 4 (a) (i) of July/Aug.-2005.

(iii) **Streak line:** It is the locus of the positions of fluid particles which have passed through a given point in space in succession.

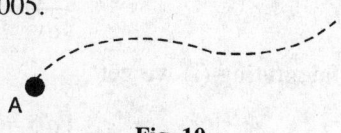

Fig. 10

Consider a space point $A(x_1 \, y_1 \, z_1)$. The various fluid particles which have passed through A in succession lie in some position at certain instant. The locus of these points constitutes the streak line.

(iv) **Stream tube:** An imaginary tabular space formed by a number of stream lines. Stream tube is a collection of stream lines which would form a very small closed passage.

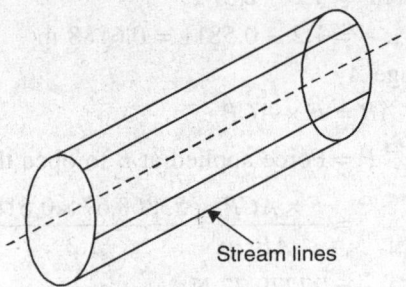

Stream lines

Fig. 11

(c) Does the stream function $\psi = 2xy$ represent a physically possible flow? If so, determine the velocity at point (1, 1) and sketch the stream function. Also determine the velocity potential function.

Solution. Yes, the stream function $\psi = 2xy$ represents a physically possible flow.

$\psi = 2xy$. The velocity components u and v are in terms of ψ are—

$$u = -\frac{\partial \psi}{\partial y} = -\frac{\partial}{\partial y}(2xy) = -2x$$

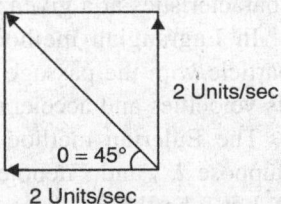

2 Units/sec

$0 = 45°$

2 Units/sec

$$v = +\frac{\partial \psi}{\partial x} = \frac{\partial}{\partial x}(2xy) = 2y$$

At the point P (1, 1)

Fig. 12

$$u = -2 \text{ units/sec}$$
$$v = 2 \text{ units/sec}$$

$\therefore$ Resultant velocity at $P(1, 1) = \sqrt{u^2 + v^2}$

$$= \sqrt{4 + 4} = 2.828 \text{ units/sec}$$

Velocity potential function ϕ

$$\frac{\partial \phi}{\partial x} = -u = -(-2x) = 2x \qquad \qquad ...(i)$$

$$\frac{\partial \phi}{\partial y} = -v = -2y \qquad \qquad ...(ii)$$

Integrating (i) we get

$$\int d\phi = \int 2x \, dx$$

or
$$\phi = \frac{2x^2}{2} + c = x^2 + c \qquad \ldots\text{(iii)}$$

where c is a constant which is independent of x but can be a function of y.
Differentiating w.r.t. y we get

$$\frac{\partial \phi}{\partial y} = \frac{\partial c}{\partial y}$$

But from (ii)
$$\frac{\partial \phi}{\partial y} = -2y$$

∴
$$\frac{\partial c}{\partial y} = -2y$$

Integrating this equation we get

$$c = \int -2y \, dy = -\frac{2y^2}{2} = -y^2$$

Substituting this value of c in equation (iii) we get

$$\phi = x^2 - y^2 \qquad \qquad \textbf{Ans.}$$

$$\tan\theta = \frac{v}{u} = \frac{2}{+2} = +1$$

∴
$$\theta = +45°$$

∴ Resultant velocity makes an angle of 45° with x axis.

$$\psi = 2xy$$

Let $\psi = 1, 2, 3$ and so on.
We have
$$xy = 1/2$$

$$xy = 1$$

$$xy = \frac{3}{2}$$

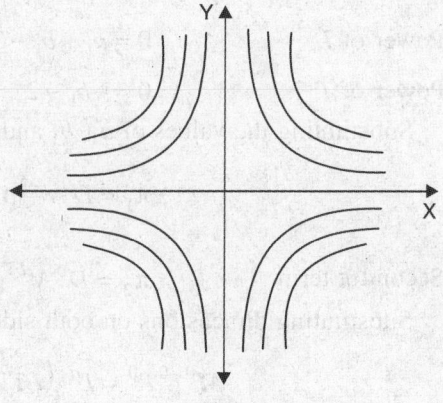

Fig. 13

Each equation is a equation of line parallel to each other.

Q.5 (a) The pressure difference ΔP in a pipe of diameter D and length L due to turbulent flow depends on the velocity V, viscosity μ, density ρ and roughness K. Using Buckingham's π theorem, obtain an expression for ΔP in the form of Darcy-Weisbach equation. (10)
Solution. ΔP is a function of D, l, V, μ, ρ, k.

$$\Delta P = f(D, l, V, \mu, \rho, k) \text{ or } f_1(\Delta, P, D, l, V, \mu, \rho, k) = 0 \qquad \ldots\text{(i)}$$

Total number of variables = 7
Dimensions of $\Delta P = ML^{-1}T^{-2}$, $D = L$, $l = L$, $V = LT^{-1}$

$$\mu = ML^{-1} T^{-1}, \rho = ML^{-3}, k = L$$

$\therefore$ Total number of fundamental dimensions = 3

$\therefore$ Number of π terms = $n - m = 7 - 3 = 4$.

Now equation (i) can be grouped in 4 π term as

$$f_1\left(\pi_1, \pi_2, \pi_3, \pi_4\right) = 0$$

Each π term contains $m + 1$ or $3 + 1 = 4$ variables. Out of four variables, there are repeating variables. Choosing D, V, e as the repeating variables, we have the four π terms as,

$$\pi_1 = D^{a_1} \cdot V^{b_1} \cdot \rho^{c_1} \cdot \Delta p$$

$$\pi_2 = D^{a_2} \cdot V^{b_2} \cdot \rho^{c_2} \cdot l$$

$$\pi_3 = D^{a_3} \cdot V^{b_3} \cdot \rho^{c_3} \cdot \mu$$

$$\pi_4 = D^{a_4} \cdot V^{b_4} \cdot \rho^{c_4} \cdot k$$

First π term $\qquad \pi_1 = D^{a_1} V^{b_1} \rho^{c_1} \Delta p$

Substituting dimensions on both sides

$$M^0 L^0 T^0 = L^{a_1}\left(LT^{-1}\right)^{b1}\left(ML^{-3}\right)^{c1}\left(ML^{-1}T^{-2}\right)$$

Equating the powers of M, L and T on both sides.

Power of M $\qquad 0 = c_1 + 1 \qquad \therefore c_1 = -1$

Power of L $\qquad 0 = a_1 + b_1 - 3c_1 - 1 \quad \therefore a_1 = -b_1 + 3c_1 + 1 = 2 - 3 + 1 = 0$

Power of T $\qquad 0 = -b_1 - 2 \qquad \therefore b_1 = -2$

Substituting the values of a_1, b_1 and c_1 in π_1

$$\pi_1 = D^0 V^{-2} \rho^{-1} \Delta P = \frac{\Delta P}{\rho V^2}$$

Second π term $\qquad \pi_2 = D^{a_2} V^{b_2} \rho^{c_2} l$

Substituting dimensions on both sides

$$M^0 L^0 T^0 = L^{a_2}\left(LT^{-1}\right)^{b_2}\left(ML^{-3}\right)^{c_2} L$$

Equating the powers of M, L and T on both sides

Power of M $\qquad 0 = c_2 \qquad\qquad\qquad \therefore c_2 = 0$

Power of L $\qquad 0 = a_2 - b_2 - 3c_2 + 1 \qquad \therefore a_2 = b_2 + 3c_2 - 1 = -1$

Power of T $\qquad 0 = -b_2 \qquad\qquad\qquad \therefore b_2 = 0$

Substituting the values of a_2, b_2 and c_2 in π_2

$$\pi_2 = D^{-1} V^0 \rho^0 l = l/D$$

Third π term $\qquad \pi_3 = D^{a_3} V^{b_3} \rho^{c_3} \mu$

Substituting dimensions on both sides

$$M^0 L^0 T^0 = L^{a_3} \left(LT^{-1}\right)^{b_3} \left(ML^{-3}\right)^{c_3} ML^{-1}T^{-1}$$

Equating the powers of M, L and T on both sides

Power of M $0 = c_3 + 1$ $\therefore c_3 = -1$

Power of L $0 = a_3 + b_3 - 3c_3 - 1$ $\therefore a_3 = -b_3 + 3c_3 + 1$

                                                             $= 1 - 3 + 1 = -1$

Power of T $0 = -b_3 - 1$ $\therefore b_3 = -1$

Substituting the values of a_3, b_3, c_3 in π_3

$$\pi_3 = D^{-1} V^{-1} \rho^{-1} \pi = \frac{\mu}{DV\rho}$$

Fourth π term $\pi_4 = D^{a_4} V^{b_4} \rho^{c_4} k$

$$M^0 L^0 T^0 = L^{a_4} \left(LT^{-1}\right)^{b_4} \left(ML^{-3}\right)^{c_4} L$$

Substituting dimensions on both sides

Power of M $0 = c_4$ $\therefore c_4 = 0$

Power of L $0 = a_4 - b_4 - 3c_4 + 1$ $\therefore a_4 = b_4 + 3c_4 - 1 = -1$

Power of T $0 = -b_4$ $\therefore b_4 = 0$

Substituting the values of a_4, b_4, c_4 in π_4

$$\pi_4 = D^{-1} V^0 \rho^0 k = \frac{K}{D}$$

Substituting the values of π_1, π_2, π_3, π_4 in equation (ii) we get

$$f_1 \left(\frac{\Delta \rho}{\rho V^2}, \frac{l}{D}, \frac{\mu}{DV\rho}, \frac{K}{D} \right) = 0$$

or $$\frac{\Delta P}{\rho V^2} = \phi \left[\frac{l}{D}, \frac{\mu}{DV\rho}, \frac{K}{D} \right]$$

From experiments, it was observed that pressure difference, ΔP is a linear function of $\dfrac{l}{D}$ and hence it is taken out of function.

$$\frac{\Delta p}{\rho V^2} = \frac{l}{D} \phi \left[\frac{\mu}{DV\rho}, \frac{K}{D} \right]$$

$$\frac{\Delta p}{\rho} = \frac{V^2 l}{D} \phi \left[\frac{\mu}{DV\rho}, \frac{K}{D} \right]$$

Dividing by g on both sides

$$\frac{\Delta p}{\rho g} = \frac{V^2 \cdot l}{gD} \phi\left[\frac{\mu}{DV\rho}\frac{K}{D}\right]$$

Now $\phi\left[\frac{\mu}{DV\rho}, \frac{K}{D}\right]$ contains two terms,

$$\frac{\mu}{DV\rho} = \frac{1}{R_e}, \quad \frac{K}{D} = \text{Roughness factor.}$$

$\therefore$ $$\phi\left[\frac{1}{R_e}, \frac{K}{D}\right] = f$$

where f is the co-efficient of fraction.

$\therefore$ $$\frac{\Delta p}{\rho g} = \frac{4f}{2}\frac{V^2 l}{gD} \qquad\qquad \because f = p\left(\frac{\mu}{DV\rho}, \frac{K}{D}\right)$$

Multiplying or dividing by any constant does not change the character of π terms.

$\therefore$ $$\frac{\Delta p}{\rho g} = hf = \frac{4flV^2}{2gD} \text{ in Daray weisback equation.}$$

(b) State Bernoulli's theorem for an incompressible flow stating all the assumptions. Write Bernoulli's equation applied between 2 points on a stream line in an ideal fluid and in a real fluid. (4)

Solution. Bernoulli's theorem for steady flow of an incompressible fluid states that the total energy at any point of the fluid is constant. The total energy consists of pressure energy. Kinetic energy and potential energy or data energy.

$\therefore$ $$\frac{P}{W} + \frac{V^2}{2g} + z = \text{constant.}$$

Assumptions
  (i) Fluid is ideal i.e., viscosity is zero.
 (ii) Flow is steady.
(iii) The flow is incompressible.
(iv) The flow is irrotational.
   Bernoulli's equation for ideal fluid.

$$\frac{p}{W} + \frac{v^2}{2g} + z = \text{constant}$$

Bernoulli's equation for real fluid

$$\frac{p_1}{W} + \frac{v_1^2}{2g} + z_1 = \frac{p_2}{W} + \frac{v_2^2}{2g} + z_2 + h_1$$

$$h_1 = \text{loss of energy between points 1 and 2}$$

(c) A pipe of 300 mm diameter conveying 300 litres/sec of water has a right angled bend in a horizontal plane. Find the resultant force exerted on the bend if the pressure at inlet and outlet of the bend are 245.55 kPa and 235.44 kPa respectively. (6)

Solution. Given

$$D = 300 \text{ mm} = 0.3 \text{ m}$$

$$A = A_1 = A_2 = \frac{\pi}{4}(0.3)^2 = 0.07068 \text{ m}^2$$

$$Q = 300 \text{ lit/sec} = 0.3 \text{ m}^3/\text{s}$$

$$V = V_1 = V_2 = \frac{Q}{A} = \frac{0.3}{0.07068} = 4.244 \text{ m/s}$$

$$\theta = 90°$$

$$P_1 = 245.25 \text{ kPa} = 245.250 \text{ N/m}^2$$

$$P_2 = 235.44 \text{ kPa} = 235440 \text{ N/m}^2$$

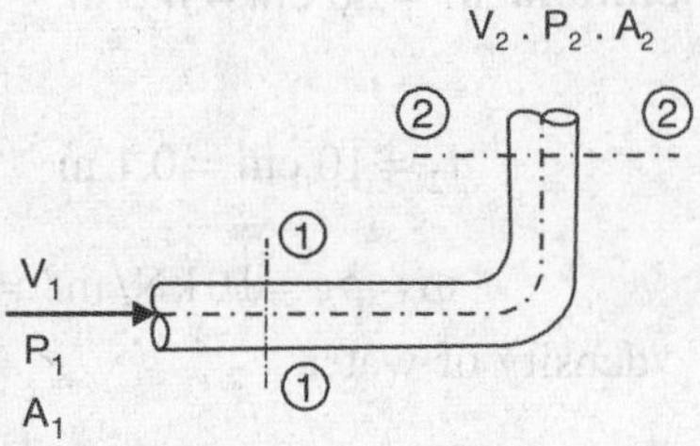

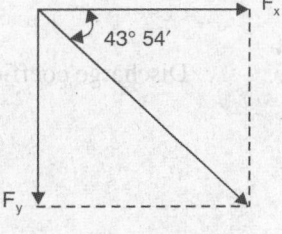

Fig. 14

Force of bend along x-axis

$$F_x = \rho Q(V_1 x - V_2 x) + (P_1 A_1)_x + (P_2 A_2)_x$$

$$= 1000 \times 0.3(4.244 - 0) + 245250 \times 0.07068 + 0$$

$$= 18607.5 \text{ N}$$

Force on bend along y axis

$$f_y = \rho Q(v_{1y} - v_{2y}) + (P_1 A_1)_y + (P_2 A_2)_y$$

$$= 1000 \times 0.3(0 - 4.244) + 0 - (1235440 \times 0.07068)$$

$$= -17914.1 \text{ N}$$

$\therefore$ Resultant force $F_R = \sqrt{F_x^2 + F_y^2}$

$$F_R = \sqrt{(18607.5)^2 + (-17914.1)^2}$$

$$= 25829.3 \text{ N}$$

$$\tan\theta = \frac{f_y}{f_x} = \frac{-17914.1}{18607.5} = -0.9627$$

$$\theta = -43°54'$$

Fig. 15

Ans.

Q.6 (a) State the working principle of pitot tube. (2)

Solution. Working principle of pitot tube: If the velocity of flow at a point becomes zero, the pressure there is increased due to the conversion of the kinetic energy into pressure energy.

(b) A horizontal venturimeter with 30 cm diameter inlet and 10 cm throat is used for measuring the flow of water through a pipeline. If the pressure in pipe is 150 kPa and the vacuum pressure at the throat is 40 cm of mercury, calculate the rate of flow. Assume 5% of differential head is lost between the pipe main and the throat section. Find the discharge coefficient. Take specific weight of water as 10 kN/m³. (10)

Solution. $d_1 = 30$ cm $= 0.3$ m $\qquad a_1 = \dfrac{\pi}{4}(0.3)^2 = 0.07069$ m^2

$\qquad d_2 = 10$ cm $= 0.1$ m $\qquad a_2 = \dfrac{\pi}{4}(0.1)^2 = 7.853 \times 10^{-3}$ m^2

$\qquad \omega = \rho g = 10\,\text{kN}/\text{m}^3 = 10 \times 10^3\,\text{N}/\text{m}^3$

density of water

$$\rho = \frac{10 \times 10^3}{9.31} = 1019.37 \text{ kg}/\text{m}^3$$

$$p_1 = 150 \text{ kpa} = 150 \times 10^3\,\text{N}/\text{m}^2$$

$$\frac{p_1}{\rho g} = \frac{150 \times 10^3}{1019.17 \times 9.81} = 15 \text{ m of water}$$

$$\frac{p_2}{\rho g} = -40 \text{ cm of mercury} = -\frac{40 \times 13.6}{100 \times 1.0194}$$

$$= -5.34 \text{ m of water}$$

$\therefore \qquad$ Differential head $= \dfrac{p_1}{\rho g} - \dfrac{p_2}{\rho g} = 15 - (-5.34)$

$$= 20.34 \text{ m of water}$$

Head lost $h_f = 4\%$ of h

$$= \frac{4}{100} \times 20.39 = 0.8136$$

$\therefore \qquad$ Discharge coefficient $c_d = \sqrt{\dfrac{h - h_f}{h}}$

$$= \sqrt{\frac{20.34 - 0.8136}{20.34}} = 0.98$$

$\therefore \qquad$ Discharge $\theta = c_d \dfrac{a_1 a_2}{\sqrt{a_1^2 - a_2^2}} \sqrt{2gh}$

$$= 0.98 \times \frac{0.07069 \times 7.853 \times 10^{-3}}{\sqrt{(0.07069)^2 - (7.853 \times 10^{-3})^2}} \times \sqrt{2 \times 9.81 \times 20.34}$$

$$= 0.15466 \text{ m}^3/\text{s.} \qquad\qquad\qquad \textbf{Ans.}$$

(c) Derive Hagen-Poiseuille equation for flow through circular pipes. (8)

Solution. Reference line. It is obtained by joining the tops of all vertical ordinates showing the sum of pressure head and kinetic head from the centre of the pipe.

Hydraulic gradient line: It is defined as the line which gives the sum of pressure head $\left(\dfrac{P}{w}\right)$ and datum head (z) of a flowing fluid in a pipe with respect to some reference line or it is the line which is obtained by joining the top of all vertical ordinates showing the pressure head (P/w) of a flowing fluid in a pipe from the centre of the pipe.

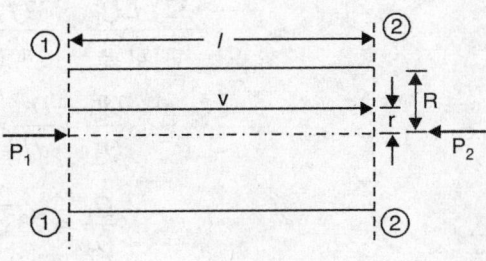

Fig. 16

(b) Two pipes of diameter D and d with equal length L are considered. If the pipes are arranged in parallel, the loss of head for either pipe when a total quantity of water flows through them is h. If the pipes are arranged in series and the same quantity Q flows through them, the loss of head is H. If $d = 0.5 D$, find the percentage of total flow through each pipe when placed in parallel and the ratio of H to h. Neglect minor losses and assume friction co-efficients to be constants. (10)

Solution.

Given $\qquad L_1 = L \quad d_1 = D$

$\qquad\qquad\qquad L_2 = L \quad d_2 = d$

Total discharge $= Q$
Head loss when pipes are arranged parallel $= h$
Head loss when pipes are arranged in series $= H$
$d = 0.5$ D and $f =$ constant
1st case $\rightarrow$ when the pipes are connected to parallel
$Q = Q_1 + Q_2$ loss of head in each pipe $= h$
For pipe AB

$$\therefore \qquad h = \frac{4 f L_1 V_1^2}{2 g d_1} \qquad V_1 = \frac{Q_1}{A_1} = \frac{Q_1}{\frac{\pi}{4} D^2} = \frac{4 Q_1}{\pi D^2} \text{ and } d_1 = D$$

$$\therefore \qquad h = \frac{4 f l \times \left(\dfrac{4 Q_1}{\pi D^2}\right)^2}{2 g D} = \frac{32 f L Q_1^2}{\pi^2 D^5 g} \qquad \ldots\text{(i)}$$

For pipe AC

$$h = \frac{32 f L Q_2^2}{\pi^2 d^5 \times g} \qquad \ldots\text{(ii)}$$

Equating (i) and (ii)

$$\frac{32 f L Q_1^2}{\pi^2 D^5 g} = \frac{32 f L Q_2^2}{\pi^2 d^5 g}$$

i.e.

$$\frac{Q_1^2}{Q_2^2} = \frac{D^5}{d^5} = \frac{D^5}{(0.5D)^5} = 32$$

$\therefore$

$$\frac{Q_1}{Q_2} = \sqrt{32} = 5.657$$

or

$$Q_1 = 5.657 Q_2$$

$\therefore$

$$Q = 5.657 Q_2 + Q_2 = 6.657 Q_2$$

$$Q_2 = \frac{Q}{6.657} = 0.15 Q$$

and

$$Q_1 = Q_1 - Q_2 = Q - 0.15Q = 0.85Q$$

Case II:

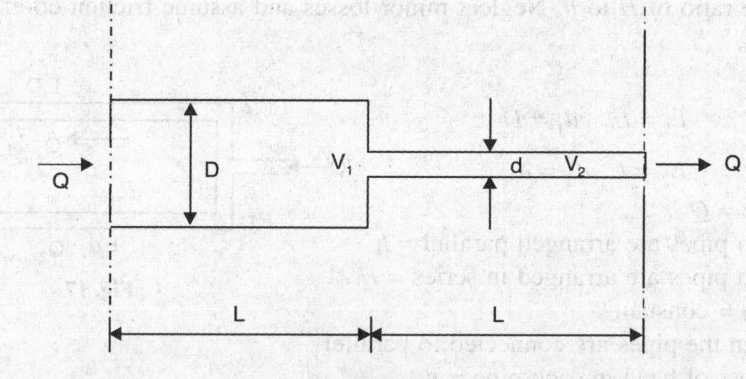

Fig. 18

When the pipes are connected in series.

Total loss = Sum of head losses in two pipes

$$H = \frac{4 f L V_1^2}{d_1 \times 2g} + \frac{4 f L V_2^2}{d^2 \times 2g}$$

$$V_1 = \frac{Q}{\dfrac{\pi}{4} D^2} = \frac{4Q}{\pi D^2}$$

$$V_2 = \frac{Q}{\dfrac{\pi}{4} d^2} = \frac{4Q}{\pi d^2}$$

$$H = \frac{4fL \times \left(\frac{4Q}{\pi D^2}\right)^2}{D \times 2g} + \frac{4fL\left(\frac{4Q}{\pi d^2}\right)^2}{d \times 2g}$$

$$= \frac{32 fLQ^2}{D^5 \pi^2 \times g} + \frac{32 f LQ^2}{d^5 \pi^2 \times g}$$

But $\dfrac{32 fL}{\pi^2 D^5 \times g} = \dfrac{h}{Q_1^2}$ from equation (*i*)

$$\frac{32 fL}{\pi^2 d^5 \times g} = \frac{h}{Q_2^2}$$

$$H = Q^2 \times \frac{h}{Q_1^2} + Q^2 \times \frac{h}{Q_2^2}$$

$$H = \left[\frac{Q^2}{Q_1} + \frac{Q^2}{Q_2}\right]h$$

$\therefore$ $\qquad \dfrac{H}{h} = \dfrac{Q^2}{(0.85Q)^2} + \dfrac{Q^2}{(0.15Q)^2}$ $\qquad\qquad \because Q_1 = 0.85Q$

$$\text{and } Q_2 = 0.15Q$$

$\therefore$ $\qquad \dfrac{H}{h} = \dfrac{1}{(0.85)^2} + \dfrac{1}{(0.15)^2}$

$$= 1.384 + 44.444$$

$$\frac{H}{h} = 45.828 \qquad\qquad\qquad \textbf{Ans.}$$

(c) Find the displacement thickness and momentum thickness for the velocity distribution in the boundary layer given by $\dfrac{u}{U} = 2\left(\dfrac{y}{\delta}\right) - \left(\dfrac{y}{\delta}\right)^2$. $\qquad\qquad$ (6)

Solution. $\qquad \dfrac{u}{U} = 2\left(\dfrac{y}{\delta}\right) - \left(\dfrac{y}{\delta}\right)^2$

(i) Displacement thickness - δ'

$$\delta' = \int_0^\delta \left(1 - \frac{u}{U}\right) dy$$

$$= \int_0^\delta \left\{ 1 - \left[2\left(\frac{y}{\delta}\right) - \left(\frac{y}{\delta}\right)^2 \right] \right\} dy$$

$$= \int_0^\delta \left\{ 1 - 2\left(\frac{y}{\delta}\right) + \left(\frac{y}{\delta}\right)^2 \right\} dy$$

$$= \left[y - \frac{2y}{2\delta} + \frac{y}{3\delta^2} \right]_0^\delta$$

$$= \delta - \frac{\delta^2}{\delta} + \frac{\delta^3}{3\delta^2} = \delta - \delta + \frac{\delta}{3} = \frac{\delta}{3} \qquad \textbf{Ans.}$$

(ii) Momentum thickness θ

$$\theta = \int_0^\delta \frac{u}{U}\left(1 - \frac{u}{U}\right) dy$$

$$= \int_0^\delta \left[\left(\frac{2y}{\delta}\right) - \left(\frac{y^2}{\delta^2}\right) \right]\left[1 - \left(\frac{2y}{\delta} - \frac{y^2}{\delta^2}\right) \right] dy$$

$$= \int_0^\delta \left[\frac{2y}{\delta} - \frac{y^2}{\delta^2} \right]\left[1 - \frac{2y}{\delta} + \frac{y^2}{\delta^2} \right] dy$$

$$= \int_0^\delta \left[\frac{2y}{\delta} - \frac{4y^2}{\delta^2} + \frac{2y^3}{\delta^3} - \frac{y^2}{\delta^2} + \frac{2y^3}{\delta^3} - \frac{y^4}{\delta^4} \right] dy$$

$$= \int_0^\delta \left[\frac{2y}{\delta} - \frac{5y^2}{\delta^2} + \frac{4y^3}{\delta^3} - \frac{y^4}{\delta^4} \right] dy$$

$$= \left[\frac{2y^2}{2\delta} - \frac{5y^3}{3\delta^2} + \frac{4y^4}{4\delta^3} - \frac{y^4}{5\delta^4} \right]_0^\delta$$

$$= \frac{\delta^2}{\delta} - \frac{5\delta^3}{3\delta^2} + \frac{\delta^4}{\delta^3} - \frac{\delta^5}{5\delta^4}$$

$$= \delta - \frac{5\delta}{3} + \delta - \frac{\delta}{5}$$

$$= \frac{15\delta - 25\delta + 15\delta - 3\delta}{15}$$

$$= \frac{2\delta}{15} \qquad \textbf{Ans.}$$

Q.8 (a) Explain the terms:

(i) friction drag (ii) pressure drug (iii) lift.(6)

Solution. (i) Friction drag and (ii) Pressure drag: Please refer Q. 8(b) of July/Aug.-2005.

(iii) **Lift:** The component of the total force (F_R) in the direction perpendicular to the direction of motion is known as lift. It is denoted by F_L. Lift force occurs only when the axis of the body is inclined to the direction of fluid flow.

(b) Describe the Mach cone and Mach angle with the help of neat sketches. (6)

Solution. When $m > 1$ the flow is supersonic flow. The projectile moves faster than the sphere of propagation of disturbance. If we draw a tangent to the different circles which represent the propagated spherical waves on both sides, we shall get a cone with vertex at *B*. This cone is known as mach cone.

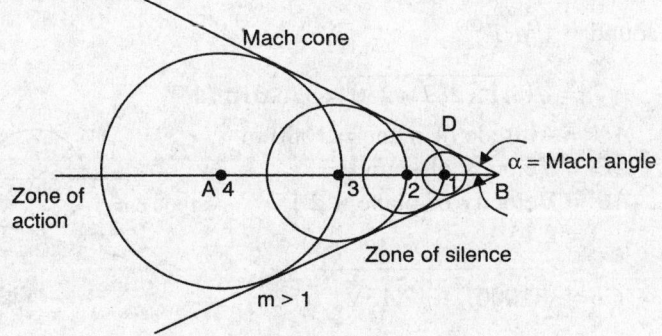

Fig. 19

Mach angle: This is defined as the half of the angle of the mach cone. Angle α is mach angle in $\Delta\ IBD$ the distance IB = velocity of projectile = *V*.

The distance ID = velocity of sound wave = *C*

$$\sin\alpha = \frac{ID}{IB} = \frac{C}{V} = \frac{1}{V/C} = \frac{1}{m}$$

(c) A passenger car with frontal projected area of 1.5 m^2 travels at 56 km/hr. Determine the power required to overcome wind resistance if the drag co-efficient of the car is 0.4. Take ρ at air = 1.2 kg/m³. (4)

Solution. $A = 1.5\ \text{m}^2$

$$u = 56\,\text{km/hr}$$

$$= \frac{56\times100}{60\times60} = 15.55\,\text{m/s}$$

$$\rho = 1.2\,\text{kg/m}^3$$

$$C_D = 0.4$$

Power required to over come wind resistance

$$F_D = C_D A \times \frac{\rho y^2}{2}$$

$$= 0.4 \times 1.5 \times \frac{1.2 \times 15.55}{2}$$

$$= 5.598 \text{ N} \qquad\qquad\qquad \textbf{Ans.}$$

(d) Calculate the velocity and Mach number of a supersonic aircraft flying at an altitude of 1000 *m* where the temperature is 280° K. Sound of the aircraft is heard 2.15 seconds after the passage of aircraft on the head of an observer. Take $\gamma = 1.41$ and $R = 287$ J/kg-K. (4)

Solution. $\quad Z \doteq 1000$ m $\qquad t = 280°$k

$$\gamma = 1.41 \qquad R = 287 \text{ J/kgk}$$

Velocity of sound $= \sqrt{\gamma R T}$

$$= \sqrt{1.41 \times 287 \times 230} = 336.61 \text{m/s}$$

AC = Altitude of plane = 1000 m
AD = Velocity of sound
AB = Velocity of plane × 2.15
$\quad = 2.15$ V

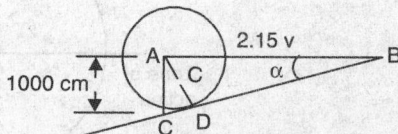

Fig. 20

$$CB = \sqrt{(1000)^2 + (2.15V)^2}$$

$$\therefore \quad \tan\alpha = \frac{1000}{2.15V}$$

$$m = \frac{V}{C} = \frac{1}{\sin\alpha} \quad \therefore V = \frac{C}{\sin\alpha} \text{ or } \sin\alpha = \frac{336.61}{V}$$

$$\therefore \quad \tan\alpha = \frac{1000}{2.15 \times \dfrac{C}{\sin\alpha}} = \frac{1000 \times \sin\alpha}{2.10 \times 336.61}$$

$$\frac{\sin\alpha}{\cos\alpha} = \frac{1000\sin\alpha}{2.15 \times 336.61}$$

$$\therefore \quad \cos\alpha = \frac{2.15 \times 336.61}{1000} = 0.7237$$

$$\sin\alpha = \sqrt{1 - \cos^2\alpha} = \sqrt{1 - (0.7237)^2} = 0.6901$$

$$V = \frac{C}{\sin\alpha} = \frac{336.61}{0.6901} = 487.77 \text{ m/s} \qquad\qquad \textbf{Ans.}$$

$$m = \frac{487.77}{336.61} = 1.449. \qquad\qquad\qquad\qquad \textbf{Ans.}$$

Fluid Mechanics
(July/Aug. 2006)
Semester - IV (ME/IP/AU/IM/MA)

Time: 3 Hours   Maximum Marks: 100

Note: 1. Answer any five full questions.
  2. Write legibly and draw neat sketches wherever required.

Q.1 (a) Recognize the following substances as fluids or solids and if fluids, classify them further. The values are obtained from isothermal tests.   (10)

Substance A	$\dfrac{dv}{dy} =$	0	1	2	3	4
	$\tau =$	0	2	4	6	8
Substance B	$\dfrac{dv}{dy} =$	0	1	2	3	4
	$\tau =$	1	2	3	4	5
Substance C	$\dfrac{dv}{dy} =$	0	0.5	1.0	1.5	2.0
	$\tau =$	0	1.0	2.5	4.0	6.0
Substance D	$\dfrac{dv}{dy} =$	0	0	0	0	0
	$\tau =$	0	0.5	1	1.5	2
Substance E	$\dfrac{dv}{dy} =$	0	1	2	3	4
	$\tau =$	0	0	0	0	0

Ans.: Substance A:

Newtonian fluid-constant of proportionality $\mu = \dfrac{\tau}{\dfrac{dv}{dy}}$ does not change with the rate of

deformation. Such fluids are represented by a straight line passing through the origin and slope of line depends upon the viscosity of the individual fluid.

Substance B: It is plastic. Initial shear stress = 1

Substance C:

Fluid—Non-Newtonian fluid. Since $\mu = \dfrac{\tau}{\dfrac{dv}{dy}}$ is 0, 2, 2.5, 2.66, and 3.00. There is non-linear

relation between the magnitude of applied shear stress and the resulting rate of deformation.

Substance D:

It is elastic solid because $\dfrac{dv}{dy} = 0$. It is represented by y axis.

Substance E:

Ideal fluid due to absence of shear stress i.e. $\tau = 0$. Such fluids are represented by x axis.

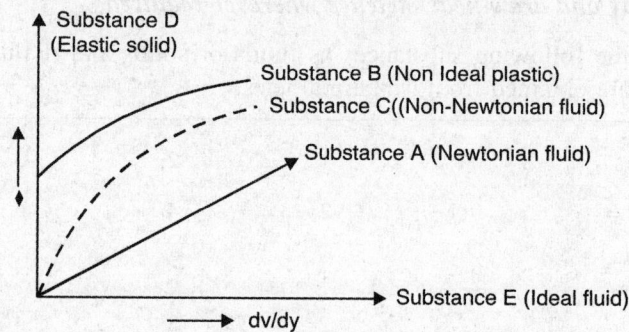

Substance D
(Elastic solid)

Substance B (Non Ideal plastic)

Substance C((Non-Newtonian fluid)

Substance A (Newtonian fluid)

Substance E (Ideal fluid)

dv/dy

Fig. 1

(b) An oil of viscosity 5 poise is used for lubrication between a shaft and sleeve. The diameter of the shaft is 25 mm and rotates at 200 r.p.m. Calculate the power lost in the oil of the bearing for a sleeve length of 100 mm. The thickness of the oil film is 1.2 mm. (5)

Ans.: Given data:

$$N = 200 \text{ r.p.m}, D = 25 \text{ mm}, r = 12.5 \text{ mm} = 0.0125 \text{ m},$$
$$l = 100 \text{ mm} = 0.1 \text{ m}, \text{ Film thickness} = 1.2 \text{ mm}$$

$$\text{Outer diameter} = 2.5 + 2 \times 1.2 = 27.4 \text{ mm}$$
$$= 0.0274 \text{ m}$$

Angular velocity of shaft

$$\omega = \frac{2\pi N}{60} = \frac{2\pi \times 200}{60} = 20.944 \text{ rad/s}$$

Linear velocity of the shaft

$$v = r\omega = 0.0125 \times 20.944$$
$$\therefore \qquad dv = 0.262 \text{ m/s}$$
$$dy = 1.2 \text{ mm}$$

$$\therefore \quad \text{Velocity gradient} = \frac{dv}{dy} = \frac{0.262}{0.0012} = 218.33 \; l/s$$

Viscosity $\mu = 5$ poise $= 0.5 \text{ Ns/m}^2$

$$\text{Shear stress developed } \tau = \mu \cdot \frac{dv}{dy}$$
$$= 0.5 \times 218.33 = 109.167 \text{ N/m}^2$$

Force acting on the shaft is
$$F = \text{Shear stress} \times \text{Area}$$
$$f = \tau \cdot \pi \cdot d \cdot l.$$
$$= 109.167 \times \pi \times 0.0274 \times 0.1$$

Torque acting on the shaft
$$T = \text{Force} \times \text{Radius of shaft}$$
$$= 0.9397 \times 0.0125$$
$$= 0.01175 \text{Nm}$$

Power required to overcome the torque
$$P = T \cdot \omega$$
$$= 0.01175 \times 20.944$$
$$= 0.246 \text{ watts.}$$

(c) Derive an expression for surface tension on liquid droplet.

Ans.:

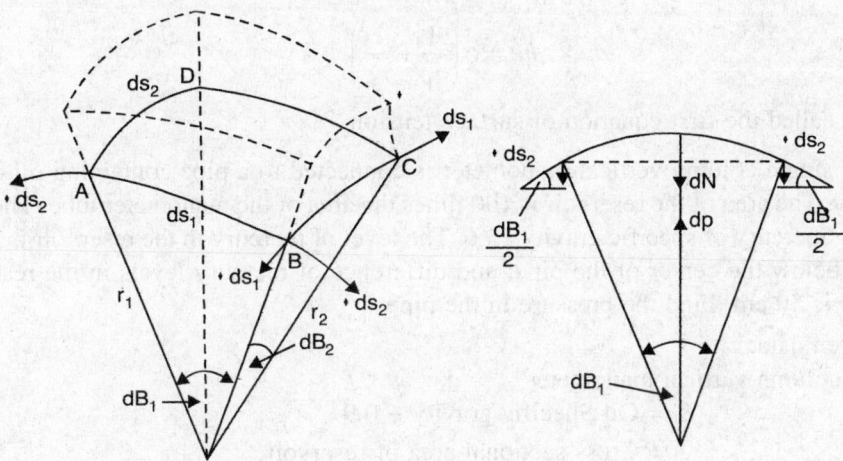

Fig. 2

Consider a small rectangular curvilinear element of surface of droplet, with side $AB = ds_1$ and $BC = ds_2$ and the radii of curvature r_1 and r_2 respectively.

Then the area of the surface $ABCD$ is

$$ds = ds_1 \cdot ds_2$$

Let the pressure on the lower side be P_1 and that on the upper side the surface be P_2. The difference on the two sides is, therefore.

$$d_p = P_1 - P_2 = \gamma h_k,$$

Where h_k is the equivalent static head due to this pressure and is called as the surface tension head.

The resultant force due to the pressure difference is $dP = d_p \cdot ds_1 \cdot ds_2$.

The net vertically downward force due to surface tension will be

$$dN = \sigma \left(2ds_2 \cdot \sin \frac{dB_1}{2} + 2 \cdot ds_1 \sin \frac{dB_1}{2} \right)$$

But for a very small element of the surface area ds and for a small angle

$$\sin dB_1 = d = \frac{ds_1}{r_1} \text{ and } dB_2 = \frac{ds_2}{r_2}$$

Hence the surface tension force is

$$dN = \sigma \left(\frac{1}{r_1} + \frac{1}{r_2} \right) ds_1 \cdot ds_2$$

Equating the two forces on the surface element for equilibrium $dP = dN$

$$dp \cdot ds_1 \cdot ds_2 = \sigma \left(\frac{1}{r_1} + \frac{1}{r_2} \right) ds_1 \cdot ds_2$$

or

$$dp = \gamma hk = \sigma \left(\frac{1}{r_1} + \frac{1}{r_2} \right)$$

This is called the first equation of surface tension.

Q.2 (a) A single column vertical manometer is connected to a pipe containing oil of specific gravity 0.9. The area of the reservoir is 100 times the area of the manometer tube. The reservoir containing mercury of specific gravity 13.6. The level of mercury in the reservoir is at a height of 30 cm below the center of the pipe; and difference of mercury levels in the reservoir and right limb is 50 cm. Find the pressure in the pipe. (10)

Ans.: Given data:

Single column vertical manometer

$$S_1 = \text{Oil Specific gravity} - 0.9$$
$$A = \text{Cross sectional area of reservoir}$$
$$A/a = 100$$
$$A = 100 \times a$$
$$a = \text{Area of right limb}$$
$$S_2 = \text{Specific gravity of mercury} = 13.6$$

$$h_1 = 30 \text{ cm} \qquad\qquad h_2 = 50 \text{ cm}$$
$$= 0.3 \text{ m} \qquad\qquad = 0.5 \text{ m}$$

$$dh = \frac{a}{A} h_2$$

$$dh = \frac{50}{100} = 0.5 \text{ cm}$$

$$= 0.005 \text{ m}$$

Pressure at the new level of Mercury in the reservoir is
$P_{oil} + 0.9\gamma (0.3 + 0.005)$

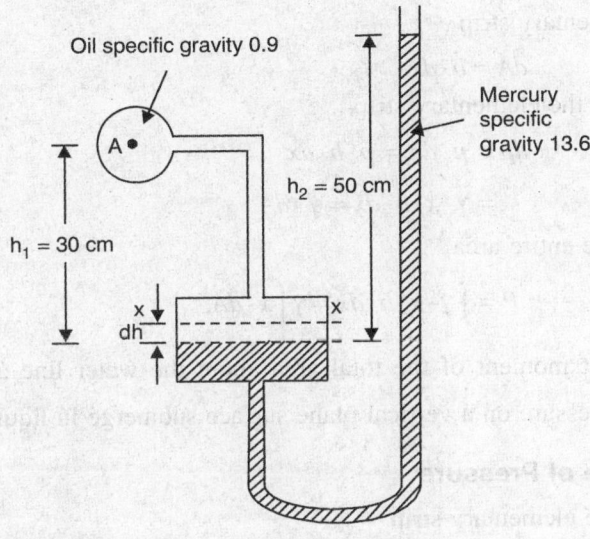

Fig. 3

Pressure above the same level in the right limb open to atmosphere
$$13.6 \, \gamma \, (0.005 + 0.5)$$

∴ Equating the two pressures.

$$P_{oil} + 0.9\gamma(0.3 + 0.005) = 13.6\gamma(0.005 + 0.5)$$

$$\frac{P_{oil}}{0.9\gamma} + 0.3 + 0.005 = \frac{13.6}{0.9}(0.005 + 0.5)$$

$$\frac{P_{oil}}{0.9\gamma} = 7.326 \text{m of oil}$$

$$P_{oil} = 7.326 \times 0.9 \times 9.81 = 64.68 \text{ m of mercury.}$$

(b) Derive an expression for total pressure force and position of centre of pressure of a vertical plane surface submerge in liquid.　　(10)

Ans.: Total pressure force of a vertical plane surface submerged in liquid.

Consider any irregular area of a plane submerged vertically in liquid.

A = Area of the given plane surface AB

$\bar{x}$ = Depth of the centre of gravity of the area AB below the free surface of water.

Consider an elementary horizontal strip of width b and thickness dx which is situated at a depth x below the free surface of water.

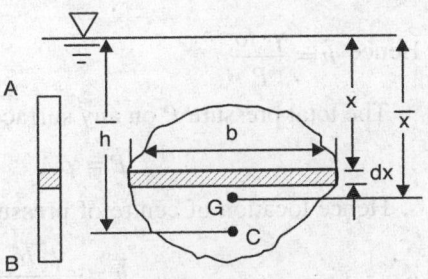

Fig. 4

Intensity of pressure on elementary strip

$$P = \gamma \cdot x.$$

Area of the elementary strip

$$dA = b \cdot dx$$

∴ Total pressure on the elementary strip

$$dp = p \cdot dA = p \cdot b \cdot dx$$

$$= \gamma \cdot x \cdot b \cdot dx = \gamma \, x dA$$

Total pressure on the entire area

$$P = \int \gamma \cdot x \cdot b \cdot dx = \gamma \int x \cdot dA.$$

$\int x \cdot dA$ is the first moment of the total area about the water line and is $= A \cdot \overline{x}$. Hence $P = \gamma \overline{x} \cdot A$. Total pressure on a vertical plane surface submerge in liquid.

Position of Centre of Pressure

Total pressure on the elementary strip

$$dp = \gamma \times dA$$

Moment of this total pressure on the elementary strip about the waterline is.

$$dM = dp \cdot x = \gamma x^2 \cdot dA$$

Total moment of all the total pressures on the elementary strips, i.e. for the whole area AB is

$$M = \int dm = \int \gamma x^2 \cdot dA$$

But $\int x^2 \cdot dA =$ Moment of inertia of the area AB about the water line $= 10$

∴ $$M = \gamma \cdot Io$$

This moment M is also equal to the moment of the total pressure P on the entire area AB about the water line. Hence $M = p \cdot \overline{h}$.

Equating the two values of the moment M. we get

$$A\overline{h} = \gamma \cdot Io \quad \text{when } h = \text{depth of the centre of pressure below the free surface of water}$$

Hence $\overline{h} = \dfrac{\gamma \cdot Io}{P}$

The total pressure P on any surface, whether regular or irregular in shape on vertical plane.

$$\dot{P} = \gamma \overline{x} A$$

Hence location of centre of pressure below water line

$$\overline{h} = \frac{\gamma \cdot Io}{\gamma \cdot \overline{x} \cdot A} = \frac{10}{A \cdot \overline{x}}.$$

Q.3 (a) Derive an analytical expression for the metacentric height of a floating body. (10)

Ans.: Please refer Q.2 (a) (i) of Jan./Feb.-2006.

(b) A rectangular pontoon is 4 m long, 3 m wide and 1.4 m high. The depth of immersion of the pontoon 1 m in sea water. If the centre of gravity is 0.7 m above the bottom of the pontoon, determine the metacentric height. Take the density of sea water as 1030 kg/m³.

Given data:

Plane of pontoon = 4 m × 3 m

Height of pontoon = 1.4 m.

OB = Height of center of buoyancy above the base of the block.

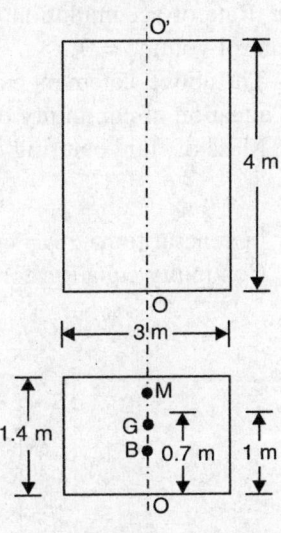

$$= \frac{1m}{2} = 0.5 \text{ m}$$

OG = Height of centre of gravity above base = 0.7 m.

M = Metacentre

$$BM = \frac{I}{\forall}$$

Where I = Moment of inertia of the water line about OO'

$$= \frac{4 \times (3)^3}{12} = 9$$

∀ = Volume of the fluid displaced

$$= L \cdot b \cdot h = 4 \times 3 \times 1 = 12 \text{ m}^3$$

Fig. 5

∴ $$BM = \frac{9}{12} = 0.75 \text{ m}$$

$$BG = OG - OB = 0.7 - 0.5 = 0.2 \text{ m}$$

Metacentric height MG = BM − BG

$$MG = 0.75 - 0.2$$

$$= 0.55 \text{ m}$$

Q.4 (a) Define continuity equation. Write its equation. Derive the continuity equation for the three dimensional flow in Cartesian co-ordinates and modify it for two and one dimensional flow. (10)

Ans.: Continuity equation

The law of conservation of mass states that mass can neither be created nor be destroyed. Conservation of mass is inherent to the definition of a closed system and can be written mathematically as $\frac{\Delta m}{\Delta t} = 0$.

Where m is the mass of the system.

i.e. Rate at which mass enters the region = Rate at which mass leaves the region + Rate of accumulation of mass in the region.

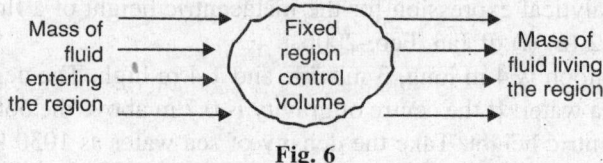

Fig. 6

or Rate of accumulation of mass in the control volume + Net rate of mass of flux from the control volume = 0.

The above statement expressed analytically in terms of velocity and density field of a flow is equation of continuity or continuity equation.

Mass of fluid entering control volume = Mass of fluid leaving the control volume

$$\rho_1 a_1 v_1 = \rho_2 a_2 v_2$$

or in general for $\rho \cdot a \cdot v$ = constant is continuity equation.

Continuity equation for the three dimensional flow in cartesian co-ordinates.

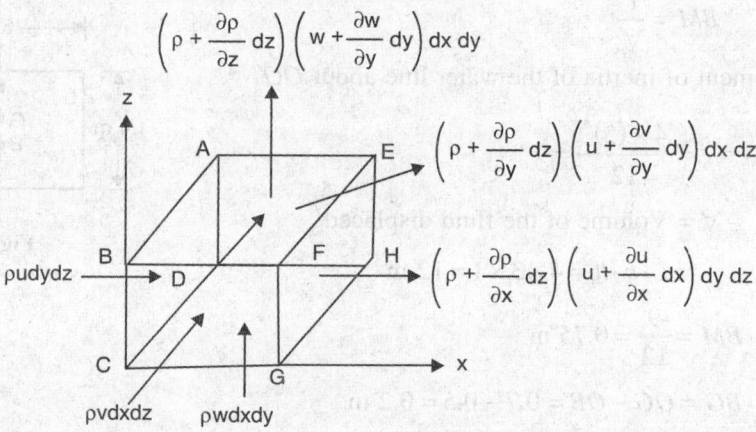

Fig. 7

Consider a rectangular parallelopiped as a control volume in a Cartesian frame of co-ordinate axes.

Let the fluid enter across the face *ABCD* with a velocity *u* and a density ρ and leaves the face *EFGH* with velocity $u + \dfrac{\partial u}{\partial x} dx$ and density $\rho + \dfrac{\partial \rho}{\partial x} \cdot dx$ respectively.

∴ As per continuity equation.

Rate of mass entering the control volume through *ABCD* = the rate of mass leaving the control volume through face *EFGH*.

∴

$$\rho \cdot u \cdot dy\, dz = \left(\rho + \frac{\partial \rho}{\partial x} \cdot dx \right)\left(u + \frac{\partial u}{\partial x} dx \right) dy\, dz$$

$$\rho \cdot u \cdot dy\, dz = \left(\rho \cdot u + \frac{\partial}{\partial x}(\rho u) dx \right) dy\, dz$$

Neglecting the higher order terms in dx.

Hence the net rate of mass of flux from the control volume in x direction

$$= \left(\rho \cdot u + \frac{\partial}{\partial x}(\rho \cdot u) dx \right) dy \, dz - \rho \cdot u \cdot dy \cdot dz$$

$$= \frac{\partial}{\partial x}(\rho \cdot u) dx \cdot dy \, dz$$

$$= \frac{\partial}{\partial x}(\rho \cdot u) d\forall \qquad \text{where } d\forall \text{ is elemental volume } y \text{ and } z$$

Similarly, the net rate of mass of flux in the y and z direction

$$= \frac{\partial}{\partial z}(\rho \cdot v) d\forall$$

$$= \frac{\partial}{\partial z}(\rho \cdot w) d\forall \quad \text{respectively}$$

The rate of accumulation of mass within the control volume is $\frac{\partial}{\partial t}(\rho \cdot d\forall) = \frac{\partial \rho}{\partial t} d\forall$. Therefore, according to the statement of conservation of mass for a control volume

$$\left\{ \frac{\partial \gamma}{\partial t} + \frac{\partial}{\partial x}(\rho u) + \frac{\partial}{\partial y}(\rho v) + \frac{\partial}{\partial z}(\rho w) \right\} d\forall = 0$$

Equation is valid irrespective of the size $d\forall$ of the control volume

$$\therefore \qquad \frac{\partial \gamma}{\partial t} + \frac{\partial}{\partial x}(\rho u) + \frac{\partial}{\partial y}(\rho v) + \frac{\partial}{\partial z}(\rho w) = 0$$

is the equation of continuity of compressible fluid in a rectangular Cartesian co-ordinate system. For two dimensional flow, taking velocity to be constant in z-direction the continuity equation becomes.

$$\frac{\partial u}{\partial x}(\rho u) + \frac{\partial}{\partial y}(\rho v) + \frac{\partial p}{\partial t} = 0$$

and for one dimensional flow taking velocity in y direction to be constant, equation becomes

$$\frac{\partial(\rho u)}{\partial x} + \frac{\partial \rho}{\partial t} = 0$$

(b) What do you mean by velocity potential function and stream function? Write their properties and relation for 2–D flow. (10)

Ans.: Stream Function: The concept of stream function is a direct consequence of the principle of continuity.

In a two-dimensional incompressible flow parallel to the x–y plane in a rectangular Cartesian co-ordinate system, the flow field is defined as

$$u = u(x, y, t)$$

$$v = v(x, y, t)$$

$$w = 0$$

∴ The equation of continuity is

$$\frac{\partial u}{\partial x} + \frac{\partial v}{\partial y} = 0 \qquad \qquad ...(1)$$

If a function $\psi(x, y, t)$ is defined in the manner

$$u = \frac{\partial \psi}{\partial y} \qquad \text{and} \qquad v = -\frac{\partial \psi}{dx}$$

So that it automatically satisfies the equation of continuity, then the function ψ is known as stream function.

For a steady flow, ψ is a function of two variables x and y only. In case of a two-dimensional irrotational flow

$$\frac{\partial v}{\partial x} - \frac{\partial u}{\partial y} = 0$$

∴

$$\frac{\partial}{\partial x}\left(-\frac{\partial \psi}{\partial x}\right) - \frac{\partial}{\partial y}\left(\frac{\partial \psi}{\partial y}\right) = 0$$

or

$$\nabla^2 \psi = -\frac{\partial^2 \psi}{\partial x^2} + \frac{\partial^2 \psi}{\partial y^2} = 0.$$

This for an irrotational flow stream function satisfies the Laplace's equation.

Velocity potential function

The flow will be irrotational when the vorticity of the fluid is absent or zero. This means that the conditions for the flow irrotational are

$$\frac{\partial v}{\partial x} = \frac{\partial u}{\partial y} \qquad \frac{\partial w}{\partial y} = \frac{\partial v}{\partial z}, \qquad \frac{\partial u}{\partial z} = \frac{\partial w}{\partial x}$$

for the two dimensional flow, the first of the three conditions $\dfrac{\partial v}{\partial x} = \dfrac{\partial u}{\partial y}$ must be satisfied. This can be mathematically stated as $udx + vdy$ must be exact or it must be possible to express this as perfect differential $udx + vdy = -d\phi$.

Here ϕ is some function of x and y and the negative sign is arbitrary $\phi = \phi(x, y)$

∴

$$d\phi = \frac{\partial \phi}{\partial x} \cdot dx + \frac{\partial \phi}{\partial y} dy$$

Hence

$$udx + vdy = \frac{\partial \phi}{\partial x} \cdot dx - \frac{\partial \phi}{\partial y} dy$$

$$\therefore \qquad u = -\frac{\partial \phi}{\partial x} \quad \text{and} \quad v = -\frac{\partial \phi}{\partial y}$$

This defines the arbitrary function ϕ, which is called as the velocity potential function.

The flow of fluid is possible when the equation of continuity is satisfied. Substituting the values of the velocity components in the equation of continuity

$$\frac{\partial u}{\partial x} + \frac{\partial v}{\partial y} = 0$$

$$\therefore \qquad \frac{\partial}{\partial x}\left(-\frac{\partial \phi}{\partial x}\right) + \frac{\partial}{\partial y}\left(-\frac{\partial \phi}{\partial y}\right) = 0$$

$$\therefore \qquad -\left[\frac{\partial^2 \phi}{\partial x^2} + \frac{\partial^2 \phi}{\partial y^2}\right] = 0$$

$$-\nabla^2 \phi = 0.$$

Relations for 2-D flow

The relation between the stream function and the velocity potential function is

$$\frac{\partial \phi}{\partial x} = \frac{\partial \psi}{\partial y}$$

and

$$\frac{\partial \phi}{\partial y} = -\frac{\partial \psi}{\partial x}$$

The potential function will exist only for the irrotational flow, but this is not the case with the stream function. The stream function will exist in cases of all the flows, whether rotational or irrotational. Therefore the relation between the stream function and the potential function will be true only in case or irrotational flow.

Q.5 (a) State Bernoulli's theorem for ideal steady flow of an incompressible fluid. Derive an expression for Bernoulli's equation from the fundamental or first principle and state the assumptions made for such a derivation. Also write its applications. (10)

Ans.: Please refer Q. 5(b) of Jan./Feb.-2006.

(b) A horizontal venturimeter with inlet and throat diameters of 32 cm and 16 cm respectively is used to measure the flow of water. The reading of differential manometer connected to inlet and throat is 22 cm of mercury. Determine the rate of flow. Take co-efficient of discharge $cd = 0.97$. (10)

Ans.: Given: $D_1 = 32$ cm $= 0.32$ m, $D_2 = 16$ cm $= 0.16$ m

$$\frac{D_2}{D_1} = \frac{16}{32} = 0.5$$

$y = 22$ cm of mercury $= 0.22$ m

For differential manometer

$$\nabla h = y\left(\frac{sm}{s.p.} - 1\right)$$

$$= 0.22\left(\frac{13.6}{1.0} - 1\right) = 2.772\,\text{m}$$

$$cd = 0.97$$

$$A_2 = \frac{\pi}{4}\left(D_2\right)^2 = \frac{\pi}{4}(0.16)^2 = 0.020\,\text{m}^2$$

$$Q = \frac{cd \cdot A_2 \sqrt{2g\Delta h}}{\sqrt{1 - \left(\dfrac{D_2}{D_1}\right)^4}}$$

$$= \frac{0.97 \times 0.020\sqrt{2 \times 9.81 \times 2.772}}{\sqrt{1 - (0.5)^4}}$$

$$= \frac{0.01350 \times 7.374}{0.9375}$$

$$= 0.1534\ \text{m}^3/\text{s}$$

Q.6 (a) What do you mean by dimensionless number? Define, explain and derive
    (i) Reynold's number   (ii) Froude's number   (iii) Euler's number
    (iv) Weber's number   (v) Mach number.   (10)

Ans.: The ratio of two forces is called as dimensionless parameter or number.

(i) **Reynold's Number:** The ratio of inertia force to the viscous force is called as Reynold's number.

$$R_e = \frac{\text{Inertia force}}{\text{viscous force}} = \frac{\text{mass} \times \text{Acceleration}}{\text{shear force} \times c/s\ \text{area}}$$

$$= \frac{(\text{Mass density} \times \text{Volume}) \times (\text{Velocity}/\text{time})}{\text{shear stress} \times c/s\ \text{Area}}$$

$$= \frac{(\text{Volume}/\text{time})\,\text{mass density} \times \text{Velocity}}{\text{shear stress} \times c/s\ \text{Area}}$$

$$= \frac{(\text{Velocity} \times c/s\ \text{Area} \times \text{mass density} \times \text{Velocity})}{\text{shear stress} \times c/s\ \text{Area}}$$

$$= \frac{V \cdot \rho \cdot V}{\tau} = \frac{V \cdot \rho \cdot V}{\mu \dfrac{du}{dy}} = \frac{V \cdot \rho \cdot V}{\mu \dfrac{V}{L}}$$

$$= \frac{\rho \cdot V \cdot L}{\mu} = \frac{VL}{\mu/\rho} = \frac{V \cdot L}{v}$$

(ii) **Froude's Number:** F_r The square root of the ratio of inertia force the gravitational force is called Froude's number.

$$F_r = \sqrt{\frac{\text{Inertia force}}{\text{Gravitational force}}}$$

$$= \sqrt{\frac{m \times a}{m \times g}} = \sqrt{\frac{a}{g}} = \sqrt{\frac{V/t}{g}}$$

$$= \sqrt{\frac{V}{t \cdot g}} = \sqrt{\frac{V}{\dfrac{L}{V} \cdot g}} = \sqrt{\frac{V^2}{L \cdot g}}$$

$$f_r = \frac{V}{\sqrt{L \cdot g}}.$$

(iii) **Euler's Number:** E_u The square root of the ratio of inertia force to the pressure force is called as Euler's ratio.

$$E_u = \sqrt{\frac{\text{Inertia force}}{\text{Pressure force}}} = \sqrt{\frac{m \times A}{P \times A}} = \sqrt{\frac{\rho \cdot \forall \cdot \dfrac{V}{T}}{P \cdot A}} = \frac{\forall}{T} = Q_2 = V \cdot A$$

$$= \sqrt{\frac{\rho \cdot V^2 \cdot A}{P \cdot A}} = \sqrt{\frac{\rho V^2}{P}}$$

$$E_u = \frac{V}{\sqrt{\dfrac{P}{r}}}$$

(iv) **Weber's Number:** W_e The square root of the ratio of the inertia force to the surface tension force.

$$W_e = \sqrt{\frac{\text{Inertia force}}{\text{Surface tension force}}} = \sqrt{\frac{\text{mass} \times \text{Acceleration}}{\text{Surface tension} \times \text{Length}}}$$

$$= \sqrt{\frac{\rho \cdot \forall \times \dfrac{V}{T}}{\sigma \cdot L}} = \sqrt{\frac{\rho \cdot V \cdot V \cdot A}{\sigma \cdot L}} \qquad \because \frac{\forall}{\tau} = Q = V \cdot A$$

$$= \sqrt{\frac{\rho \cdot V^2 \cdot L^2}{\sigma \cdot L}} \qquad \because A = L^2$$

$$W_e = \frac{V}{\sqrt{\dfrac{\sigma}{\rho \cdot L}}}.$$

(v) **Mach Number:** M_n The square root of the ratio of inertia force to the elastic force is called as Mach's number.

$$M_n = \sqrt{\frac{\text{Inertia force}}{\text{Elastic force}}} = \sqrt{\frac{\text{mass} \times \text{Acceleration}}{\text{Elastic stress} \times \text{Area}}}$$

$$= \sqrt{\frac{\rho \cdot \forall \times V/T}{K \cdot A}} = \sqrt{\frac{\rho \cdot V \cdot A \cdot V}{K \cdot A}} \qquad \because \frac{\forall}{T} = Q = A \cdot V.$$

$$= \sqrt{\frac{\rho \cdot V^2}{K}} = \frac{V}{\sqrt{K/\rho}}.$$

(b) Derive an expression for the velocity of sound wave for the compressible fluid and adiabatic process. (6)

Ans.: In a compressible of fluid displaced mass compresses and increases the density of the neighbouring mass. A disturbance in the form of an elastic wave or a pressure wave travels through the medium. If the amplitude of the elastic wave is infinitesimal, it is termed as acoustic wave or sound wave.

Velocity of sound wave through the air is known as sonic velocity or velocity of sound.

Consider motionless air has density ρ, pressure p and temperature T. Sound wave is created by some source, i.e. blast and travels with velocity cm/s. Thus sound wave is a thin region of disturbance air across which the pressure, temperature and density change slightly.

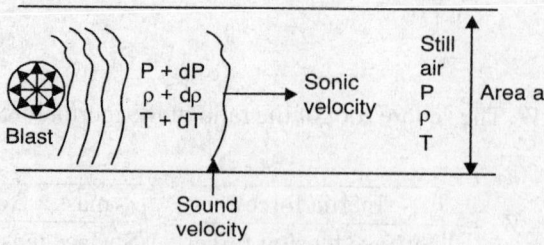

Fig. 8

Let dp, dT and $d\rho$ are the change in pressure, temperature and density of air respectively due to passing of sound wave through it. The air behind the wave is moving away from the wave with velocity $c + dc$.

Let the point 1 and 2 be ahead of and behind the sound wave respectively.

Applying the continuity equation

$$\rho a_1 C = (\rho + d\rho) a_2 + (C + dC)$$

$a_1 = a_2$ = area of steam tube through which the waves are travelling.

$\therefore$

$$\rho C = (\rho + d\rho)(C + dC)$$

$$\rho C = (\rho C + C d\rho + \rho dC + d\rho \cdot dC)$$

$d\rho \cdot dc$ is very small $\therefore$ Neglected

$\therefore$

$$0 = C d\rho + \rho \cdot dC$$

$$C d\rho = -\rho \cdot dC$$

$$C = -\rho \frac{dC}{d\rho} \qquad \qquad \dots(1)$$

Applying the momentum equation by replacing v by c.

$$dP = -\rho \cdot C \cdot dC$$

or

$$dC = -\frac{1}{\rho \cdot C} dP.$$

Substitute dC in equation (1) we get,

$$C = -\frac{r}{d\rho}\left(-\frac{dP}{\rho \cdot C}\right) = \frac{dP}{C \cdot d\rho}$$

$$C^2 = \frac{dp}{d\rho} \quad \text{or} \quad c = \sqrt{\frac{dp}{d\rho}}$$

Sonic velocity involves no heat addition and the effect of friction is negligible. Hence the flow through a sound wave is isentropic.

$\therefore$ Velocity of sound $C = \sqrt{\left(\dfrac{dp}{d\rho}\right)_{isentropic}}$

For isentropic flow we know that,

$$\frac{P_2}{P_1} = \left(\frac{\rho_2}{\rho_1}\right)^\gamma \text{ in } \frac{P_3}{\rho_2^\gamma} = \frac{P_1}{\rho_1^\gamma} = n \text{ constant.}$$

$$\frac{P}{\rho^\gamma} = n \quad \text{or} \quad P = n\rho^\gamma$$

and
$$\rho\left(\frac{dp}{d\rho}\right)_{isentropic} = \frac{d}{d\rho}(n\rho^\gamma) \quad \therefore \quad \left(\frac{dP}{d\rho}\right)_{isentropic} = n\gamma\rho^{\gamma-1}$$

Substituting the value of n is equation we get
$$\left(\frac{dp}{d\rho}\right)_{isentropic} = \frac{P}{\rho^\gamma} \cdot \gamma \cdot \rho^{\gamma-1} = \frac{\gamma P}{\rho}$$

$\therefore$
$$C = \sqrt{\left(\frac{dp}{dr}\right)_{isentropic}} = \sqrt{\frac{\gamma P}{\rho}}$$

However for a perfect gas P and ρ are related through the equation of state $P = \rho R T$

$\therefore$
$$C = \sqrt{\gamma RT}$$

(c) Find the sonic velocity for the following fluids:
   (i) Crude oil of specific gravity 0.85 and bulk modulus of 150000 N/cm^2.
   (ii) Mercury having a bulk modulus of 2600000 N/cm^2. (4)

Ans.: Given:

(i) Specific gravity of oil = 0.35 ρ oil = 0.35×10^3 kg/m^3
$$E = 150000 \text{ N/cm}^2 = 150,000 \times 10^4 \text{ N/m}^2$$
$$C = \sqrt{\frac{E}{\rho}} = \sqrt{\frac{150,000 \times 10^{14}}{0.85 \times 10^3}}$$
$$= 1328.42 \text{ m/s}$$

(ii) $\rho = 13.6 \times 10^3$ kg/m^3
$$E = 2600.000 \text{ N/cm}^2 \qquad\qquad = 2600000 \times 10^4 \text{ N/m}^2$$
$$C = \sqrt{\frac{E}{\rho}} \qquad\qquad = \sqrt{\frac{2600000}{13.6 \times 10^3}}$$
$$= 191.176 \text{ m/s}$$

Q.7 (a) The water is flowing at a velocity of 4 m/s through a pipe of diameter 35 cm and length 55 m. Find the head lost due to friction using (i) Darcy equation and (ii) Chezy equation. Take value of Chezy's constant = 60 and kinematic viscosity γ for water = 0.01 stoke. (10)

Ans.: Given: $V = 4$ m/s, $D = 35$ cm = 0.35 m
$l = 55$ m, $\gamma = 0.01$ stoke.

(i) Head loss due to friction using darcy equation
$$h_f = \frac{fLV^2}{2gD}$$

$$R_e = \frac{dv}{\gamma} = \frac{35 \times 400}{0.01} = 1400000$$

Using the relation $f = 0.0032 + \dfrac{0.221}{R_e^{0.237}}$

$$f = 0.0032 + \frac{0.221}{(1400000)^{0.237}} = 0.0109$$

By Darcy equation

$$h_f = \frac{0.0109 \times 55 \times (4)^2}{2 \times 3.81 \times 0.35} = 1.3997 \text{ m}$$

(ii) Chezy's equation, $v = C\sqrt{ms}$

$$h_f = \frac{4V^2 L}{dC^2} \qquad C = 60$$

$$= \frac{4 \times (4)^2 \times 55}{0.35 \times (60)^2} = 2.7936 \text{ m}$$

(b) Prove that the velocity distribution for a viscous flow between two parallel plates, when both plates are fixed across a section is parabolic in nature. Also prove that maximum velocity is equal to one and a half times the average velocity. (10)

Ans.: Consider two parallel plates fixed across a section 1.1 and 2.2 placed at distance $2b$ from each other or y distance from the centre. A viscous incompressible fluid of viscosity μ is flowing between the two plates. Consider section 1.1 and 2.2 l units apart.

Let P_1 and P_2 be the pressure intensities at theses sections Velocity of the fluid is different distance from the centre of the plates.

Let the velocity be u at the distance r from the centre.

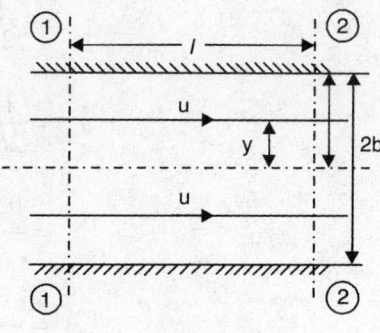

Fig. 9

$\therefore$ Velocity gradient $= -\dfrac{du}{dy}$

$C - ve$ sign indicates as r increases v decreases.

$\therefore$ shear stress intensity $= -\mu \dfrac{du}{dy}$

and $\qquad \dfrac{dp}{dx} = \mu \dfrac{d^2 u}{dy^2}$

The boundary conditions are at $y = b$, $u = 0$

and $\qquad\qquad y = -b$, $u = 0$

from equation (1) we can write

$$\frac{du}{dy} = \frac{1}{\mu}\frac{dp}{dx}y + C_1$$

$$u = \frac{1}{2\mu}\frac{dp}{dx}y^2 + C_1 y + C_2$$

Applying the boundary conditions, the constants are evaluated as $C_1 = 0$.

$$C_2 = -\frac{1}{\mu}\frac{dp}{dx}\frac{b^2}{2}$$

∴ Solution is

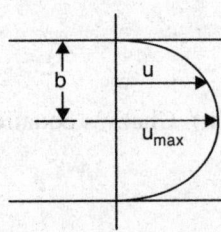

Fig. 10

$$u = -\frac{1}{2\mu}\frac{dp}{dx}\left(b^2 - y^2\right)$$

Which implies that the velocity profile is parabolic. Hence proved.
Velocity is maximum at $b = 0$

∴ $$u_{max} = -\frac{b^2}{2\mu}\frac{dp}{dx}$$

Average velocity $u_{av} = \dfrac{Q}{2b} = \dfrac{\text{flow rate}}{\text{flow area}}$

$$= \frac{1}{2b}\int_{-b}^{b} u\, dy$$

$$u_{av} = \frac{1}{2b}\int_{0}^{b} u\, dy = \frac{1}{b}\int_{0}^{b} -\frac{1}{2\mu}\frac{dp}{dx}\left(b^2 - y^2\right) dy$$

$$= -\frac{1}{2\mu}\frac{dp}{dx}\frac{1}{b}\left\{\left[b^2 y\right]_0^b - \left(\frac{y^3}{3}\right)_0^b\right\}$$

$$= -\frac{1}{2\mu}\frac{dp}{dx}\frac{2}{3}b^2$$

∴ So $$\frac{u_{av}}{u_{max}} = \frac{2}{3} \quad \text{or} \quad U_{max} = \frac{3}{2}U_{av}.$$

Thus it is proved that maximum velocity is equal to one and a half time the average velocity.

Q.8 (a) Define drag force and lift force. Also derive their expressions. (10)

Ans.: Drag: The component of total force in the direction parallel to the direction of motion, when the body is moving or fluid is flowing over stationary body is known as drag.

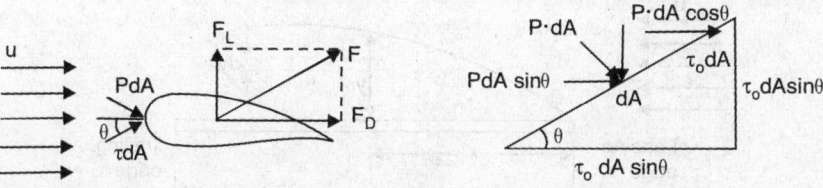

Fig. 11

Lift: The component of total force in the direction perpendicular to the direction of motion is known as lift.

Consider a body held stationary in a stream of real fluid moving at uniform velocity u. The force acting at any point on the small element dA of the surface of the body are resolved in to two component, i.e. shear force τdA and pressure force $P \cdot dA$ acting along the tangential direction and normal direction to the surface.

The sum of component of the shear force in the direction of flow of fluid is called as friction drag F_{D_f} or skin drag of shear drag.

$\therefore$ Friction drag $F_{D_f} = \int_A \tau \cdot dA \cdot \cos\theta$.

Similarly the sum of component of the pressure forces in the direction of the fluid motion is called as pressure drag $F_{D_P} = \int_A P \cdot dA \cdot \sin\theta$.

The total drag $f_D = F_{Df} + F_{DP}$

$\therefore$
$$F_D = \int_A \tau \cdot dA \cdot \cos\theta + \int_A P \cdot dA \cdot \sin\theta$$

$$= C_D \frac{1}{2} \rho u^2 \cdot A$$

And the lift on the body is given by the summation of the components of shear and pressure force acting over the body in the direction perpendicular to the direction of fluid motion.

$$F_L = \int_A \tau \cdot dA \cdot \sin\theta + \int_A P \cdot dA \cdot \cos\theta$$

$$= C_L \cdot \frac{1}{2} \rho u^2 \cdot A$$

C_D = Coefficient of drag

C_L = Coefficient of lift.

(b) Derive an expression for displacement thickness and momentum thickness of a flow over a plate. (10)

Ans.: Displacement thickness:

Consider a fluid moving with a velocity u approaching flat plate at rest. At a section distant x from the leading edge, let δ be the thickness of the boundary layer. At this section the velocity varies from zero at the plate to u at a distance δ from the plate.

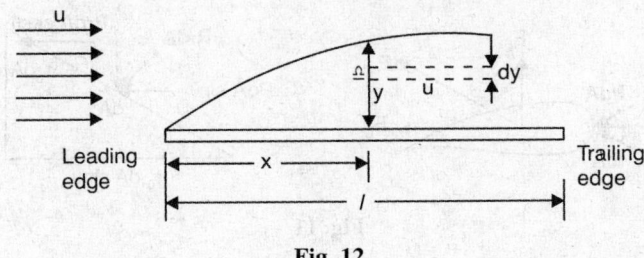

Fig. 12

Consider unit width of the plate. Consider an elemental strip $(1 \times dy)$ distant y from the plate. Let u be the velocity at this level.

Mass flowing per second thought the elemental strip $= \rho \cdot u \cdot dy$.

If the plate had not been present the mass flowing per second through the above elemental strip would have been $\rho \cdot U \cdot dy$.

∴ Reduction in mass flowing per second through the elemental strip $= \rho(U - u)\,dy$.

∴ Total reduction in mass flowing per second due to the plate

$$= \int\limits_0^\delta \rho(U - u)\,dy$$

Let the above quantity be $= \rho \cdot U \cdot d^*$

∴
$$\rho U d^* = \int\limits_0^\delta \rho(U - u)\,dy$$

$$d^* = \int\limits_0^\delta \frac{\rho(U - u)\,dy}{\rho u}$$

$$= \int\limits_0^\delta \left(1 - \frac{u}{U}\right)dy$$

δ^* is called displacement thickness.

Momentum Thickness θ^*: Again consider the flow through an elemental strip of area $(1 \cdot dy)$ distant y from the boundary.

Mass flowing per second through the elemental strip $= \rho \cdot u \cdot dy$

Let us consider the above quantity of the fluid momentum of this quantity.

$$= (\rho \cdot u \cdot dy)u$$

$$= \rho \cdot u^2 \cdot dy$$

Momentum of this quantity in the absence of the boundary layer $= (\rho \cdot u \cdot dy)\,U$.

Loss of momentum per second

$$= (\rho \cdot u \cdot dy)U - \rho u^2 dy$$

$$= \rho \cdot u(U - u)\,dy$$

∴ Total loss of momentum per second

$$= \int_0^\infty \rho \cdot u(U - u)\, dy$$

Let the above quantity be $= \rho \cdot u^2 \cdot \theta^*$.

∴

$$\rho \cdot U^2 \cdot \theta^* = \int_0^\delta \rho \cdot u \cdot (U - u)\, dy$$

∴

$$\theta^* = \int_0^\delta \frac{u}{U}\left(1 - \frac{u}{U}\right) dy$$

The momentum thickness θ^* may be visualized as the depth of flow with uniform velocity U, so as to have a momentum per second equal to the loss and momentum per second due to boundary layer. For a depth of flow θ^* with a velocity U, momentum per second per unit width $= (\rho \cdot \theta^* \cdot U)\, U = \rho U^2\, \theta^*$. The depth θ^* is called momentum thickness.

Fluid Mechanics
Jan./Feb. 2005
Semester - IV (ME/IP/AU/IM/MA)

Time: 3 Hours   Total Marks 100

Note:
1. *Answer any five full questions.*
2. *Draw neat sketches (using pencil) wherever necessary.*
3. *Missing data may be suitably assumed.*

Q.1 (a) Give technical reasons for the following: (2)
  (i) Certain insects are able to walk on the surface of water.
  (ii) Viscosity of liquids decrease on heating whereas that of gases increase.

Ans.: (i) Surface tension of the water (0.073 N/m) maintains the free water surface as a film or membrane in tension. Hence very small loads like some insects, needle put horizontally, etc. can be easily supported.

  (ii) Due to heating, cohesive forces between molecules of liquid get reduced. Viscosity of liquid is predominantly due to cohesive forces as exchange of molecular momentum is negligibly small. Hence viscosity of liquid reduces.

  In gases viscosity is due to exchange of molecular momentum (as cohesive forces are negligible). As gas molecules get more heat, 'More exchange' leads to increase in viscosity of gas.

  (b) Define capillarity. Derive an expression for the capillary rise.   (2 + 4)

Ans.: When a small diameter tube is partially dipped in liquid, it is found that liquid level inside tube is either raised or lowered as compared to liquid level outside the tube. This effect is known as capillarity, which is due to surface tension and adhesion/cohesion properties of liquid.

  Consider a tube of diameter d dipped in a liquid of specific weight γ and capillary rise in tube be 'h' as shown in Fig. 1. There is apparent rise of liquid level at the contact with tube due to predominance of adhesion in liquids like water, kerosene, oils, etc.

Height of small cylindrical liquid element in the tube $W = \gamma \times$ volume of element.

$$\therefore \qquad W = \gamma\left(\frac{\pi}{4}\right)d^2 h$$

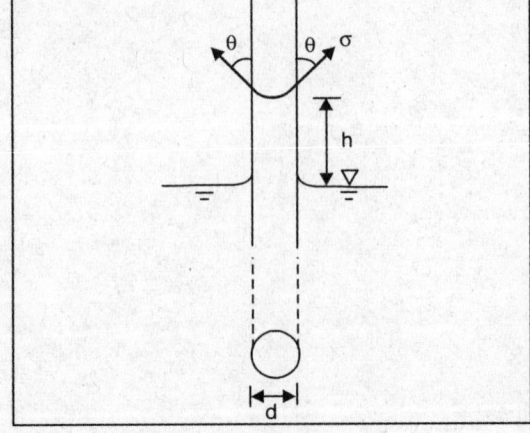

Upward vertical force due to surface tension $= (\sigma\cos\theta)(\pi d)$

Equating these forces in vertical direction, we get h.

$$\gamma\left(\frac{\pi}{4}\right)d^2 h = (\sigma\cos\theta)(\pi d)$$

$$\therefore \qquad \boxed{h = \frac{4\sigma\cos\theta}{\gamma d}}$$

Fig. 1

(c) At a certain point in castor oil film, the shear stress in 0.2 N/m² and the velocity gradient is 0.216 s⁻¹. If the mass density is 959.42 kg/m³, find the kinematic viscosity of the oil.   (4)

Ans.: Newton's law of viscosity is $\tau = \mu \dfrac{du}{dy}$. Given $\tau = 0.2$ Pa, $\dfrac{du}{dy} = 0.216$ and $\rho = 959.42$

$\therefore$ $$\mu = \frac{0.2}{0.216} = 0.926 \text{ Ns}/\text{m}^2$$

Now $\quad v = \dfrac{\mu}{\rho}$. Hence $\boxed{v = 965.166 \times 10^{-6} \, \text{m}^2/\text{s}}$

(d) State Newton's law of viscosity. A *U*-tube is made up of two capillaries of bore 1 mm and 2 mm respectively. The tube is held vertically and is partially filled with liquid of surface tension 0.05 N/m and zero contact angle. Calculate the mass density of the liquid if the estimated difference in the level of two meniscii is 12.5 mm.   (2+6)

Ans.: Newton's law of viscosity state that for certain liquids, shear stress between two layers in directly proportional to the rate of shear deformation, (i.e. velocity gradient).

$\therefore$ $$\tau \propto \frac{du}{dy} \quad \text{or} \quad \tau = \mu \frac{du}{dy}$$

Formula for capillary rise for liquid with $\theta = 0°$ is given as

$$h = \frac{4\sigma}{\gamma d}$$

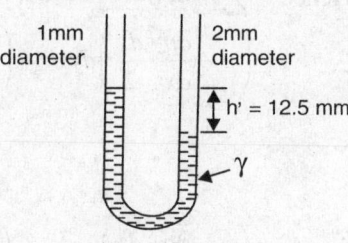

Hence $h' = \dfrac{4\sigma}{\gamma}\left[\dfrac{1}{d_1} - \dfrac{1}{d_2}\right]$ as shown in the Fig. 2.

h' is given in the problem.

$$\left(12.5 \times 10^{-3}\right) = \frac{4(0.05)}{\gamma}\left[\frac{1}{1 \times 10^{-3}} - \frac{1}{2 \times 10^{-3}}\right]$$

$\therefore$ $\quad \gamma = 8000$ N/m³

But $\quad \gamma = \rho g$. Hence $P = \dfrac{\gamma}{g} = \dfrac{8000}{9.81}$

$$\boxed{\therefore \rho = 815.494 \text{ kg}/\text{m}^3}$$

Fig. 2

Q.2 (a) Draw a rectangular parallelopid element of a fluid at rest, indicating the pressures on the faces. For the element derive the hydrostatic equation in the form $p = \gamma h$, where p is the pressure intensity at a depth h from a liquid surface of specific weight γ.   (10)

Ans.: Assume pressure at the center of a rectangular parallelopiped shaped element be p and increases in the x, y and z directions. Let rates of increase be $\dfrac{\partial p}{\partial x}, \dfrac{\partial p}{\partial y}$ and $\dfrac{\partial p}{\partial z}$ respectively. For size of the element $dx \times dy \times dz$, forces acting (including weight) will be as shown in Fig. 3, with origin at the center of gravity 0.

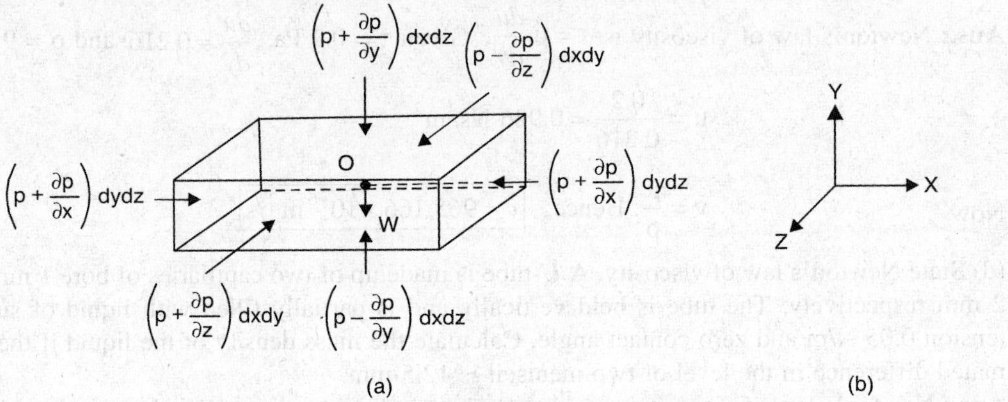

(a)   (b)

Fig. 3

$$W = \gamma \times \text{volume} = \gamma \left(dx\, dy\, dz \right)$$

Since the element is in static liquid mass, it also must be in equilibrium.

Hence $\sum F_x = 0$

$$\therefore \quad \left(p - \frac{\partial p}{\partial x} \cdot \frac{dx}{2} \right) dy\, dz - \left(p + \frac{\partial p}{\partial x} \cdot \frac{dx}{2} \right) dy\, dz = 0$$

$$\therefore \quad -\frac{\partial p}{\partial x} \cdot dx \cdot dy \cdot dz = 0 \quad \text{or} \quad \frac{\partial p}{\partial x} = 0 \quad\quad \text{...(i)}$$

$$\sum F_z = 0$$

$$\therefore \quad \left(p - \frac{\partial p}{\partial z} \cdot \frac{dz}{2} \right) dx\, dy - \left(p + \frac{\partial p}{\partial z} \cdot \frac{dz}{2} \right) dx\, dy = 0$$

$$\therefore \quad -\frac{\partial p}{\partial z} \cdot dx\, dy\, dz = 0$$

$$\therefore \quad \frac{\partial p}{\partial z} = 0 \quad\quad \text{...(ii)}$$

Thus we observe that pressure intensity does not vary in the horizontal (x and z) directions. Finally $\Sigma F_y = 0$.

$$\therefore \quad \left(p - \frac{\partial p}{\partial x} \cdot \frac{dy}{2} \right) dx\, dy - \left(p + \frac{\partial p}{\partial y} \cdot \frac{dy}{2} \right) dx\, dz - r\, dx\, dy\, dz = 0$$

$$\therefore \quad -\frac{\partial p}{\partial y} \cdot dx\, dy\, dz - r\, dx\, dy\, dz = 0$$

$$\text{or} \quad \frac{\partial p}{\partial y} = -\gamma \quad\quad \text{...(iii)}$$

This can be rewritten as $\frac{\partial p}{\partial y} = -\gamma$ [as equations (i) and (ii) already proved that pressure varies only in vertical direction]. In words we state the equation (iii) as —'rate of change of pressure is negative upward and proportional to specific weight of liquid γ'. If depth of a point in liquid is measure from free surface as h, we write $\frac{\partial p}{\partial \dot{g}h} = \gamma$ (negative sign vanishes as h is measured downward). Recalling that gauge pressure on free surface of liquid is zero and integrating, we get

$$\therefore p = \gamma h$$

(b) A certain fluid of specific gravity 0.8 flows upwards through a vertical pipe A and B are two points on the pipe, B being 0.3 m higher than A. A U-tube mercury manometer is connected at gage points A and B. If the difference of pressure between A and B is 0.18 N/m², find the reading shown by the manometer. (6)

Ans.: Sketch of the setup as shown in Fig. 4 assuming monometric reading x. Assume datum $X–Y$.

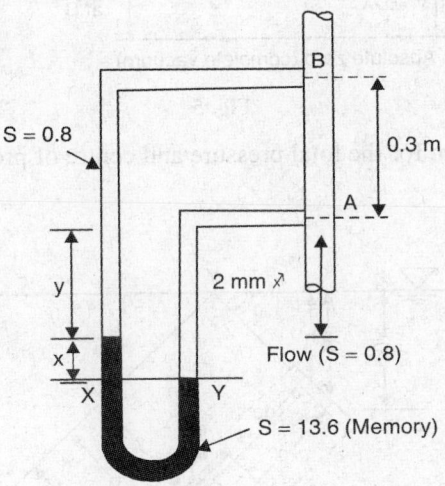

Fig. 4

$\therefore$ Pressure above X = Pressure above Y

$\therefore \qquad P_B + (0.8\gamma)(0.3) + (0.8\gamma)\,y + (13.6\gamma)\,x = p_A + (0.8\gamma)(x+y)$

Assume $\qquad P_A > P_B$ Hence $P_A - P_B = 0.18$ N/m² (given)

$\therefore \qquad 0.8\,\gamma\,(0.3) + 13.6\,\gamma x = 0.8\,\gamma\,x + 0.18$

Put $\qquad \gamma = 9810$ N/m³ (for water)

$\therefore \qquad x(13.6 - 0.8)9810 = 0.18 - (0.8)\,(0.3)\,(9810)$

$\qquad (9810)\,(12.8\,x) = 0.18 - 2354.4$

Thus we find that pressure difference given is not correct.

Let us assume
$$P_A - P_B = 18 \text{ kN/m}^2$$
$$\therefore \quad (9810)(12.8)\, x = 18000 - 2354.4$$
$$\therefore \quad x = 0.125\text{m or } \boxed{x = 124.6 \text{ mm}}.$$

(c) Define the terms gauge pressure, vacuum pressure and absolute pressure. Indicate their relative positions on a chart. (3+1)

Ans.: A pressure measured with reference to local atmospheric pressure (i.e. atom. pr. at a point or station) is called gauge pressure. It can be positive or negative as shown in Fig. 5. Negative gauge pressure is also called 'vacuum' or 'suction' pressure. Absolute pressure is the press measured from absolute zero line (hypothetical datum below which is not a single gas molecule can exist).

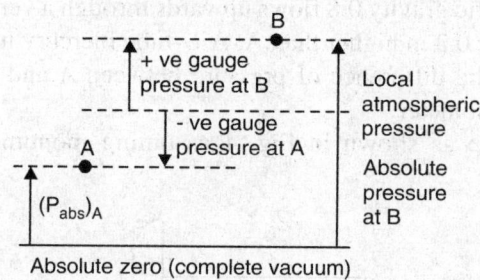

Fig. 5

Q.3 (a) Derive an expression for the total pressure and centre of pressure for an incline surface immersed in a liquid.

Ans.:

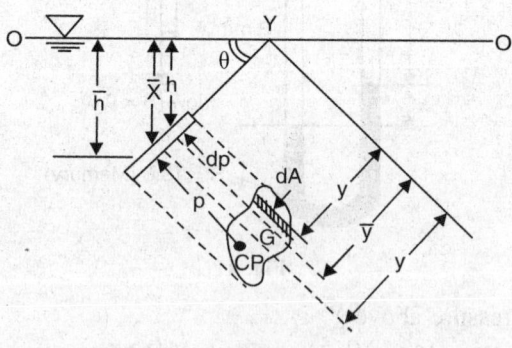

Fig. 6

Consider an arbitrary shaped plane surface inclined at angle θ with free liquid surface '$O - O$' having are A. Let Q and CP be the positions of centroid and centre pressure, their depth from $O - O$ are $\bar{x}$ and $\bar{h}$ respectively as shown in Fig. 6. Since pressure intensity varies on the surface from point to point according to depth, consider infinitely small elemental horizontal strip of area dA at depth 'h'. Total pressure for strip = dP.

For simplicity, consider an axis YY which is *ar* right angles to the plane of the surface.

$\therefore$ $\qquad$ $pdA = \gamma\,hdA$, but $y = \sin\theta = h$

$\therefore$ $\qquad$ $dp = \gamma y \sin\theta\,dA$

$\therefore$ $\qquad$ $p = \int dP = \gamma \sin\theta \int ydA$

From definition of centroid, $\int ydA = A\bar{y}$

$\qquad$ $p = \gamma \sin\theta A\bar{y}$

again $\bar{y}\sin\theta = \bar{x}$

$\therefore$ $\qquad$ $p = \gamma\,A\bar{x}$

This gives total pressure on the surface to locate, center of pressure, consider moments of all dP about YY and apply Varignon's Theorem of moments.

$\therefore$ $\qquad$ $py_1 = \int dp \times y$

$\therefore$ $\qquad$ $\left(\gamma A\bar{x}\right)\left(\dfrac{\bar{h}}{\sin\theta}\right) = \int\left(\gamma y \sin\theta\,dA\right)y$ since $y_1\sin\theta = \bar{h}$

Now, $\int y^2 dA = I_y$ by definition of moment of inertia.

Also by parallel axis theorem, $I_y = I_G + A\bar{y}^2$

$\therefore$ $\qquad$ $\left(A\bar{x}\right)\left(\dfrac{\bar{h}}{\sin\theta}\right) = \sin\theta\left(I_G + A\bar{y}^2\right)$ put $\bar{y} = \dfrac{\bar{x}}{\sin}\theta$

$\therefore$ $\qquad$ $\left(A\bar{x}\right)\left(\dfrac{\bar{h}}{\sin\theta}\right) = \sin\theta\left(I_G + A\dfrac{\bar{x}^2}{\sin^2\theta}\right)$

$\therefore$ $\qquad$ $\bar{h} = \bar{x} + \dfrac{I_G \sin^2\theta}{A\bar{x}}$. This gives depth of C.P.

(b) Describe the experimental method of determining the metacentric height of a floating object. (6)

Ans.: A small weight 'w' initially placed on the axis of symmetry is shifted through distance 'x' to tilt the body of weight 'W'.

θ can be measured with a protractor scale or other arrangement can be used to measure tan θ directly. 'W' and 'w' are known.

$\therefore$ $\qquad$ $\overline{GM} = \dfrac{wx}{W\tan\theta}$

Sometimes a small plum bob is used for obtaining θ and a small weight with bob can be used. On the slots or grooves made to place this jockey weight, the weight can attached as that x is known.

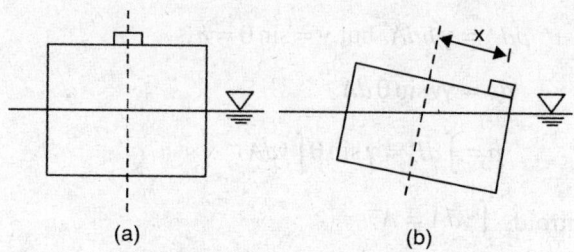

(a) (b)

Fig. 7

(c) A wooden cylinder having a specific gravity of 0.6 is required to float in an oil specific gravity 0.8. If the diameter of the cylinder is d and length L, show that L do not exceed $0.817\, d$ for the cylinder to float with its longitudinal axis vertical.

Ans.: For the floating cylinder $F_B = W$ gives

$$\gamma_{oil}\,(Ah) = \gamma_{cyl}\,(AL)$$

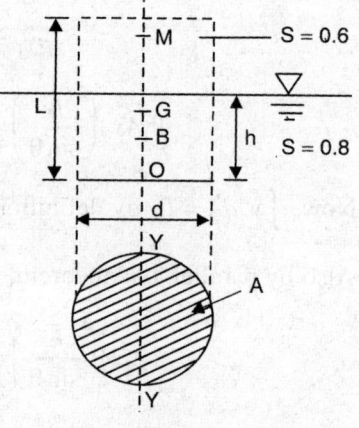

Fig. 8

$$\therefore \qquad 0.8\gamma h = 0.6\gamma L \quad \text{or} \quad h = \left(\frac{3}{4}\right)L$$

Assume points O, B, G and M on the axis.

$$\therefore \qquad OB = \frac{h}{2} = \left(\frac{3}{8}\right)L$$

and

$$OG = \frac{L}{2}$$

$$\therefore \qquad BG = OG - OB\left(\frac{1}{8}\right)L$$

Now $BM = \dfrac{I}{V}$ where $I = \dfrac{\pi}{64}d^4$ and

$$V = Ah = \frac{\pi}{4}d^2\left(\frac{3}{4}\right)L$$

$$\therefore \qquad BM = \frac{\dfrac{\pi}{64}(d)^4}{\left(\dfrac{\pi}{4}\right)(d^2)\left(\dfrac{3}{4}\right)L} = \frac{(d^2)}{4(3L)}$$

or

$$BM = \frac{d^2}{12L}$$

For stable equilibrium; $BM \geq BG$

$$\therefore \qquad \frac{d^2}{12L} \geq \frac{L}{8}$$

$$\therefore \qquad \frac{L^2}{2} \le \frac{d^2}{3}$$

$$\therefore \qquad L^2 \le 0.666\,d^2$$

i.e. $\qquad L \le \sqrt{0.666}\,d$ or $\boxed{L \le 0.816\,d}$

Q.4 (a) Define $\hfill$ (8)
      (i) Stream line $\qquad\qquad$ (ii) Streak line
      (iii) Path line $\qquad\qquad$ (iv) Velocity potential

Ans.: (i) **Stream line** is defined as the line along which flow occurs. It is the line of constant stream function ψ. It can also be defined as the line is the flow field, tangent on which always gives velocity vector.

    (ii) **Streak line** is the line showing instantaneous positions of all fluid particles which have passed through a reference point or section. Smoke line or dyeline are examples of streak line.

    (iii) **Path line** is the line traced by a fluid particle in motion. Thus it is the locus of a single particle over a period of time.

    (iv) **Velocity potential** ϕ is the scalar function of co-ordinates x and y such that $\dfrac{\partial \phi}{\partial x} = -u$

and $\dfrac{\partial \phi}{\partial y} = -v$ i.e. Its partial derivative gives velocity in opposite direction along the co-ordinate

axis. It also means potential or ability of a particle to flow. (flow occurs from higher potential to lower potential).

(b) In a flow the stream function is given by $\psi = 3x^2 - 3y^2$. Find the velocity potential function.
$\hfill$ (8)

Ans.: $\psi = 3x^2 - 3y^2$ and gives $\dfrac{\partial \psi}{\partial x} = 9x^2$ and $\dfrac{\partial \psi}{\partial y} = -6y$

Be definition of ψ we have $\dfrac{\partial \psi}{\partial x} = v$ and $\dfrac{\partial \psi}{\partial y} = -u$

Similarly $\qquad\qquad \dfrac{\partial \phi}{\partial x} = -u$ and $\dfrac{\partial \phi}{\partial y} = -v$

$$\therefore \qquad\qquad \frac{\partial \phi}{\partial x} = -6y$$

Integrating w.r.t. 'x' we get $\phi = -6xy + f(y)$ $\hfill$ ...(i)

Differentiating this with respect to y we get $\dfrac{\partial \phi}{\partial y} = -6x + f'(y)$

But $\qquad\qquad\qquad \dfrac{\partial \phi}{\partial y} = -v = -9x^2$

$$\therefore \qquad -9x^2 = -6x + f'(y)$$

or
$$f'(y) = 6x - 9x^2$$

Integrating with respect to y we get $f(y) = 6xy - 9x^2 y$

(neglecting numerical constant of integration)

Putting $f(y)$ in equation (i) we get

$$\phi = -6xy + 6xy - 9x^2 y \quad \text{or} \quad \boxed{\phi = -9x^2 y}$$

(c) Distinguish between
 (i) Steady flow and unsteady flow
 (ii) Laminar flow and Turbulent flow.

Ans.: (i)

Steady flow	Unsteady flow
1. Flow parameters like velocity, pressure do not vary with time.	1. Flow parameters change from time to time in the flow.
2. Mathematically $\dfrac{\partial v}{\partial t} = 0, \dfrac{\partial p}{\partial t} = 0,$ etc. represents the steady flow	2. Mathematically, $\dfrac{\partial v}{\partial t} \neq 0, \dfrac{\partial p}{\partial t} \neq 0,$ etc.
3. e.g. flow through pipe under constant head.	3. e.g. flow under varying head (either increasing or decreasing head)

(ii) Please refer Q. 8(c)(ii) of July/Aug-2004.

Q.5 (a) Assuming that the rate of discharge Q of a hydraulic machine is dependent upon the mass density ρ of the fluid, speed of machine N, diameter of the impeller pressure p an viscosity μ, show using Buckingham's π theorem that it can be represented by

$$Q = ND^3 \phi \left[\frac{9H}{N^2 D^2}, \frac{v}{ND^2} \right]$$

H being the head and v the kinematic viscosity of the fluid. $\hspace{2cm}$ (8)

Ans.: Let, $Q = f(\rho, N, D, p, \mu)$

Hence total number of quantities $n = 6$

Now $\quad Q \equiv L^3 T^{-1}, \rho \equiv ML^{-3}, N \equiv T^{-1}, D \equiv L, p \equiv ML^{-1}T^{-2}$ and $\mu \equiv ML^{-1}T^{-1}$

Hence total fundamental dimensions involved $m = 3$

$\therefore$ No. of π terms $n - m = 3$

Let ρ, N, D be repeating variables.

$$\pi_1 = \rho^{a_1} N^{b_1} D^{c_1} Q$$

$$\pi_2 = \rho^{a_2} N^{b_2} D^{c_2} p$$

$$\pi_3 = \rho^{a_3} N^{b_3} D^{c_3} \mu \quad \text{where } f_1(\pi_1, \pi_2, \pi_3) = 0 \qquad \text{...(i)}$$

First π term: Put M, L, T dimensions on both sides.

$$\therefore \qquad M^0 L^0 T^0 \equiv [ML^{-3}]^{a_1} [T^{-1}]^{b_1} [L]^{c_1} [L^3 T^{-1}]$$

Equate powers of M, T and L for dimensional homogenity.

$$\therefore \qquad 0 = a_1$$

$$0 = -b_1 - 1 \text{ or } b_1 = -1$$

$$0 = -3a_1 + c_1 + 3. \text{ Hence } c_1 = -3$$

$$\therefore \qquad \pi_1 = \rho^0 N^{-1} D^{-3} Q$$

$$\therefore \qquad \pi_1 = \frac{Q}{ND^3}$$

Second π term: $M^0 L^0 T^0 \equiv [ML^{-3}]^{a_2} [T^{-1}]^{b_2} [L]^{c_2} [ML^{-1}T^{-2}]$

$$\therefore \qquad 0 = a_2 + 1 \quad \text{or} \quad a_2 = -1$$

$$0 = -b_2 - 2 \quad \text{or} \quad b_2 = -2$$

$$0 = -3a_2 + c_2 - 1 \quad \text{or} \quad c_2 = -2$$

$$\therefore \qquad \pi_2 = \frac{p}{\rho N^2 D^2}$$

Now from fundamentals of statics, we know that $p = \rho gH$

$$\therefore \qquad \pi_2 = \frac{gH}{N^2 D^2}$$

Third π term: $\qquad M^0 L^0 T^0 = [ML^{-3}]^{a_3} [T^{-1}]^{b_3} [L]^{c_3} [ML^{-1}T^{-1}]$

$$\therefore \qquad 0 = a_3 + 1 \text{ or } a_3 = -1$$

$$0 = -b_3 - 1 \text{ or } b_3 = -1$$

$$0 = -3a_3 + c_3 - 1 \text{ or } c_3 = -2$$

$$\therefore \qquad \pi_3 = \frac{\mu}{\rho ND^2} \text{ Using } \frac{\mu}{\rho} = v \text{ (kinematic viscosity) we get}$$

$$\pi_3 = \frac{v}{ND^2}$$

Putting values of these π terms in equation (i) we get

$$f_i \left(\frac{Q}{ND^3}, \frac{gH}{N^2 D^2}, \frac{v}{ND^2} \right) = 0$$

or

$$\frac{Q}{ND^3} = \phi\left(\frac{gH}{N^2D^2}, \frac{v}{ND^2}\right)$$

i.e.

$$\boxed{Q = ND^3\phi\left(\frac{gH}{N^2D^2}, \frac{v}{ND^2}\right)}$$

(b) Derive Euler's equation along a streamline and reduce it to Bernoulli's equation.
Ans.: Please refer Q. 4(c) of July/Aug.-2004.

(c) A 0.25 m diameter pipe carries an oil of SG 0.8 at the rate of 120 *l/s* and the pressure a point *A* is 19.62 kN/m². If the point *A* is 3.5 m above the datum line, calculate total energy at point *A* in *m* of oil.
Ans:

$$Q = 120 l/s = 120 \times 10^{-3}\,m^3/s \text{ and } A = \frac{\pi}{4}(0.25)^2 = 0.049\,m^2.$$

∴ $V = \frac{Q}{A} = 2.445$ m/s. Now total head $H = z + \frac{V^2}{2g} + \frac{p}{\gamma}$

Where velocity head $= \frac{V^2}{2g} = 0.305$ m

Pressure head $= \frac{p}{\gamma} = \frac{19.62}{0.8 \times 9.81} = 2.5$ m

Datum head = 3.5 m

Hence total head $\boxed{H = 6.305 \text{ m}}$

Q.6 (a) The inlet and throat diameters of a vertically mounted venturimeter are 300 mm and 100 mm respectively. The throat is below the inlet at a distance of 100 mm. The mass density of the liquid is 900 kg/m³. The pressure intensity at the inlet is 140 kPa while at the throat is 80 kPa. Calculate the flow rate. Assume that 2% of the differential head is lost between the inlet and the throat. (8)

Ans.: $\rho = 900$ kg/m³ means $s = 0.9$

$$\frac{p_1}{\gamma} = \frac{140}{0.9 \times 9.81} = 15.857 \text{ m}$$

$$\frac{p_2}{\gamma} = \frac{80}{0.9 \times 9.81} = 9.061 \text{ m}$$

∴ Differential head = 15.857 − 9.061
$$= 6.796 \text{ m}$$

∴ Loss = 0.02 × 6.796 = 0.136 m

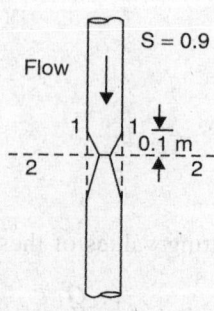

Fig. 9

Now $A_1 V_1 = A_2 V_2$ gives

$$V_2 = \left(\frac{A_1}{A_2}\right) V_1 = \left(\frac{300}{100}\right)^2 V_1$$

$\therefore$
$$V_2 = 9 V_1$$

Apply Bernoulli's equation between inlet $(1-1)$ and throat $(2-2)$—considering loss.

$\therefore$
$$\frac{p_1}{\gamma} + z_1 + \frac{V_1^2}{2g} = \frac{p_2}{\gamma} + z_2 + \frac{V_2^2}{2g} + h_L$$

$\therefore$
$$\frac{V_2^2 - V_1^2}{2g} = \frac{p_1}{\gamma} - \frac{p_2}{\gamma} + z_1 - z_2 - h_L$$

$\therefore$
$$V_2^2 - V_1^2 = (19.62)(15.857 - 9.061 + 0.1 - 0 - 0.136)$$

$\therefore$
$$(9V_1)^2 - V_1^2 = 132.631$$

$\therefore$
$$V_1 = 1.288 \ \text{m/s}$$

$\therefore$
$$Q = A_1 V_1 = \frac{\pi}{4}(0.3)^2(1.288)$$

$\therefore$
$$\boxed{Q = 0.091 \ \text{m}^3/\text{s}}$$

or
$$\boxed{Q = 91.01 \ \text{R.P.S.}}$$

(b) Derive the Darcy-Weisbach equation for the loss of head due to friction in a pipe. (8)

Ans.: Consider portion of fluid between sections $1-1$ and $2-2$ as shown $\tau_0 =$ wall shear stress.

For steady, uniform flow through pipe, acceleration is zero Hence $\Sigma F = 0$ along length of pipe in the direction of flow.

$\therefore$
$$p_1\left(\frac{\pi}{4}D^2\right) - p_2\left(\frac{\pi}{4}D^2\right) \tau_0 (\pi D L) + W \sin\theta = 0$$

But
$$W = \gamma\left(\frac{\pi}{4}D^2 L\right) \ \text{and} \ \sin\theta = \frac{z_1 - z_2}{L}$$

$$p_1 - p_2 - \tau_0\left(\frac{4L}{D}\right) + \gamma L\left(\frac{z_1 - z_2}{L}\right) = 0$$

or
$$\left(\frac{p_1}{\gamma} - \frac{p_2}{\gamma}\right) + (z_1 - z_2) = \frac{4\tau_0 L}{\gamma D} \qquad \text{...(i)}$$

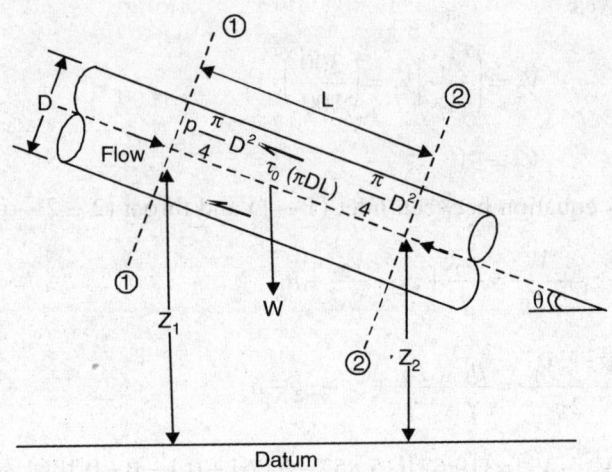

Fig. 10

But Darcy-Weisbach friction factor $f = \dfrac{4\tau_0}{\dfrac{1}{2}\rho V^2}$ and $\rho g = \rho$

Hence $4\tau_0 = \dfrac{1}{2} f \left(\dfrac{\gamma}{g}\right) V^2$ and L.H.S. of equation (i) is frictional loss.

$\therefore \qquad \boxed{h_f = \dfrac{flV^2}{2gD}}$. The required Darcy-Weisbach equation.

(c) What are hydraulic gradient and total energy lines? (4)

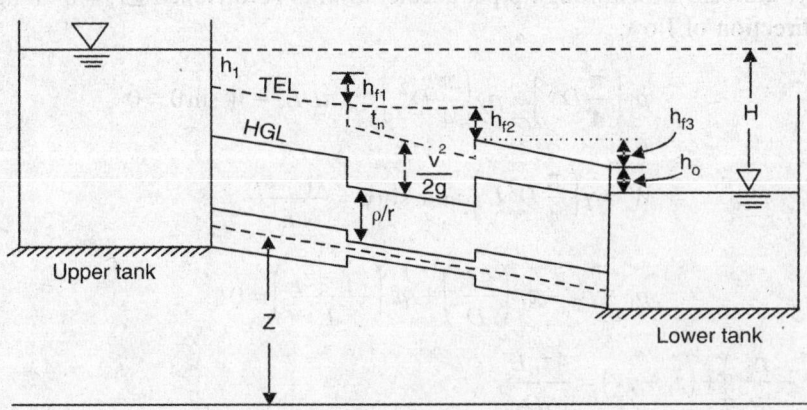

Fig. 11

Ans.: Hydraulic Grade Line (HGL) is the line obtained by joining all piezometric heads along the given pipe. If pipe is horizontal, it becomes pressure head line only as datum head becomes zero. It always slopes along flow direction, except in case of sudden expansion. It is usually above pipe axis, except in case of syphon.

Total Energy Line (TEL) also called total head line is obtained by joining total heads above datum. It is at a distance of $\dfrac{V^2}{2g}$ (Velocity Head) above the HGL and parallel to HGL. Considering all losses, there will be sudden drop in TEL showing loss of head due to sudden expansion/contraction and entry or exit loss as shown in Fig. 11.

Q.7 (a) Water is supplied to a town having a population of 1 lakh from a reservoir 6 km away from the town and it is stipulated that half of the daily supply of 15 liters per head should be delivered in 8 hours. What should be the diameter of the supply pipe. The loss of head due to friction in the pipe line is 12 m. Take Chezy's constant as 45. (8)

Ans.: Note: Chezy's formula is actually for open channel flow.

$$Q_{total} = (1 \text{ lakh population}) \times (15 \text{ liters per head})/\text{time}$$

$\therefore$
$$Q = \frac{1.5 \times 10^6 \text{ litres}}{8 \text{ hours}} = \frac{1.5 \times 10^3 \text{ m}^3}{8 \times 3600 \text{ s}}$$

$\therefore$
$$Q = 0.0521 \text{ m}^3/\text{s}$$

$$A = \frac{\pi}{4} d^2 \text{ and } p = \pi d \text{ ....wetted perimeter}$$

$\therefore$
$$R = \frac{A}{p} = \frac{d}{4}$$

Chezy's formula is $V = C\sqrt{RS}$

Now
$$s = \frac{12}{6000} = 2 \times 10^{-3} \text{ and } V = \frac{Q}{A}$$

$\therefore$
$$\left[\frac{0.0521}{\left(\frac{\pi}{4}\right)d^2} \right]^2 = (45)^2 \left(\frac{d}{4}\right)\left(2 \times 10^{-3}\right) \quad \text{...squaring both sides of Chezy's formula}$$

$\therefore$
$$\left(\frac{\pi}{4}\right)^2 d^4 \left(\frac{d}{4}\right) = \frac{(0.0521)^2}{(45)^2 \left(2 \times 10^{-3}\right)}$$

or
$$d = 0.337 \text{ m say } \boxed{340 \text{ mm diameter}}$$

(b) Define Reynold's number. What is its significance? (4)

Ans.: Reynolds number is the ratio of inertia force and viscous force.

$$\therefore \qquad R_e = \frac{F_1}{F_V} = \frac{\rho L^2 V^2}{\mu V L}$$

$$\therefore \qquad R_e = \frac{\rho V L}{\mu} = \frac{V L}{\nu}$$

It shows relative importance or predominance of inertia and viscous forces. It is used therefore to differentiate flows as Laminar and turbulent. If $F_i >> F_v$, particles move randomly and flow B turbulent (for pipes, $R_e > 4000$) if $F_i > F_v$, viscous forces dominate and flow is orderly as particles move in some particular layers (laminae) without mixing.

(c) A supersonic plane travels at 1.8 Mach at an altitude of 20 km above the ground. How far ahead the plane will be when one hears the sonic boom on the ground? (8)

Ans.: Semi-vertex angle α of Mach cone moving with the plane is given as

$$\alpha = \sin^{-1}\left(\frac{1}{M_a}\right)$$

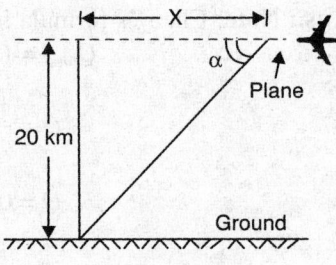

Given $\qquad M_a = 1.8$

$\therefore \qquad \alpha = 33.75°$

From geometry of figure,

$$\tan \alpha = \frac{20}{x}$$

$$\therefore \qquad \boxed{x = 29.933 \text{ km}}$$

Fig. 12

(When a person is in zone of disturbance he will hear the boom.)

Q.8 (a) Experiments were conducted in a wind tunnel with a wind speed of 50 km/h on a flat plate of size 2 m long and 1 m wide. Specific weight of air is 11.282 N/m³. The plate is kept at such an angle that the coefficients of lift and drag are 0.75 and 0.15 respectively. Determine:

  (i) Lift force

 (ii) Drag force

(iii) Resultant force

(iv) Power excited by air stream on the plate. (10)

Ans.:

$$F_L = \frac{1}{2}\rho A V^2 C_L$$

$$\therefore \qquad F_L = \frac{1}{2}\left(\frac{11.282}{9.81}\right)(2 \times 1)\left(\frac{50 \times 5}{18}\right)^2 (0.75)$$

$$\therefore \qquad \boxed{F_L = 166.385 \text{ N}}$$

$$F_D = \frac{1}{2}\rho A V^2 C_D$$

$$\therefore \qquad F_D = \frac{1}{2}\left(\frac{11.282}{9.81}\right)(2 \times 1)\left(\frac{50 \times 5}{18}\right)^2 (0.15)$$

$$\therefore \qquad \boxed{F_D = 33.277 \text{ N}}$$

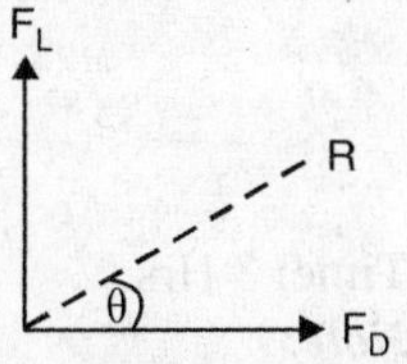

$$\therefore \qquad R = \sqrt{\left(F_L\right)^2 + \left(F_D\right)^2} = 169.68 \text{ N}$$

$$\theta = \tan^{-1}\left(\frac{F_L}{F_D}\right) = 78.69° \text{ as shown.}$$

Power $P = F_D \times V$

$$P = (33.277)\left(\frac{50 \times 5}{18}\right) \text{ watts}$$

$$\therefore \qquad \boxed{P = 462.18 \text{ watts}}$$

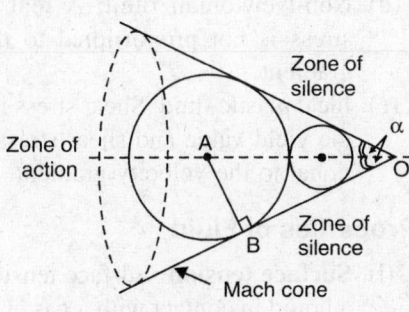

Fig. 13

(b) Distinguish between friction drag and pressure drag. (4)

Ans.: Please refer Q. 3(b) (i) of July/Aug.-2004.

(c) Sketch the nature of propagation of disturbance in compressible flow when Mach number is more than one and hence define Mach angle and Mach cone. (6)

Ans.: As $M_a > 1$, $V > C$

Hence object (projectile) travels faster than the elastic or pressure waves as shown in Fig. 14.

From object O if tangents are drawn to the spheres formed by elastic waves at different times, a can is generated having its axis as direction of motion of projectile (object).

Note that distance $OA = vt$ and $AB = ct$

$$\therefore \qquad \sin \alpha = \frac{AB}{OA} = \frac{ct}{vt}. \text{ Hence } \alpha = \sin^{-1}\left(\frac{1}{M_a}\right)$$

This semivertex angle α is called Mach angle.

Fig. 14

Fluid Mechanics
(July/August 2005)
Semester - IV (ME/IP/AU/IM/MA)

Time: 3 Hrs.   Max. Marks: 100

Note:
   1. *Answer any five full questions.*
   2. *Answer should be precise and to the point.*

Q.1 (a) Define a fluid and explain its following properties.
(i) Surface tension (ii) Viscosity (iii) Vapour pressure (iv) Bulk modulus.   (10)

Solution. Fluid: It is a substance which deform or yield continuously due to application of shear stress no matter how small it is.

The molecules in a fluid can move more freely within the fluid mass and therefore it do not possess any rigidity of form.

Types of Fluids

(a) Ideal fluid: Incompressible and having no viscosity (Imaginary fluids).
(b) Real fluid: Fluids which possess viscosity.
(c) Newtonian fluid: A real fluid in which shear stress is directly proportional to velocity gradient.
(d) Non-Newtonian fluid: A real fluid shear stress is not proportional to the velocity gradient.
(e) Ideal plastic fluid: Shear stress is more than the yield value and shear stress to proportional to the velocity gradient.

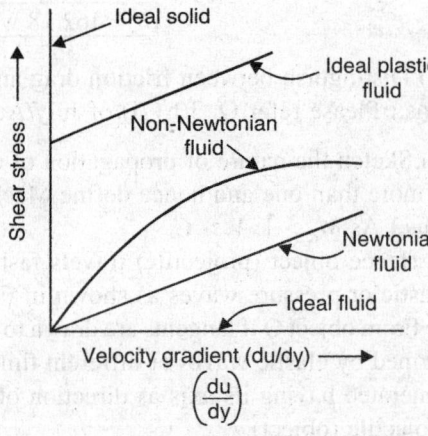

Fig. 1

Properties of Fluid

(i) **Surface tension:** Surface tension is defined as the tensile force acting on the surface of a liquid in contact with a gas on the surface between two immersible liquids such that the contact surface behaves like a membrane under tension.

It is denoted by σ and expressed as kgf/m in MKS units and N/m in SI units.

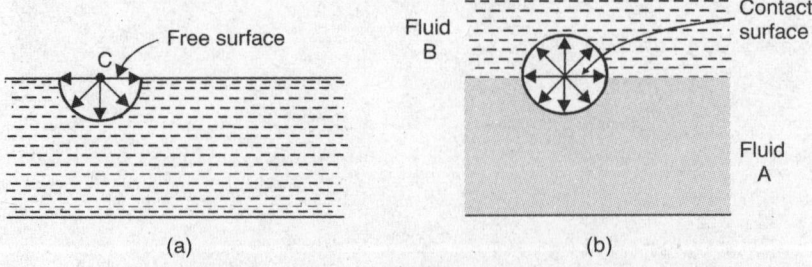

(a)   (b)

Fig. 2

(ii) **Viscosity:** It is the property of fluid which offers resistance to the movement of one layer of fluid over another adjacent layer of the fluid.

When two layers of a fluid, a distance '*dy*' apart move one over the other at different velocities. Say *u* and *u* + *du*, the viscosity together with relative velocity causes a shear stress acting between the fluid layers.

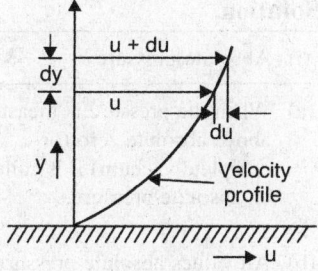

This shear stress is proportional to the rate of change of velocity with respect to *y*. It is denoted by τ called Tau.

$$\tau \propto \frac{du}{dy} \quad \text{or} \quad \tau = \mu \frac{du}{dy}$$

Fig. 3

μ (mu) is the constant of proportionality and is known as the coefficient of dynamic viscosity or only viscosity.

$$\frac{du}{dy} = \text{Rate of shear strain or velocity gradient}$$

∴

$$\mu = \frac{\tau}{\left(\dfrac{du}{dy}\right)} = \text{Shear stress required to produce unit rate of shear strain}$$

its unit is $\dfrac{\text{kg f sec}}{\text{m}^2}$ in MKS $= \dfrac{\text{Ns}}{\text{m}^2}$ is S.I. = 98.1 Poise

(iii) **Vapour pressure:** Consider a liquid which is confined in a closed vessel. When vapourisation of liquid takes place, the molecules escapes from the force surface of the liquid. These vapour molecules get accumulated in the space between the free liquid surface and top of the vessel, which exert a pressure on the liquid surface. This pressure is known as vapour pressure of the liquid.

Vapour pressure is dependent on temperature. A liquid having very high vapour pressure will evaporate more easily than a liquid having low vapour pressure. When the vapour pressure of a liquid is slightly greater than the pressure impressed on the surface the liquid will boil.

(iv) **Bulk modulus:** Bulk modulus of elasticity is a measure of the incremental change in pressure *dp*, which takes place when a volume *v* of the fluid is changed by incremental amount *dv*.

It is denoted by *K* and it is inverse of compressibility of a fluid.

$$K = \frac{\text{Stress}}{\text{Strain}} = -\frac{dp}{(dv/v)} = \frac{\text{Change in pressure}}{\left(\dfrac{\text{Change in volume}}{\text{Original volume}}\right)}$$

Rise in pressure always causes a decrease in volume *dv* is always negative and the minus sign is included in the equation to give a positive value of *K*.

It is expressed in N/m^2 is SI units. Bulk modulus of elasticity of a fluid is not constant it increases with increase is pressure.

(b) Clearly distinguish between

    (i) Absolute pressure (ii) Gauge pressure (iii) Gauge vacuum.   (6)

Solution.

(i) Absolute pressure	(ii) Gauge pressure	(iii) Gauge vacuum
(a) When the pressure is measured above absolute zero (or complete vacuum) it is called an absolute pressure.	When pressure is measured either above or below atmospheric pressure as datum, it is called as gauge pressure.	If the pressure of a fluid is below atmospheric pressure (or suction pressure) is known as gauge vacuum and also vacuum pressure.
(b) All values absolute pressure are positive.	Gauge pressure are positive if they are above that of atmospheric and negative if they are vacuum pressure.	It is negative.

Absolute pressure = Atmospheric pressure + Gauge pressure

Absolute pressure = Atmospheric pressure + Vacuum pressure

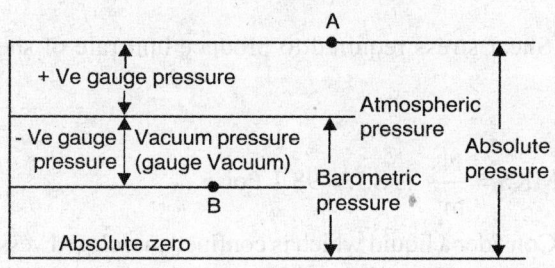

Fig. 4

(c) Two horizontal plates are placed 1.25 cms apart and the intermediate space is far with an oil viscosity 14 poise. Calculate the shear stress in the oil if the upper plate moved with a velocity of 2.5 m/s.

Solution. Given:

$$y = 1.25 \text{ cms} = 0.0125 \text{ m}$$

$$v = 2.5 \text{ m/s}$$

$$\mu = 14 \text{ poise} = 1.4 \frac{Ns}{m^2}$$

Shear stress $\tau = \mu \dfrac{v}{y}$

$$= 1.4 \times \frac{2.5}{0.0125} = 280 \frac{N}{m^2}$$

Q.2 (a) State and prove Pascal's law.

Solution. Pascal's law: It states that the pressure or intensity of pressure at a point in static fluid is equal in all directions.

Proof: Consider the fluid element of very small dimensions *dx*, *dy* and *ds* wedge shape in a fluid mass at rest with width of the element parallel to the plane the paper is unity.

Let P_x, P_y and P_z are the pressure or intensity of pressure acting on the face *AB*, *AC* and *BC* respectively.

Let $\angle ABC = \theta$

Then the forces acting on the element are—
1. Pressure forces normal to the surfaces.
2. Weight of element in the vertical direction.

∴ Forces on the face $AB = P_x \times$ Area of face *AB*
$$= P_x dy \, 1$$

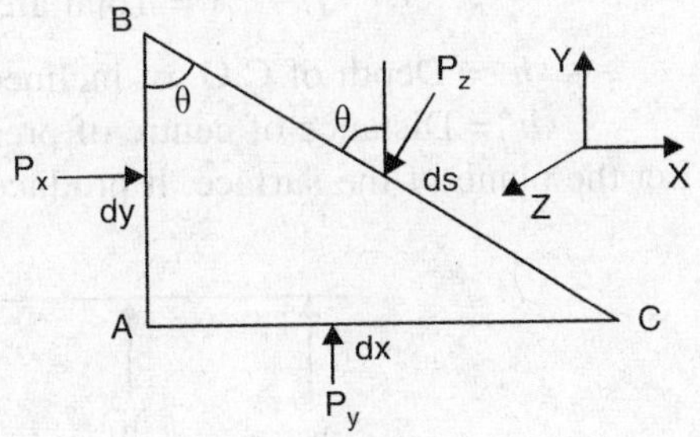

Similarly forces on the face $AC = P_y \, dx \, 1$
and on face $BC = P_z \, ds \, 1$

Weight of element = (mass of element) × g
$$= \text{volume} \times \rho \times g \qquad \rho = \text{density of fluid}$$

$$= \left(\frac{AB \times AC \times 1}{2} \right) \times \rho \times g$$

Fig. 5

Resolving the forces in X-direction.

$$P_x \times dy \times 1 - P_z \,(ds \times 1) \times \sin(90° - \theta) = 0$$

$$P_x dy - P_z ds \cos\theta = 0$$

But from the Fig. 5 $\qquad ds \cos\theta = AB = dy$

∴ $\qquad\qquad P_x \, dy - P_z \, dy = 0$

or $\qquad\qquad P_x = P_z$ ...(1)

Similarly, resolving the forces in *Y* direction, we get

$$P_y dx - P_z ds \cos(90° - \theta) - \frac{dx\,dy}{2} \times \rho \times g = 0$$

$$P_y dx - P_z ds \sin\theta - \frac{dx\,dy}{2}\rho \times g = 0$$

But $ds \sin\theta = dx$ and also the element is very small and hence weight is neglected.

∴ $\qquad\qquad P_y dx - P_z dx = 0$

or $\qquad\qquad P_y = P_z$ ...(2)

∴ From equation (1) and (2) we have $P_x = P_y = P_z$

The above equation shows that the pressure at any point in *x*, *y* and *z* directions is equal. Hence Pascal's law is proved.

(b) Derive an expression for the total pressure force and the depth of centre of pressure for an inclined surface submerged in water.

Solution. Consider a plane surface of arbitrary shape immersed in a liquid in such a way that the plane of the surface makes an angle θ with the free surface of the liquid.

$$A = \text{Total area of inclined surface}$$

$\bar{h}$ = Depth of C.G. of inclined area from free surface

h' = Distance of centre of pressure from free surface of liquid.

Let the plane of the surface, it produced meet the free liquid surface at 0.

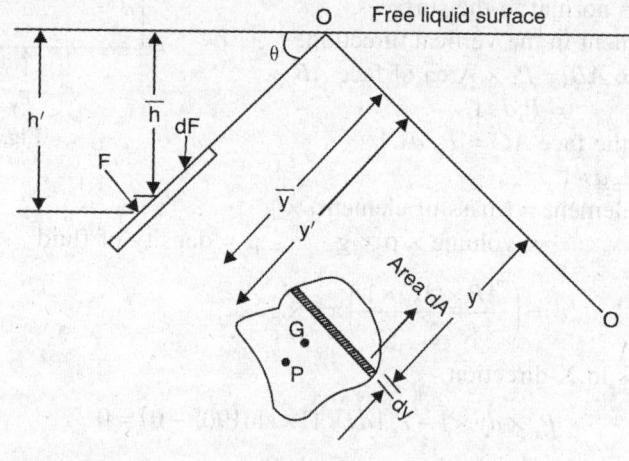

Fig. 6

0–0 is the axis perpendicular to the plane of the surface.

Let $\bar{y}$ = distance of the C.G. of the inclined surface from 0–0

y' = distance of the centre of pressure from 0–0.

Consider a small strip of area dA at a depth 'h' from free surface and at a distance y from the axis 0–0.

Pressure intensity on the strip $P = \rho \cdot g \cdot h$

∴ Pressure force, df on the strip $df = P \times \text{Area of strip} = \rho \cdot g \cdot h \times dA$

Total pressure force on the whole area.

$$F = \int df = \int \rho \cdot g \cdot h \cdot dA$$

But

$$\frac{h}{y} = \frac{\bar{h}}{\bar{y}} = \frac{h'}{y'} = \sin\theta$$

∴

$$h = y\sin\theta$$

∴

$$F = \int \rho \cdot g \cdot y \cdot \sin\theta\, dA$$

$$= \rho \cdot g \cdot \sin\theta \int y\, dA$$

$$= \rho \cdot g \cdot \sin\theta \cdot \bar{y} \cdot A \qquad \because \int y\, dA = \bar{y}A$$

$$\boxed{F = \rho \cdot g \cdot A\bar{h}} \qquad \because \bar{h} = \bar{y}\sin\theta$$

Depth of centre of pressure:

Pressure force on the strip $df = \rho \cdot g \cdot h\, dA$

$$= \rho \cdot g \cdot y \sin\theta\, dA$$

Moment of the forces, df about axis 0–0.

$$= df \times y = \rho \cdot g \cdot y \sin\theta\, dA \times y = \rho \cdot g \cdot y^2 \sin\theta\, dA$$

Sum of moments of all such forces about 0–0.

$$= \int \rho \cdot g \cdot y^2 \sin\theta \cdot dA = \rho \cdot g \cdot \sin\theta \int y^2\, dA$$

But $\int y^2 dA$ = moment of inertia of the surface about 0–0 = I_0.

∴ Sum of moments of all forces about 0–0 = $\rho \cdot g \cdot \sin\theta\, I_0$ ...(a)

Moment of the total force, F about 0–0 is $= f \times y'$ ...(b)

Equating equation (a) and (b)

$$F \times y' = \rho \cdot g \cdot \sin\theta I_0$$

$$y' = \frac{\rho \cdot g \cdot \sin\theta \cdot I_0}{F}$$

Now $y' = \dfrac{h'}{\sin\theta}$, $F = \rho \cdot g \cdot A\bar{h}$ and $I_0 = I_G + A\bar{y}^2$ (by theorem of parallel axis)

∴

$$\frac{h'}{\sin\theta} = \frac{\rho \cdot g \cdot \sin\theta}{\rho \cdot g \cdot A\bar{h}}\left[I_G + A\bar{y}^2\right]$$

$$h' = \frac{\sin^2\theta}{A\bar{h}}\left[I_G + A\bar{y}^2\right]$$

But

$$\frac{\bar{h}}{\bar{y}} = \sin\theta \quad \text{or} \quad \bar{y} = \frac{\bar{h}}{\sin\theta}$$

∴

$$h' = \frac{\sin^2\theta}{A\bar{h}}\left[I_G + A \times \frac{\bar{h}^2}{\sin^2\theta}\right]$$

or

$$\boxed{h' = \frac{I_G \sin^2\theta}{A\bar{h}} + \bar{h}}.$$

$h' > \bar{h}$. ∴ C.P. is always below the C.G. of an immersed surface.

(c) Define centre of Buoyancy and Metacentre for a floating body.

Solution. Centre of Buoyancy: It is defined as the point through which the force of buoyancy is supposed to act. If vertical buoyancy force is equal to the weight of the fluid displaced by the body, the centre of buoyancy will be the centre of gravity of the liquid displaced.

Metacentre for a floating body: It is the point about which a body start oscillating when body is tilted by a small angle and it is also the line of action of the force of buoyancy will meet the normal axis of the body when the body is given small angular displacement.

(d) Measurements of pressure at the base and top of a mountain are 74 cm and 60 cm mercury respectively. Calculate the height of the mountain if air has a specific weight of 1.22 kg/mt³. (4)

Solution. Pressure at base of mountain.

$$P_1 = 74 \text{ cm of } H_g = \frac{74}{100} \times 13.6 \times 1000 \times 9.81 \text{ N/m}^2$$

$$= 98727.84 \text{ N/m}^2$$

Pressure at top of mountain $P_2 = \rho \cdot g h_2$

$$= 13.6 \times 10^3 \times 9.81 \times \frac{60}{100} = 80049.60 \text{ N/m}^2$$

Density of air = 1.22 kg/m³

Let h = height of the mountain from the base

$$P_2 = P_1 - \rho \cdot g \cdot h$$

$$h = \frac{P_1 - P_2}{\rho \cdot g} = \frac{98727.84 - 80049.60}{1.22 \times 9.81}$$

$$= 1560.65 \text{ m}$$

Q.3 (a) Clearly explain how Metacentric-height is determined experimentally. (6)
Solution. Determination of metacentric-height experimentally. Take a floating vessel with known centre of gravity. Let w_1 = known weight placed over the centre of the vessel.

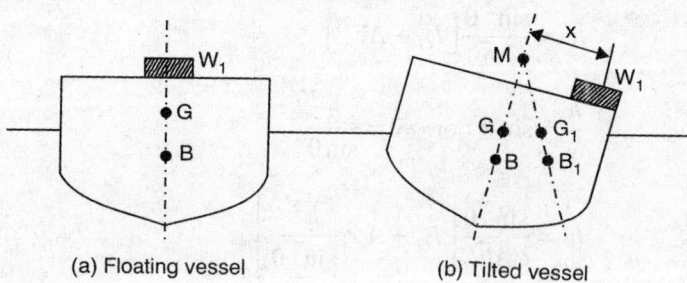

(a) Floating vessel   (b) Tilted vessel

Fig. 7

W = Weight of vessel including w_1
G = Centre of gravity
B = Centre of buoyancy.

The weight w_1 is moved across the vessel towards right through a distance x. The vessel is tilted. The angle of heel θ is measured by means of a plumb line and a contractor attached to the vessel. The new centre of gravity of the vessel will shift to and buoyancy to β_1 as the vessel tilted.

Under equilibrium, the moment caused by the movement of the load w_1 through distance are must be equal to the moment caused by the shift of the centre of gravity G to G_1.

∴ The moment due to change of G

$$= GG_1 \times W = W \times Gm \tan\theta$$

The moment due to movement of $w_1 = w_1 \times x$

∴ $$w_1 x = WGm \tan\theta$$

∴ $$Gm = \frac{w_1 x}{W \tan\theta} = \text{metacentric height}$$

(b) A triangular plate of 2 mt base and height of 3 mt is immersed in water, with its plane making an angle of 60° with free water surface as shown in Fig. 8. Determine the hydrostatic force and the depth of centre of pressure when the apex is 5 mts below the water surface.

(10)

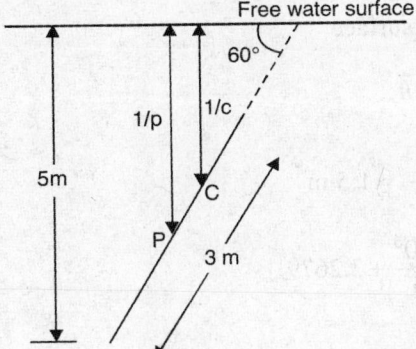
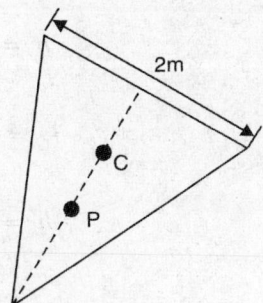

Fig. 8

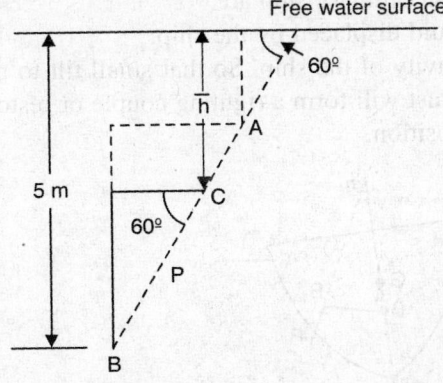
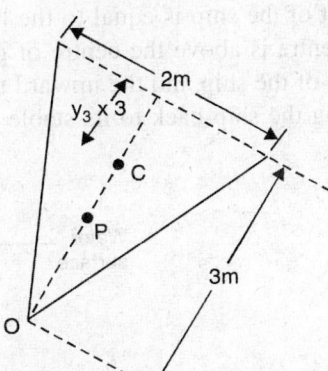

Fig. 9

$$\text{Area of plate} = \frac{b \times h}{2} = \frac{2 \times 3}{2} = 3\,\text{m}^2$$

$$\theta = 60°$$

$$\overline{h} = 5 - CB \sin 60° \qquad \left(CB = \frac{2}{3} \times 3\text{m} \right)$$

$$= 5 - \frac{2}{3} \times 3 \times \sin 60°$$

$$= 3.2679\,\text{m}$$

(i) Hydrostatic force $F = \rho \cdot g \cdot A\overline{h}$

$$= 1000 \times 9.81 \times 3 \times 3.2679$$

$$= 36175.75\,\text{N}$$

(ii) Centre of pressure depth below the water surface

$$h' = \frac{I_G \sin^2 \theta}{A\overline{h}} + \overline{h}$$

$$I_G = \frac{bh^3}{36} = \frac{2 \times 3^3}{36} = 1.5\,\text{m}^4$$

$\therefore$
$$h' = \frac{1.5 \times \sin^2 60°}{3 \times 32679} + 3.2679$$

$$= 3.3826\,\text{m} \qquad\qquad\qquad \textbf{Ans.}$$

(c) Briefly explain what is the basic criteria for the stability of (i) an ordinary ship (ii) a submarine. (4)

Solution. (i) Basic criteria for stability of an ordinary ship are:
1. Weight of the ship is equal to the liquid displaced by the ship.
2. Metacentre is above the centre of gravity of the ship. So that small tilt to the ship, the weight of the ship and the upward thrust will form a righting couple or restoring couple to being the ship back to its stable position.

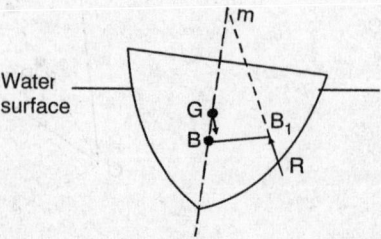

Fig. 10

(ii) Basic criteria for stability of a submarine is
1. Weight of submarine is equal to the buoyancy force.
2. Centre of gravity is below the position of buoyancy force.

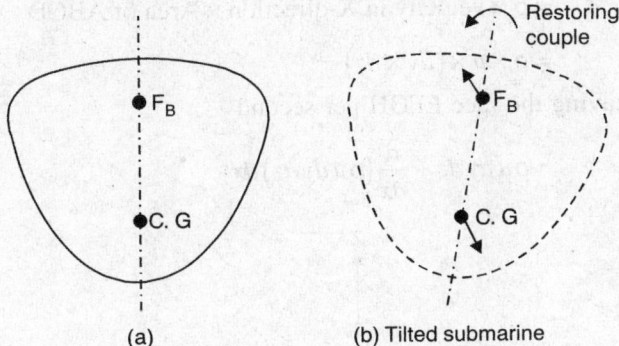

Fig. 11

Q. 4 (a) Define: (i) Stream line (ii) Path line (iii) Rotational flow (iv) Compressible flow. (4)

Solution. (i) Stream line: A stream line is a line which shows the direction of the velocity of the fluid at each point along the line. Fluid particles lying on a stream line at an instant move along the stream line.

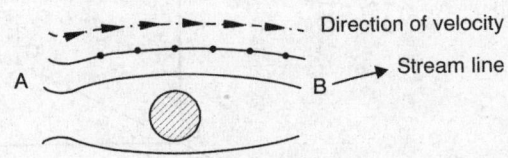

Fig. 12

(ii) Path line: Path traced by a single fluid particle only.

Fig. 13

(iii) Rotational flow: A particle which is moving along a stream line, rotates about its own axis also, them the particle is said to have rotational flow.

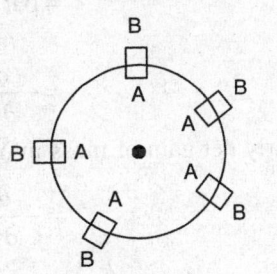

(iv) Compressible flow: A flow in which the density of the fluid changes from point to point is called a compressible flow.

Fig. 14

(b) Clearly derive the continuity equation in three dimensions in the differential form and write the same for steady incompressible flow. (8)

Solution. Continuity equation in three dimensions in the differential form.

Consider a fluid element of lengths dx, dy and dz in the direction of x, y and z. Let u, v and w are the inlet velocity components in x, y, z directions respectively. Mass of fluid entering the face ABCD per second.

$$= \rho \times \text{velocity in X-direction} \times \text{Area of ABCD}$$

$$= \rho \times u \times (dy \times dz)$$

The mass of fluid leaving the face EFGH per second

$$= \rho u \, dy \, dz + \frac{\partial}{\partial x}(\rho u \, dy \, dz) dx$$

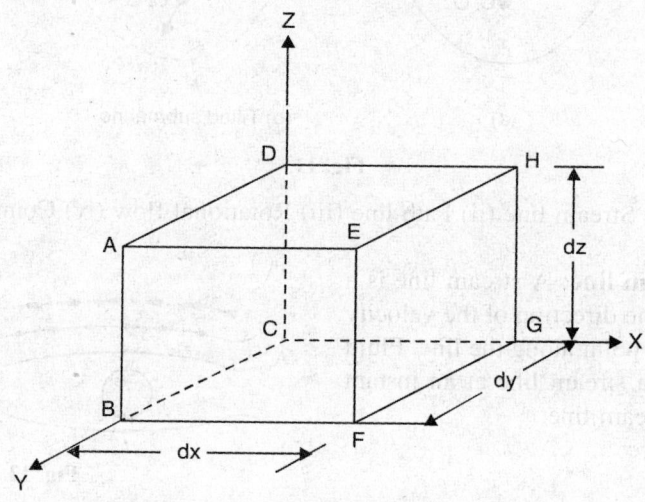

Fig. 15

∴   Gain of mass in X-direction.

$$= (\text{Mass through } ABCD - \text{Mass through } EFGH) \text{ per second}$$

$$= \rho \, u \, dy \, dz - \rho u \, dy \, dz - \frac{\partial}{\partial x}(\rho \cdot u \cdot dy \, dz) dx$$

$$= -\frac{\partial}{\partial x}(\rho \cdot u \partial y \, dz) dx = -\frac{\partial}{\partial x}(\rho \cdot u) dx \, dy \, dz$$

Similarly net gain of mass in Y-direction

$$= -\frac{\partial}{\partial y}(\rho \cdot v) dx \, dy \, dz$$

and in Z-direction $= -\dfrac{\partial}{\partial z}(\rho \cdot w) dx \, dy \, dz$

∴   Net gain of masses $= -\left[\dfrac{\partial}{\partial x}(\rho \cdot u) + \dfrac{\partial}{\partial y}(\rho \cdot x) + \dfrac{\partial}{\partial z}(\rho \cdot w)\right] dx \, dy \, dz$ ...(1)

Since the mass is neither created nor destroyed in the fluid element, the net increase of mass per unit time in the fluid element must be equal to the rate of increase of mass of fluid in the element.

But the mass of fluid in the element is $\rho \cdot dx \cdot dy \cdot dz$ and its rate of increase with time is

$$\frac{\partial}{\partial t}(\rho \cdot dx \cdot dy \cdot dz) \text{ or } \frac{\partial}{\partial t}(dx \cdot dy \cdot dz)$$

Equating equation (1) and (2)

$$-\left[\frac{\partial}{\partial x}(\rho \cdot u) + \frac{\partial}{\partial y}(\rho \cdot v) + \frac{\partial}{\partial z}(\rho \cdot w)\right] dx\, dy\, dz = \frac{\partial \rho}{\partial t} dx\, dy\, dz$$

$$\frac{\partial \rho}{\partial t} + \frac{\partial}{\partial x}(\rho \cdot u) + \frac{\partial}{\partial y}(\rho \cdot v) + \frac{\partial}{\partial z}(\rho \cdot w) = 0$$

Equation (3) to the continuity equation in the differential form.

For steady in compressible flow. ρ = constant, $\dfrac{\partial \rho}{\partial t} = 0$

$\therefore$ Equation becomes $\dfrac{\partial u}{\partial x} + \dfrac{\partial v}{\partial y} + \dfrac{\partial w}{\partial z} = 0$

(c) Clearly define convective and local acceleration. Find the acceleration components at the point (1, 1, 1) in a flow field described by

$$u = 2x^2 + 2y, \ v = -2xy + 3y^2 + 3yz, \ w = -\frac{3}{2}z^2 + 2xz - 9y^2z. \tag{8}$$

Solution. Convective acceleration: It is the rate of change of velocity due to the change of position of fluid particles in a fluid flow.

$$a_x = u \cdot \frac{\partial u}{\partial x} + v \frac{\partial u}{\partial y} + w \frac{\partial u}{\partial z} \qquad \text{Acceleration in } x \text{ direction}$$

$$a_y = u \cdot \frac{\partial v}{\partial x} + v \frac{\partial v}{\partial y} + w \frac{\partial v}{\partial z} \qquad \text{Acceleration in } y \text{ direction}$$

$$a_z = u \cdot \frac{\partial w}{\partial x} + v \frac{\partial w}{\partial y} + w \frac{\partial w}{\partial z} \qquad \text{Acceleration in } z \text{ direction}$$

Local acceleration: It is the rate of increase of velocity with respect to time at a given point in a flow field. $\dfrac{\partial u}{\partial t}, \dfrac{\partial v}{\partial t}, \dfrac{\partial w}{\partial t}$ are local acceleration in x, y and z direction.

Total acceleration is sum of convective and local acceleration.

Given point (1, 1, 1), i.e. $x = 1$ $y = 1$ $z = 1$

$$u = 2x^2 + 2y \qquad\qquad v = -2xy + 3y^2 + 3yz \qquad\qquad w = -\frac{3}{2}z^2 + 2xz - 9y^2z$$

$$= 4 \qquad\qquad = -2 + 3 + 3 \qquad\qquad = -\frac{3}{2} + 2 - 9$$

$$\qquad\qquad\qquad = 4 \qquad\qquad = -\frac{3 + 4 - 18}{2}$$

$$\qquad\qquad\qquad\qquad\qquad = -\frac{17}{2}$$

$$\frac{\partial u}{\partial x} = 4x = 4 \qquad\qquad \frac{\partial v}{\partial x} = -2y = 2 \qquad\qquad \frac{\partial w}{\partial x} = 2z = 2$$

$$\frac{\partial u}{\partial y} = 2 \qquad\qquad \frac{\partial v}{\partial y} = -2x + 6y + 3z \qquad\qquad \frac{\partial w}{\partial x} = -18yz = -18$$

$$\qquad\qquad\qquad = -2 + 6 + 3 = 7$$

$$\frac{\partial u}{\partial z} = 0 \qquad\qquad \frac{\partial v}{\partial z} = 3y = 3 \qquad\qquad \frac{\partial w}{\partial z} = -3z + 2x - 9y^2 = -10$$

For steady flow $\dfrac{\partial u}{\partial t}, \dfrac{\partial v}{\partial t}, \dfrac{\partial w}{\partial t} = 0$

Acceleration components at (1, 1, 1)

$$a_x = u\frac{\partial u}{\partial x} + v\frac{\partial u}{\partial y} + w\frac{\partial u}{\partial z} + \frac{\partial u}{\partial t}$$

$$= 4 \times 4 + 4 \times 2 + -\frac{17}{2} \times 0 + 0 = 24 \text{ units}$$

$$a_y = u\frac{\partial v}{\partial x} + v\frac{\partial v}{\partial y} + w\frac{\partial v}{\partial z} + \frac{\partial v}{\partial t}$$

$$= 4 \times (-2) + 4(7) + -\frac{17}{2}(3) + 0$$

$$= -8 + 28 - \frac{51}{2} = -\frac{11}{2} \text{ units}$$

$$a_z = u\frac{\partial w}{\partial x} + v\frac{\partial w}{\partial y} + w\frac{\partial w}{\partial z} + \frac{\partial w}{\partial t}$$

$$= 4 \times 2 + 4 \times (-18) + -\frac{17}{2} \times (-10) + 0$$

$$= 8 - 72 + \frac{170}{2} = 21 \text{ units}$$

$\therefore$ $\qquad$ Acceleration $= a_x i + a_y j + a_z k$

$$= 24i - \frac{11}{2} j + 21k \qquad \dots \textbf{Ans}$$

Resultant acceleration $= \sqrt{(24)^2 + \left(-\frac{11}{2}\right)^2 + (21)^2}$

$$= 32.36 \text{ units} \qquad \dots \textbf{Ans}$$

Q.5 (a) State Buckingham π theorem and show that the velocity through circular orifice is

given by $V = \sqrt{2gH} \; \Phi \left\{ \dfrac{D}{H} \dfrac{\mu}{\rho VH} \right\}$.

Where H is the head causing flow D the diameter of the orifice, μ is the absolute viscosity, ρ the fluid of the mass density of the fluid and g the acceleration due to gravity. (10)

Solution. Buckingham π theorem: If there are n variables (independent and depends variables) in a physical phenomenon and if these variables contain m fundamental dimensions (M, L, T) then the variables are arranged into $(n - m)$ dimensionless form. Each term is called π term.

$\qquad \qquad H =$ Head causing the flow,

$\qquad \qquad D =$ diameter of orifice,

$\qquad \qquad \mu =$ Absolute viscosity,

$\qquad \qquad \rho =$ fluid mass density,

$\qquad \qquad g =$ Acceleration due to gravity

$\qquad \qquad V =$ is a function of H, D, μ, ρ and g.

$\qquad \qquad V = f(H, D, u, \rho, g)$ or $f_1(V, H, D, \mu, \rho, g) = 0 \qquad \dots (1)$

Total variables $n = 6$.

Writing dimensions of each variables.

$$V = LT^{-1},\ H = L,\ D = L,\ \mu = mL^{-1}T^{-1},\ \rho = mL^{-3},\ g = LT^{-2}$$

Fundamental dimensions $m = 3$.

$\therefore$ Number of π terms $= n - m = 6 - 3 = 3$

$\therefore \qquad \qquad f_1 = (\pi_1, \pi_2, \pi_3) = 0$

Each π term contains $m + 1$ variables where $m = 3$ and is also equal to repeating variables.

Here V is a dependent variable and hence should not be selected as repeating variable.

Choosing H, g, ρ as repeating variable.

We get three π terms as

$$\pi_1 = H^{a_1} \cdot g^{b_1} \cdot \rho^{c_1} \cdot V$$

$$\pi_2 = H^{a_2} \cdot g^{b_2} \cdot \rho^{c_2} \cdot D$$

$$\pi_3 = H^{a_3} \cdot g^{b_3} \cdot \rho^{c_3} \cdot \mu$$

First π term $\qquad \qquad \pi_1 = H^{a_1} \cdot g^{b_1} \cdot \rho^{c_1} \cdot V$

Substituting dimensions on both sides.

$$M^0 L^0 T^0 = L^{a_1} \left(LT^{-2}\right)^{b_1} \left(mL^{-3}\right)^{c_1} \left(LT^{-1}\right)$$

Equating the powers of M, L, T on both sides.

Power of M $0 = c_1$

Power of L $0 = a_1 + b_1 - 3c_1 + 1$ $a_1 = -b_1 + 3c_1 - 1$

Power of T $0 = -2b_1 - 1$ $b_1 = -1/2$

$\therefore$ $a_1 = +\dfrac{1}{2} + 0 - 1 = -1/2$

Substituting the values of a_1, L_1 and c_1 in π_1.

$$\pi_1 = H^{-1/2} g^{-1/2} \rho^0 V$$

$$= \frac{V}{\sqrt{gH}}$$

Second π term $\pi_2 = H^{a_2} \cdot g^{b_2} \cdot \rho^{c_2} \cdot D$

Substituting the dimensions on both sides.

$$M^0 L^0 T^0 = L^{a_2} \left(LT^{-2}\right)^{b_2} \left(ML^{-3}\right)^{c_2} L$$

Equating the powers to M, L, T.

Power of M $0 = c_2$

Power of L $0 = a_2 + b_2 - 3c_2 + 1$

Power of T $\cdot 0 = -2b_2$ $b_2 = 0$

$\therefore$ $a_2 = -b_2 + 3c_2 - 1 = 0 + 0 - 1 = -1$

Substituting the values of a_2, b_2, c_2 in π_2

$$\pi_2 = H^{-1} g^0 \rho^0 D$$

$$= \frac{D}{H}$$

Third π term $\pi_3 = H^{a_3} \cdot g^{b_3} \cdot \rho^{c_3} \cdot \mu$

Substituting the dimensions on both sides

$$m^0 L^0 T^0 = L^{a_3} \left(LT^{-2}\right)^{b_3} \left(ML^{-3}\right)^{c_3} mL^{-1}T^{-1}$$

Equating the powers to M, L, T on both sides

Power of M $0 = c_3 + 1$ $c_3 = -1$

Power of L $0 = a_3 + b_3 - 3c_3 = -1$

Power of T $\qquad\qquad 0 = -2b_3 - 1 \quad \text{or} \quad b_3 = -1/2$

and $\qquad\qquad\qquad a_3 = -b_3 + 3c_3 + 1 = 1/2 - 3 + 1 = -3/2$

Substituting the values of a_3, b_3 and c_3 in π_3.

$$\pi_3 = H^{-3/2} \cdot g^{-1/2} \cdot \rho^{-1} \cdot \mu$$

$$= \frac{\mu}{\rho \cdot H^{3/2}\sqrt{g}} = \frac{\mu V}{H\rho V\sqrt{gH}} \qquad\qquad \text{(multiply and divide by } V\text{)}$$

$$= \frac{\mu}{H\rho V}\pi_1 \qquad \because \frac{V}{\sqrt{gH}} = \pi_1$$

Substituting the values of π_1, π_2 and π_3 in $f_1(\pi_1, \pi_2, \pi_3) = 0$

$$f_1\left(\frac{V}{\sqrt{gH}}, \frac{D}{H}, \pi_1 \cdot \frac{\mu}{\rho V H}\right) = 0$$

or $\qquad\qquad \dfrac{V}{\sqrt{gH}} = \phi\left[\dfrac{D}{H}, \pi_1 \cdot \dfrac{\mu}{\rho V H}\right]$

$$V = \sqrt{2gH}\,\phi\left[\frac{D}{H}, \frac{\mu}{\rho \cdot V \cdot H}\right] \qquad\qquad \text{... Hence proved}$$

Multiplying by a constant does not change the character of π terms.

(b) Derive the Eulers equation of continuity for an ideal frictionless fluid. (8)

Solution. Consider the motion of a fluid element along a stream line in which flow is taking place in s direction.

Consider a cylindrical element of cross-section dA and length ds.

The forces acting on the cylindrical element are:

(i) Pressure force $P\,dA$ in the direction of flow.

(ii) Pressure force $\left(P + \dfrac{\partial p}{\partial s}ds\right)dA$ opposite to the direction of flow.

(iii) Weight of element $\rho \cdot g \cdot dA \cdot ds$.

Let θ = Angle between the direction of flow and the line of action of the weight of element.

The resultant force on the fluid element in the direction of 's' must be equal to the mass of fluid element x acceleration in the direction 's'.

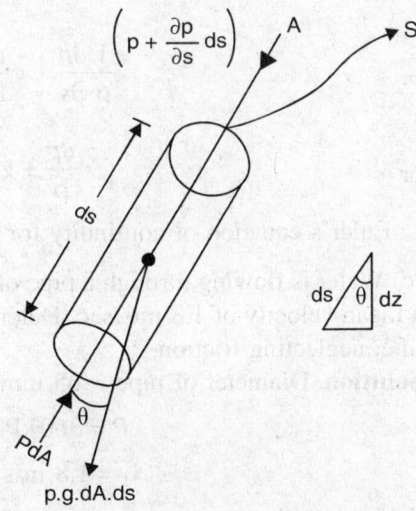

Fig. 16

$\therefore \qquad P \cdot dA - \left[P + \dfrac{\partial p}{\partial s} ds \right] dA - \rho \cdot g \cdot dA \cdot ds \cdot \cos\theta = \rho \cdot dA \cdot ds \cdot x \cdot a_s \qquad \qquad \ldots(1)$

$a_s = \text{Acceleration in the direction of } s = \dfrac{dv}{dt}$

$= \dfrac{\partial v}{\partial s} \dfrac{ds}{dt} + \dfrac{\partial v}{\partial t} = \dfrac{v \partial v}{\partial s} + \dfrac{\partial v}{\partial t} \qquad \because \dfrac{ds}{dt} = v$

If flow is steady $\dfrac{\partial v}{\partial t} = 0$

$\therefore \qquad a_s = \dfrac{v \partial v}{\partial s}$

Substituting the value of a_s in equation (1) and simplifying we get

$$-\frac{\partial p}{\partial s} ds\, dA - \rho g\, dA ds \cos\theta = \rho \cdot dA \cdot ds \times \frac{v \partial v}{\partial s}$$

Divide by $\rho \cdot ds \cdot dA$

$$-\frac{\partial p}{\rho ds} - g\cos\theta = \frac{v \partial v}{\partial s}$$

or

$$\frac{\partial p}{\rho \partial s} + g\cos\theta + v\frac{\partial v}{\partial s} = 0$$

But

$$\cos\theta = \frac{dz}{ds}$$

$\therefore$

$$\frac{1}{\rho}\frac{\partial p}{\partial s} + g\frac{dz}{ds} + v\frac{\partial v}{\partial s} = 0$$

or

$$\frac{\partial p}{\rho} + g\, dz + v\, dv = 0$$

Euler's equation of continuity for ideal frictionless fluid.

(c) Water is flowing through a pipe of 65 mm diameter under a gauge pressure of 360 kPa and a mean velocity of 1.8 mts/sec. Determine the total head if the pipe is 7 mts above the datum line, neglecting friction.

Solution. Diameter of pipe = 65 mm = 0.065 m

$$P = 360\,\text{kPa} = 360 \times 10^3\,\text{N/m}^2$$

$$v = 1.8\,\text{m/s} \quad z = 7\,\text{m}$$

$$\text{Total head} = \frac{p}{\rho g} + \frac{v^2}{2g} + z$$

$$= \frac{360 \times 10^3}{1000 \times 9.81} + \frac{(1.8)^2}{2 \times 9.81} + 7$$

$$= 36.697 + 0.165 + 7$$

$$= 43.86 \text{ m} \qquad\qquad\qquad\qquad\qquad\qquad \textbf{Ans.}$$

Q.6 (a) Name the instruments in your hydraulics laboratory that are similar to voltmeter ammeter and briefly explain the reasons for the same.

Solution. Instrument similar to voltmeter are manometers. Voltmeters measures voltage in the electric circuit, manometers measures the pressure in the hydraulic pipelines. From pressure difference the direction of flow can be identified similar voltmeter. Also high voltage and high pressure points can be detected in electric circuit and hydraulic pipelines respectively.

Instruments similar to ammeter are flow measuring devices e.g., Orifice measure Venturimeter, etc. Ammeter measures current in the electric circuit, similarly, measuring devices in hydraulic systems measures discharge through the pipelines.

(b) Clearly derive the expression for the theoretical discharge through a venturimeter Mention any assumption made.

Solution. Theoretical discharge through a venturimeter.

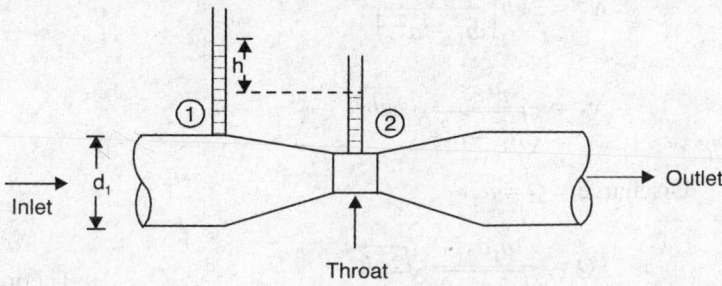

Fig. 17

Consider a venturimeter fitted in a horizontal pipe through which a fluid flowing.

d_1 = diameter at inlet or at section (1)

p_1 = pressure at section (1)

v_1 = velocity of fluid at section (1)

$a = \dfrac{\pi}{4} d_1^2$ = Area at section (1)

and d_2, p_2, v_2, a_2 are corresponding values at section (2).

Applying Bernoulli's equation at section (1) and (2) we get

$$\frac{p_1}{\rho g} + \frac{v_1^2}{2g} + z_1 = \frac{p_2}{\rho g} + \frac{v_2^2}{2g} + z_2$$

As pipe is horizontal $z_1 = z_2$

$$\frac{p_1 - p_2}{\rho g} = \frac{v_2^2}{2g} - \frac{v_1^2}{2g} \quad \text{But} \quad \frac{p_1 - p_2}{\rho g} = h$$

$$\therefore \qquad h = \frac{v_2^2}{2g} - \frac{v_1^2}{2g}$$

Applying continuity equation at section (1) and (2)

$$a_1 v_1 = a_2 v_2 \quad \text{or} \quad v_1 = \frac{a_2 v_2}{a_1}$$

$$\therefore \qquad h = \frac{v_2^2}{2g} - \frac{\left(\dfrac{a_2 v_2}{a_1}\right)^2}{2g} = \frac{v_2^2}{2g}\left(1 - \frac{a_2^2}{a_1^2}\right)$$

$$\therefore \qquad h = \frac{v_2^2}{2g}\left(\frac{a_1^2 - a_2^2}{a_1^2}\right)$$

$$v_2^2 = 2gh\left(\frac{a_1^2}{a_1^2 - a_2^2}\right)$$

$$v_2 = \frac{a_1}{\sqrt{a_1^2 - a_2^2}}\sqrt{2gh}$$

$$\text{Discharge} = Q = a_2 v_2$$

$$\therefore \qquad Q = \frac{a_1 a_2}{\sqrt{a_1^2 - a_2^2}}\sqrt{2gh} \qquad\qquad \text{...Theoretical discharge}$$

It is assumed that theoretical discharge is under ideal conditions.

(c) A horizontal circular pipe is of 50 mm diameter and 750 mm long maintains water flow rate of 0.03 mt³/min. Calculate the head loss due to friction and the power required to maintain the flow, if $\mu = 1.14 \times 10^{-3}$ W.S/mt^2 and $f = 0.008$. (8)

Solution. $d = 50$ mm $= 0.50$ m

$$a = \frac{\pi}{4}(d)^2 = \frac{\pi}{4}(0.05)^2 = 1.963 \times 10^{-3}\,\text{m}^2$$

$$Q = 0.03 \text{ m}^3/\text{min} = \frac{0.03}{60} = 5 \times 10^{-4}\,\text{m}^3/\text{s}$$

$$v = \frac{Q}{A} = \frac{5 \times 10^{-4}}{1.963 \times 10^{-3}} = 0.2547\,\text{m/s}$$

$$l = 750 \text{ mm} = 0.75 \text{ m}$$

$$f = 0.008$$

$$\mu = 1.14 \times 10^{-3} \text{ w.s.}/\text{m}^2 = 1.14 \, \frac{\text{N.s}}{\text{m}^2}$$

Let

$$\rho = 1000 \, \text{kg}/\text{m}^3$$

Loss of head due to friction $h_f = \dfrac{4 f l v^2}{2 g d}$

$$h_f = \frac{4 \times 0.008 \times 0.75 \times (0.2547)^2}{2 \times 9.81 \times 0.05}$$

$$= 1.5871 \times 10^{-3} \, \text{m}$$

Power required $= \dfrac{\rho \cdot g \cdot Q \cdot h_f}{1000} \, \text{kW}$

$$= \frac{1000 \times 3.81 \times 0.03 \times 1.5871 \times 10^{-3}}{1000}$$

$$= 4.67 \times 10^{-4} \, \text{kW} \qquad \qquad \textbf{...Ans.}$$

Q.7 (a) Define Reynold's number and clearly distinguish between Laminar and Turbulent flows.

Solution. Reynold's number: It is defined as the ratio of inertia force of a flowing fluid and the viscous force of the fluid. $R_e = \dfrac{\rho \cdot v \cdot d}{\mu}$

S.. No.	Laminar flow	Turbulent flow
(i)	Laminar flow is a type of flow in which the fluid particles move in layer.	Turbulent flow in the most common type of flow that occurs in nature.
(ii)	There is no transportation of fluid particles from one layer to another.	There is general mixing up of the fluid particles in motion.
(iii)	The fluid particles in any layer move along well defined paths or stream lines. Hence the flow is also called stream lines flow, also viscous flow.	There is continuous collision between fluid particles involving transference of momentum between them.
(iv)	This type of flow can exist only at low velocities.	The flows eddy currents and the velocity of flow changes in direction and magnitude from point to point.
(v)	When the Reynolds number is less than 2000, the flow is laminar.	When the Reynolds number is greater than 2800 the flow is turbulent.

(b) Oil is to be transported from a tanker to the shore at a rate of 0.006 m³/sec using a pipe of 32 cm diameter for a distance of 20 kms. If oil has viscosity $\mu = 0.1$ nM/sec² and density $\rho = 900$ kg/mt³. Calculate the power necessary to maintain flow. (8)

Solution. $Q = 0.006 \text{ m}^3/\text{s}$ $d = 32 \text{ cm} = 0.32 \text{ cm}$

$$l = 20 \text{ kms} = 20 \times 10^3 \text{ m}$$
$$\mu = 0.1 \text{ Ns/m}^2$$
$$\rho = 900 \text{ kg/m}^3$$
$$A = \frac{\pi}{4}(d)^2 = \frac{\pi}{4}(0.32)^2 = 0.0804 \text{ m}^2$$
$$V = \frac{Q}{A} = \frac{0.006}{0.0804} = 0.0746 \text{ m/s}$$

Reynold number $R_e = \dfrac{\rho V d}{\mu}$

$$= \frac{900 \times 0.0746 \times 0.32}{0.1}$$
$$= 214.859$$

Co-efficient of friction $f = \dfrac{16}{R_e}$

$$= 0.07447$$

∴ Loss of head due to friction $= \dfrac{4fLV^2}{2gd}$

$$= \frac{4 \times 0.07447 \times 20 \times 10^3 \times (0.0746)^2}{2 \times 9.81 \times 0.32}$$
$$= 5.2806 \text{ m of oil.}$$

Power necessary to maintain flow $= \dfrac{\rho \cdot g \cdot Q \cdot h_f}{1000}$

$$= \frac{900 \times 9.81 \times 0.006 \times 5.2306}{1000}$$
$$= 0.2797 \text{ kW.}$$

(c) Explain briefly what is meant by sonic velocity and Mach number. (2)

Solution. Sonic velocity: Velocity of sound in the form of disturbances is called sonic velocity.

Mach number: It is square root of the ratio of the inertia force of a flowing fluid to the elastic force.

It is also ratio of velocity of fluid or body moving in fluid to the velocity of sound in the fluid.

(d) Compute the velocity of a bullet fired in still air, with a mach angle of 30°. Take $R = 287.14$ J/kg°k and $\gamma = 1.4$. Assume air temperature to be 15°C. (4)

Solution. $\alpha = 30°$ $R = 287.14$ J/kg°k $\gamma = 1.4$

$$t = 15°C = 15 + 273 = 288°k$$

Velocity of sound $c = \sqrt{\gamma RT}$

$$= \sqrt{1.4 \times 287.14 \times 288} = 340.25 \text{ m/s}$$

$$\sin \alpha = \frac{c}{V} \quad \therefore V = \frac{c}{\sin \alpha}$$

$$= \frac{340.25}{\sin 30°}$$

$$V = 680.50 \text{ m/s}$$

Velocity of bullet in still air = 680.50 m/s.

Q.8 (a) Briefly explain what is meant by a boundary layer and hence define,

(i) Displacement thickness (ii) Momentum thickness (iii) Energy thickness. (10)

Solution. Boundary layer: When a real fluid flows past a solid body, the fluid particles adhere to the boundary and condition of no slip occurs. That is velocity of fluid close to the boundary of solid body will be same as that of boundary. If boundary is stationary, velocity of fluid at the boundary will be zero and velocity will be higher away from the boundary and velocity gradient $\frac{du}{dy}$ will exist.

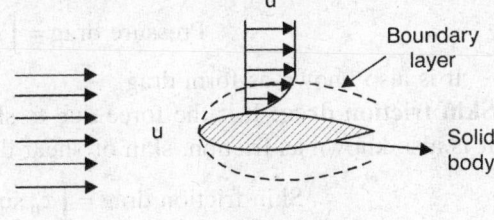

Fig. 18

The velocity of fluid increases from zero velocity on the stationary boundary to free stream velocity u of the fluid in the direction normal to the boundary.

Narrow region in the vicinity of solid boundary where variation of velocity from zero to u takes place, is called boundary layer.

(i) **Displacement thickness** (δ): It is defined as the distance, measured perpendicular to the boundary of the solid body, by which the boundary should be displaced to compensate for the reduction in flow rate on account of boundary layer formation. It is denoted by δ. It is also defined as "the

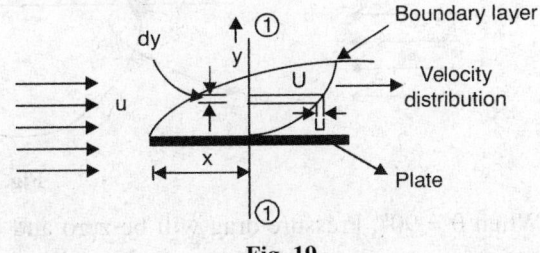

Fig. 19

distance, perpendicular to the boundary by which the free stream is displaced due to the formation of boundary layer".

$$\delta = \int_0^\delta \left(1 - \frac{u}{U}\right) dy$$

(ii) **Momentum thickness** (θ): Momentum thickness is defined as the distance, measured perpendicular to the boundary of the solid body by which the boundary should be displaced to compensate for the reduction in momentum of the flowing fluid on account of boundary layer formation. It is denoted by θ.

$$\theta = \int_0^\delta \frac{u}{U}\left[1 - \frac{u}{U}\right] dy$$

(iii) **Energy thickness** (δ''): It is defined as the distance measured perpendicular to the boundary of the solid body, by which the boundary should be displaced to compensate for the reduction in kinetic energy of the flowing fluid on account of boundary layer formation. It is denoted by δ''

$$\delta'' = \int_0^\delta \frac{u}{U}\left[1 - \frac{u^2}{U^2}\right] dy$$

(b) Clearly distinguish between pressure drag and skin friction drag. (4)
Solution. Pressure drag: It is the force due to pressure in the direction of fluid motion.

$$\text{Pressure drag} = \int P \cos\theta\, dA$$

It is also known as form drag.
Skin friction drag: It is the force due to shear stress in the direction of fluid motion. It is also known as friction, skin or shear drag.

$$\text{Skin friction drag} = \int \tau_0 \sin\theta\, dA$$

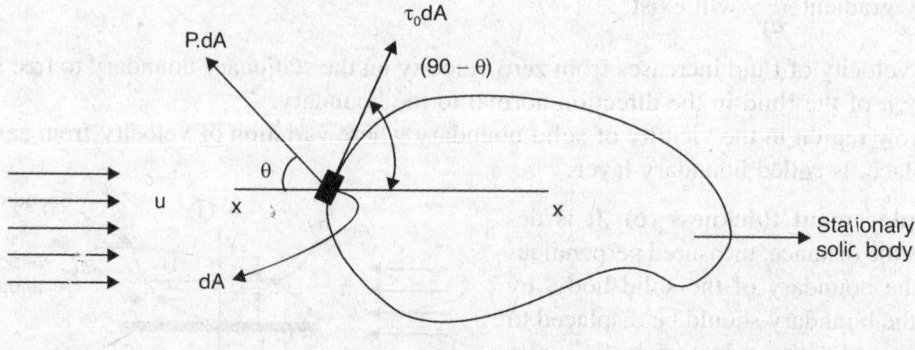

Fig. 20

When $\theta = 90°$. Pressure drag will be zero and total drag will be only friction drag.

When $\theta = 0$ or area parallel to U, skin function drag will be zero and total drag due to the pressure difference between the upstream and downstream side of the plate.

If the plate is held at an angle with the direction of flow, both the drag will exist and total drag will be equal to the sum of pressure drag and friction drag.

Please refer Q. 3 (b) of July/Aug.-2004.

(c) A flat plate of 2 mts × 2 mts moves with a velocity 50 km/hr in air of specific weight of 1.15 kg/mt^3. If the co-efficients of lift and drag are 0.75 and 0.15 respective calculate

(i) Drag force (ii) Lift force and (iii) Resultant force. (6)

Solution. Area of plate $A = 2 \times 2 = 4$ m^2

Velocity of the plate = 50 km/hr

$$= \frac{50 \times 100}{60 \times 60} = 13.39 \text{ m/s}$$

$$\rho = 1.15 \text{ kg/m}^3$$

$$C_D = 0.15$$

$$C_L = 0.75$$

(i) Drag force $\quad F_D = C_D \times A \times \dfrac{\rho V^2}{2}$

$$= 0.15 \times 4 \times \frac{1.15 \times (13.89)^2}{2} \text{N}$$

$$= 66.56 \text{ N}$$

(ii) Lift force $\quad F_L = C_L A \times \dfrac{\rho v^2}{2}$

$$= 0.75 \times 4 \times \frac{1.15 \times (13.89)^2}{2}$$

$$= 332.81 \text{ N}$$

(iii) Resultant force $\quad F_R = \sqrt{F_D^2 + F_L^2}$

$$= \sqrt{(66.56)^2 + (332.81)^2}$$

$$= 339.40 \text{ N}.$$

Fluid Mechanics
July/August 2004
Semester - IV (ME/IP/AU/IM/MA)

Time: 3 Hours   Total Marks: 100

Note: (i) Answer any Five full questions.

Q.1 (a) With the help of a neat plot (stress Vs. rate of strain) show the characteristic behaviour for the following materials.

(i) Elastic solid (ii) Ideal fluid (iii) Newtonian fluid (iv) Ideal plastic (v) Dilatant fluid (vi) Pseudo plastic fluid. (6)

Ans.:

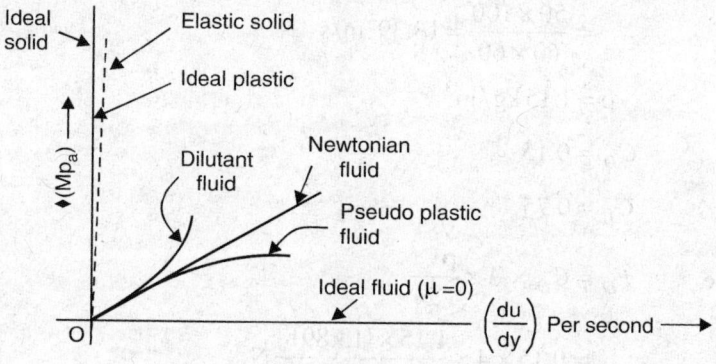

Fig. 1

(b) Determine the pressure difference $P_A - P_B$ for the system shown below. (4)

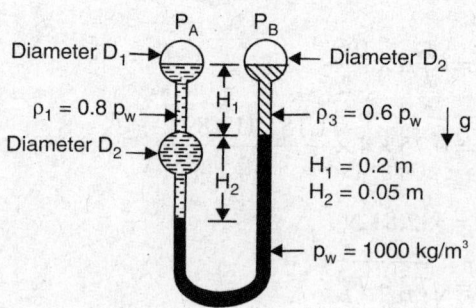

Fig. 2

Ans. Assume datum XY as shown in Fig. 2(a). For static equilibrium, pressure at X must be equal to pressure at Y. Using $p = \rho g h$ we write gauge equation.

$\therefore \qquad p_A + (0.8\rho_w)(0.25)g = p_B + (0.6\rho_w)(0.2)g + (\rho_w)(0.05)g$

$\therefore \qquad p_A - p_B = \rho_w\left[(0.6)(0.2) + (0.05) - (0.8)(0.25)\right]g$

$\therefore \qquad \boxed{p_A - p_B = -294.3 \text{ N/m}^2} \qquad i.e., \ p_A < p_B$

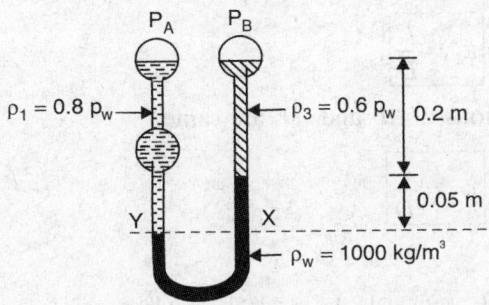

Fig. 2(a)

(c) The distance travelled by a golf ball in still air, L, is known to be a function of the following:

$$L = f\left(V_0, \rho, D, d, \mu, \omega, m\right)$$

Where,

V_0 is the initial velocity of the ball,

d is the diameter of the dimples;

ω is the angular speed of the ball;

m is the mass of the ball

D is the diameter of the ball

ρ is the density of air

μ is viscosity of air and

(i) Using ρ, V_0 and D as repeating variables find all the relevant π groups. (8)

Ans.: Let $L = f\left(V_0, \rho, D, d, \mu, \omega, m\right)$

$\therefore$ Total variables $n = 8$

Number of basic dimensions involved $m = 3$

$\therefore$ No. of π terms $n - m = 5$

Let ρ, V_0 and D be repeating variables.

$$\therefore \quad \pi_1 = \rho^{a_1} V_0^{b_1} D^{c_1} L \qquad \mu \equiv ML^{-1}T^{-1}$$
$$\pi_2 = \rho^{a_2} V_0^{b_2} D^{c_2} d \qquad \omega \equiv T^{-1}$$
$$\pi_3 = \rho^{a_3} V_0^{b_3} D^{c_3} \mu \qquad m \equiv M$$
$$\pi_4 = \rho^{a_4} V_0^{b_4} D^{c_4} \omega \qquad \rho \equiv ML^{-3}$$
$$\pi_5 = \rho^{a_5} V_0^{b_5} D^{c_5} m \qquad V_0 \equiv LT^{-1}$$

First π term: $\left[M^0 L^0 T^0\right] = \left[ML^{-3}\right]^{a_1} \left[LT^{-1}\right]^{b_1} [L]^{c_1} [L]$

Equate powers of M, T and L for dimensional homogeneity.

$\therefore$ $0 = a_1$ or $a_1 = 0$

$0 = -b_1$ or $b_1 = 0$

$0 = -3a_1 + b_1 + c_1 + 1$ Hence $c_1 = -1$

$$\therefore \qquad \boxed{\pi_1 = \frac{L}{D}}$$

Is second π term, dimensions of 'd' and 'D' are same.

Hence

$$\boxed{\pi_2 = \frac{d}{D}}$$

Third π term:

$$[M^0 L^0 T^0] = [ML^{-3}]^{a_3} [LT^{-1}]^{b_3} [L]^{c_3} [ML^{-1}T^{-1}]$$

$$\therefore \qquad 0 = a_3 + 1 \quad \text{or} \quad a_3 = -1$$

$$0 = -b_3 - 1 \quad \text{or} \quad b_3 = -1$$

$$0 = -3a_3 + b_3 + c_3 - 1 \quad \text{or} \quad c_3 = -1$$

$$\therefore \qquad \pi_3 = \rho^{-1} V_0^{-1} D^{-1} \mu$$

$$\therefore \qquad \boxed{\pi_3 = \frac{\mu}{\rho V_0 D}}$$

Fourth π term:

$$[M^0 L^0 T^0] = [ML^{-3}]^{a_4} [LT^{-1}]^{b_4} [L]^{c_4} [T^{-1}]$$

$$\therefore \qquad 0 = a_4$$

$$0 = -b_4 - 1 \quad \text{or} \quad b_4 = -1$$

$$0 = -3a_4 + b_4 + c_4$$

$$\therefore \qquad c_4 = -b_4 = 1$$

$$\therefore \qquad \pi_4 = \rho^0 V_0^{-1} D^1 \omega$$

or

$$\boxed{\pi_4 = \frac{D\omega}{V_0}}$$

Fifth π term:

$$[M^0 L^0 T^0] = [ML^{-3}]^{a_5} [LT^{-1}]^{b_5} [L]^{c_5} [M]$$

$$\therefore \qquad 0 = a_5 + 1 \quad \text{or} \quad a_5 = -1$$

$$0 = -b_5 \quad \text{or} \quad b_5 = 0$$

$$0 = -3a_5 - b_5 + c_5 \quad \text{Hence} \quad c_5 = -3$$

$$\therefore \qquad \pi_5 = \rho^{-1} V_0^0 D^{-3} m$$

$$\therefore \qquad \boxed{\pi_5 = \frac{\mu}{\rho D^3}}$$

(ii) For completely submerged body, Reynolds model law is applicable.

i.e. $(R_e)_m = (R_e)_p$ .... 'm' stands for model and 'p' for prototype.

$$\therefore \qquad \left(\frac{V_0 D}{\nu}\right)_m = \left(\frac{V_0 D}{\nu}\right)_p$$

As fluid is same, $\nu_m = \nu_p$. Given $D_m = 2D_p$

$$\therefore \qquad \boxed{\frac{(V_0)_m}{(V_0)_p} = \frac{D_p}{D_m} = 0.5}.$$

Q.2 (a) State and prove Pascal's law of pressure. (6)

Ans.: Pascal's law states that the pressure intensity at a point in static liquid mass acts equally in all directions.

Consider very small wedge shaped liquid element of size as shown in Fig. 3. Let P_x, P_y and P_s be pressure intensities at right angles to vertical, horizontal and inclined faces of element respectively. Since the liquid and the element are in equilibrium $\Sigma f_x = 0$ for forces shown on element.

$$\therefore \qquad p_x\,dy\,dz - p_s\,ds\,dz \sin\theta = 0 \qquad \text{But } \sin\theta = \frac{dy}{ds}$$

$$\therefore \qquad p_x\,dy\,dz - p_s\,ds\,dz = 0 \qquad \text{or } p_x = p_s \qquad ...(\text{i})$$

$$\Sigma f_x = 0 \text{ gives } p_y\,dx\,dz - W - p_s\,ds\,dz \cos\theta = 0$$

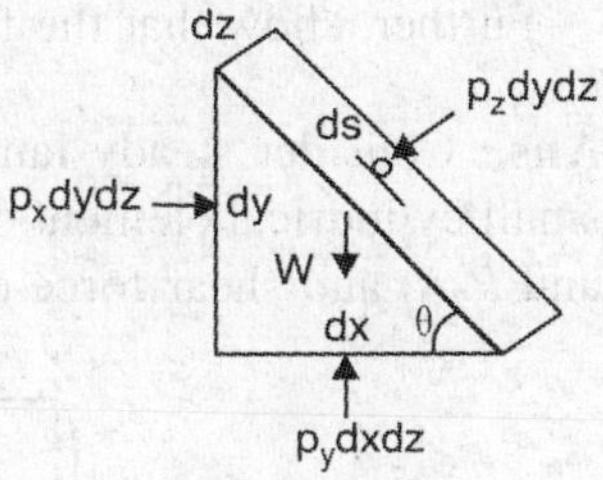

Also $\quad W = \left(\frac{1}{2}dx\,dy\,dz\right)\gamma$ and $\cos\theta = \frac{dx}{ds}$

Fig. 3

$$\therefore \qquad p_y\,dx\,dz - \frac{\gamma}{2}dx\,dy\,dz - p_s\,dx\,dz = 0$$

When wedge shaped element is reduced to a point ($dx \to 0$, $dy \to 0$, $dz \to 0$), we can neglect middle term involving weight as Pascal's law is applicable at a point only.

$$\therefore \qquad p_y = p_s \qquad ...(\text{ii})$$

From equations (i) and (ii), we conclude that pressure intensity is same in all directions.

(b) Given $V = (xy + 2zt)i + (2y^2 + xyt)j + (12xy)k$ where, x, y and z are in meters and t in seconds determine a_x the x component of the acceleration of the fluid particle at $(1, 1, 1)$ at $t = 1s$.

Ans.: $V = (xy + 2zt)i + (2y^2 + xyt)j + (12xy)k$

$$\therefore \qquad u = xy + 2zt \qquad\qquad v = 2y^2 + xyt \qquad\qquad w = 12xy$$

$$\frac{\partial u}{\partial x} = y \qquad\qquad\qquad \frac{\partial v}{\partial x} = yt \qquad\qquad\qquad \frac{\partial w}{\partial x} = 12y$$

$$\frac{\partial u}{\partial y} = x \qquad\qquad \frac{\partial v}{\partial y} = 4y + xt \qquad\qquad \frac{\partial w}{\partial y} = 12x$$

$$\frac{\partial u}{\partial z} = 2t \qquad\qquad \frac{\partial v}{\partial z} = 0 \qquad\qquad \frac{\partial w}{\partial z} = 0$$

$$\frac{\partial u}{\partial t} = 2z \qquad\qquad \frac{\partial v}{\partial t} = xy \qquad\qquad \frac{\partial w}{\partial t} = 0$$

Now $a_x = \dfrac{\partial u}{\partial x} + v\dfrac{\partial u}{\partial y} + w\dfrac{\partial u}{\partial z} + \dfrac{\partial u}{\partial t}$ and $x = y = z = 1\mathrm{m}, \quad t = 1s$ given

$\therefore \qquad a_x = (1+2)(1) + (2+1)(1) + (12)(2) + (2)$

$\therefore \qquad \boxed{a_x = 32\,\mathrm{m/s^2}}$

(c) Starting from an appropriate control volume derive the expression for the velocity distribution for steady, laminar, fully developed flow of an incompressible fluid in a circular pipe.

Further, show that the friction factor $f = \dfrac{2gDh_f}{LV^2}$ is 64/Re for this flow. (10)

Ans.: Consider steady laminar flow through a horizontal pipe of radius R. Forces acting on small cylindrical element of length dx and radius r are—pressure forces on circular faces (P_1A and P_2A) and shear force on curved surface ($\tau \cdot 2\pi r\, dx$).

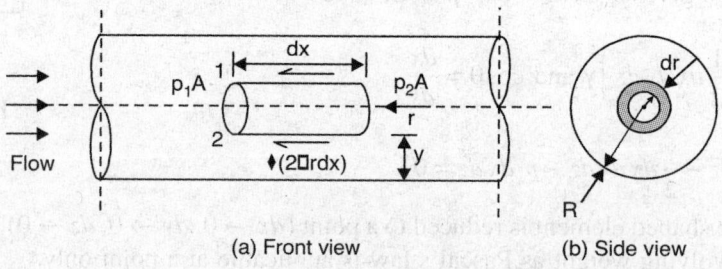

(a) Front view   (b) Side view

Fig. 4

Apply Newton's second law of motion.
(acceleration = 0 for steady uniform flow)

$\therefore \qquad p_1A - p_2A - \tau(2\pi r\,dx) = 0$

put $\qquad\qquad\qquad\qquad A = \pi r^2$

And $p_2 = p_1 + \left(\dfrac{\partial p}{\partial x}\right)dx$ by assuming increase in pressure downstream.

$\therefore \qquad dx\left(-\dfrac{\partial p}{\partial x}\right)\pi r^2 - \tau(2\pi r dx) = 0$

or
$$\tau = \left(\frac{-\partial p}{\partial x}\right)\left(\frac{r}{2}\right)$$

Now $\tau = \mu \dfrac{du}{dy}$ according to Newtons' law of viscosity and $y = R - r$.

$\therefore$
$$dy = -dr$$

$\therefore$
$$\tau = \left(\frac{-\partial p}{\partial x}\right)\left(\frac{r}{2}\right) = -\mu\left(\frac{du}{dr}\right)$$

$\therefore$
$$\frac{du}{dr} = \left(-\frac{1}{\mu}\right)\left(\frac{-rp}{\partial x}\right)\left(\frac{r}{2}\right)$$

Integrating with respect to r, we get,
$$u = \left(-\frac{1}{\mu}\right)\left(\frac{-\partial p}{\partial x}\right)\left(\frac{r^2}{4}\right) + C$$

At $r = R$, $u = 0$ (no slip condition)

$\therefore$
$$0 = \left(\frac{-1}{\mu}\right)\left(\frac{-\partial p}{\partial x}\right)\left(\frac{R^2}{4}\right) + C$$

or
$$C = \left(\frac{1}{\mu}\right)\left(\frac{-\partial p}{\partial x}\right)\left(\frac{R^2}{4}\right)$$

$\therefore$
$$u = \left(\frac{1}{4\mu}\right)\left(\frac{-\partial p}{\partial x}\right)\left(R^2 - r^2\right) \qquad \text{...point velocity}$$

This gives velocity variation (parabolic) with $u_{max} = \left(\dfrac{1}{4\mu}\right)\left(\dfrac{-\partial p}{\partial x}\right)R^2$ along axis.

For mean velocity, consider discharge through elemental ring (refer side view) of thickness dr at radial distance r.

$\therefore$
$$dQ = \text{vel. at ring} \times \text{area of ring.}$$

$$= \left(\frac{1}{4\mu}\right)\left(\frac{-\partial p}{\partial x}\right)\left(R^2 - r^2\right)\left(2\pi r\, dr\right)$$

$$\therefore \qquad Q = \int_0^R dQ = \left(\frac{\pi}{8\mu}\right)\left(\frac{-\partial p}{\partial x}\right)R^4$$

$$\therefore \qquad \text{Mean velocity } \bar{u} = \frac{Q}{A}\frac{\left(\frac{\pi}{8\mu}\right)\left(\frac{-\partial p}{\partial x}\right)R^4}{\pi R^2}$$

$$\bar{u} = \left(\frac{1}{8\mu}\right)\left(\frac{-\partial p}{\partial x}\right)R^2$$

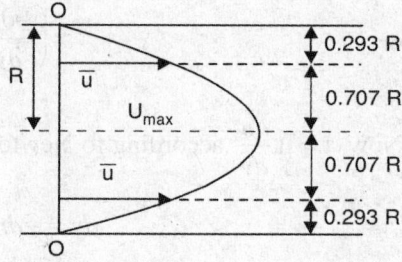

Velocity variation in pipe (Laminar flow)

Fig. 5

We know that Darcy-Weisbach equation for frictional

loss is $h_f = \dfrac{fLV^2}{2gD}$

where V = mean velocity.

For laminar flow, $\qquad V = \left(\dfrac{1}{8\mu}\right)\left(\dfrac{p_1 - p_2}{L}\right)R^2$

or $\qquad V = \left(\dfrac{p_1 - p_2}{\gamma}\right)\left(\dfrac{\gamma}{8\mu L}\right)\left(\dfrac{D}{2}\right)^2$

$\therefore \qquad V = h_f\left(\dfrac{\rho g D^2}{32\mu L}\right)$

Hence $\qquad h_f = \dfrac{32\mu VL}{\rho g D^2} = \dfrac{flV^2}{2gD}$

$\therefore \qquad f = \dfrac{64\mu}{\rho VD}$

Now $\qquad \dfrac{\rho VD}{\mu} = \text{Reynolds number } R_e$

$\therefore \qquad f = \dfrac{64}{R_e}$

Q.3 (a) Determine the minimum force F, required to keep the gate closed, in the Fig. 6 below. The gate is a square of 0.5 m side and hinged in the middle as shown. The centre of the gate is 1 m below the water surface.

(8)

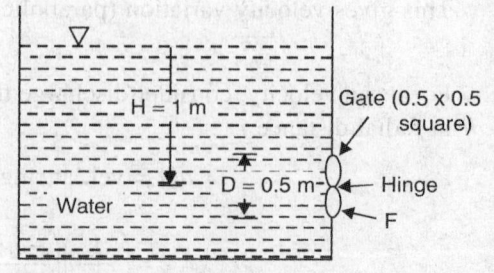

Fig. 6

Ans.:

Water force $P = \gamma A \bar{x}$

$\therefore \qquad P = (9.81)(0.5)^2 (1)$

$\therefore \qquad P = 2453$ kN

Depth of center of pressure $\bar{h} = \bar{x} + \dfrac{I_G}{A\bar{x}}$

$\therefore \qquad \bar{h} = 1 + \dfrac{\dfrac{(0.5)^4}{12}}{(0.5)^2 (1)}$

$\therefore \qquad \bar{h} = 1.021$ m

For equilibrium of gate, $\sum M_0 = 0$

$\therefore \qquad P(\bar{h} - \bar{x}) - F(0.25) = 0$

$\therefore \qquad \boxed{F = 0.204 \text{ kN}}$

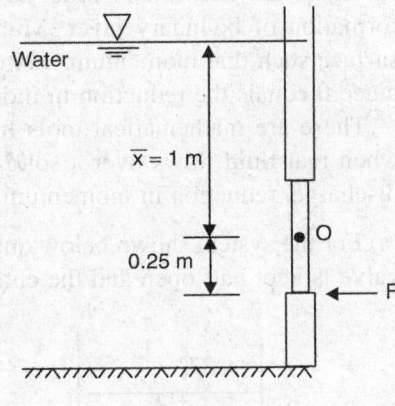

Fig. 6(a)

(b) Distinguish between the following:

    (i) Form drag and skin friction drag

    (ii) Lift force and drag force

    (iii) The physical significance of displacement thickness and momentum thickness   (6)

(i) Form Drag	*Skin Friction Drag*
1. This depends on shape and orientation of body with respect to flow direction.	1. Depends on shape of body. Also called friction drag.
2. This is usually large: $(F_D)_p$	2. Its magnitude is comparatively small: $(F_D)_{fr}$
3. If $(F_D)_p$ is very large, body experiences large resistance to flow/motion. Hence it is a bluff body.	3. For bluff body it is still smaller.
4. It is due to pressure difference between upstream and downstream sides on body.	4. It is due to friction between flowing fluid and surface of body.

(ii) Drag Force	*Lift Force*
1. It is the component of resultant force on the immersed body in the direction of flow.	1. It is the component at right angles to the direction of flow.
2. It consists of friction drag, pressure drag, form drag, deformation drag etc.	2. There are no types of lift.
3. Usually its magnitude is very high.	3. Its magnitude is relatively very small.
4. Given as $F_D = \dfrac{1}{2}\rho A V^2 \, C_D$	4. Formula is $F_L = \dfrac{1}{2}\rho A V^2 \, C_L$
5. Drag is maximum for a bluff body and minimum for Streamlined body.	5. Lift is maximum for unsymmetrical with proper angie of attach and zero for symmetrical aerofoil with zero angle of attack.

(iii) Displacement thickness is the distance through which the solid surface of body would have to be shifted outward in order to compensate for the reduction in discharge owing to the formation of boundary layer. Momentum thickness θ is the distance from the actual solid surface such that momentum flux corresponding to free stream velocity U_∞ through this distance θ equals the reduction in momentum due to for motion of boundary layer.

These are mathematical tools for interpreting the effects of formation of boundary layer when real fluid flows over a solid surface at rest. These effects are reduction in total actual discharge, reduction in momentum and reduction in kinetic energy of flowing fluid.

(c) For the system shown below qualitatively sketch the hydraulic and energy grade lines. The valve is kept half open and the entry loss into the pipe is negligible.

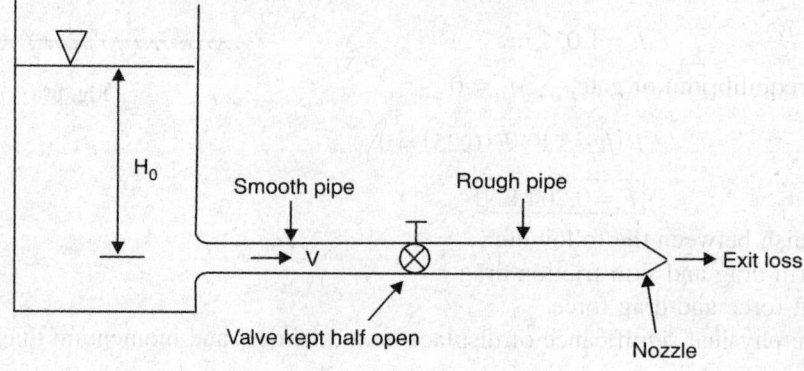

Fig. 7

Ans.:

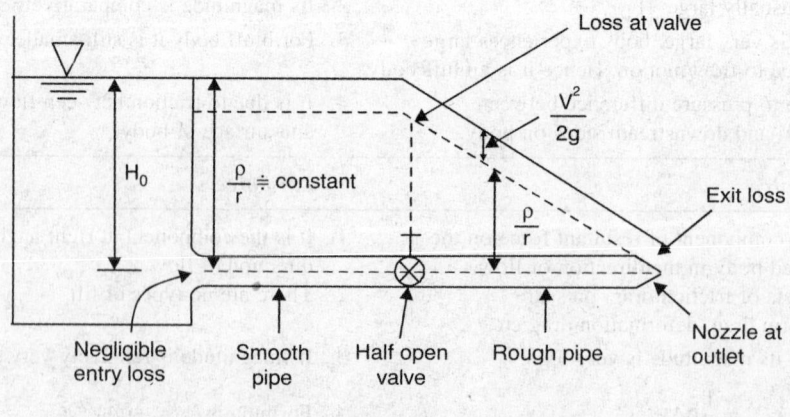

Fig. 7(a)

Q.4 (a) Calculate the horizontal and vertical forces, due to the gauge pressure of water on the cylindrical portion of the tank shown below. The radius of the cylinder, R is 2m, the level of water in the tank, H is 5m and the width of the tank, W is 5m. The tank is open at the top.

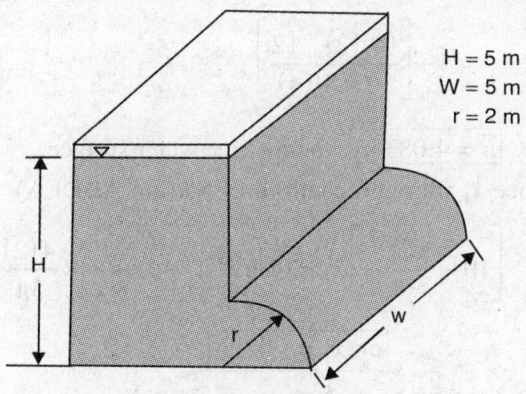

H = 5 m
W = 5 m
r = 2 m

Fig. 8

Ans.:

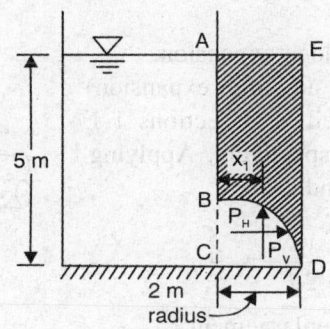

Fig. 8(a)

Projection of curved portion on vertical plane $= (BC) \times 5 m^2$

$\therefore$ Horizontal component of force on curved portion of tank $= rA\bar{x}$

$\therefore$ $$P_H = (9.81)(2 \times 5)(5-1)$$

or $$\boxed{P_H = 392.4 \text{ kN}} \rightarrow$$

Now vertical component P_v = weight of water (imagined) in portion ABDEA.

$\therefore$ $P_v = r \times$ volume ABDEA $= r \times 5 \times$ are ABDEA

$\therefore$ $$P_v = (9.81)(5)\left[(2 \times 5) - \frac{\pi(2)^2}{4}\right]$$

$\therefore$ $$\boxed{P_v = 336.405 \text{ kN}} \uparrow$$

P_H acts at depth $\bar{h} = \bar{x} + \dfrac{I_G}{A\bar{x}}$

$$= 4 + \left[\frac{(5)(2)^3/12}{(2 \times 5)(4)} \right]$$

or $\boxed{h = 4.083 \text{ m}}$ below free water surface.

P_v acts at distance x_1 (through centroid of portion ABDEA)

$$\therefore \qquad \left[10 - \frac{\pi(2)^2}{4} \right] x_1 = (2 \times 5)(1) - \frac{\pi(2)^2}{4} \times \frac{4(2)}{3\pi}$$

$$\boxed{\text{or } x_1 = 1.07 \text{ m}}$$

(b) Derive the expression for the energy loss due to a sudden expansion in a pipe from area A_1 to area $A_2(>A_1)$, in terms of the inlet dynamic pressure $\frac{1}{2}\rho V_1^2$. Clearly state all the assumptions made.

Ans.: Consider flow through sudden expansion.

Consider two sections (before and after expansion)

Pressure intensities and velocities at sections 1-1 and 2-2 are p_1, v_1 and p_2, v_2 respectively. Applying Bernoulli's theorem at sec 1-1 and 2-2.

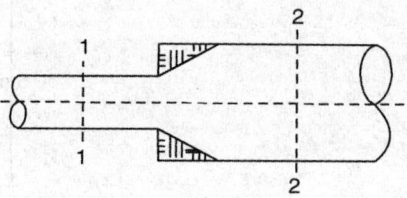

Fig. 9

$$\frac{p_1}{\gamma} + \frac{v_1^2}{2g} + z_1 = \frac{p_2}{\gamma} + \frac{v_2^2}{2g} + z_2 + h_e$$

Where 'h' is loss due to sudden enlargement.

$$h = \left(\frac{p_1 - p_2}{\gamma} \right) + \left(\frac{v_1^2 - v_2^2}{2g} \right)$$

If pipe is horizontal on control volume section 1-1 - 2-2

Net force acting in flow direction:

$$= p_1 A_1 + p_1 (A_2 - A_1) - p_2 A_2$$

$$= A_2 (p_1 - p_2)$$

By momentum principle

$$A_1 (p_1 - p_2) = \rho Q (v_2 - v_1)$$

By continuity equation $A_1 v_1 = A_2 v_2 = Q$

$$p_1 - p_2 = \rho \frac{Q}{A_2} (v_2 - v_1) = \frac{\gamma}{g} v_2 (v_2 - v_1)$$

$$\frac{p_1 - p_2}{2} = \frac{(v_1 - v_2)^2}{2g} \qquad p = \frac{(v_1 - v_2)^2}{2g}$$

(c) Derive Bernoulli's equation using an infinitesimal stream tube. Clearly state all the assumptions made. (6)

Ans.: Let a stream tube be along streamline S in the steady, ideal flow. Assume the pressure intensities be p and $p + \frac{\partial p}{\partial s}$ at the two ends of cylindrical element of height $W = \rho g \, dA \, ds$.

(This is the gravity force) where dA = cross sectional area of element. Apply Newton's second law of motion in the direction of flow.

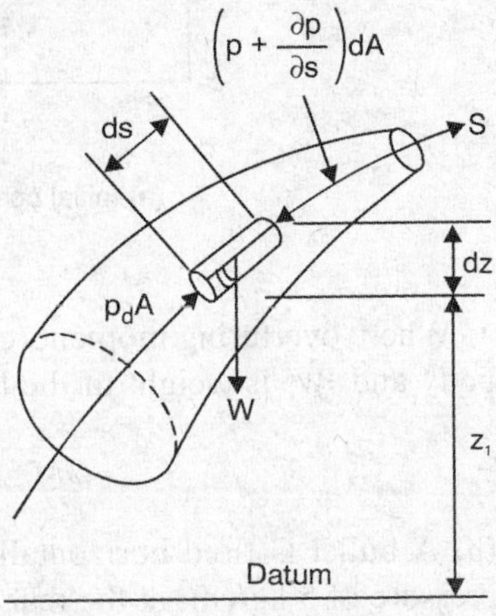

$$\therefore \quad p dA - \left(p + \frac{\partial p}{\partial s} ds \right) dA - W \cos \theta = \left(\rho \cdot dA \, ds \right) a_s$$

Now $a_s = v \cdot \frac{\partial v}{\partial s}$ for steady flow and $\cos \theta = \frac{dz}{ds}$

$$\therefore \quad -\frac{\partial p}{\partial s} dA \, ds - \rho g \, dA \, ds \left(\frac{dz}{ds} \right) - \rho \, dA \, ds \left(v \frac{\partial v}{\partial s} \right) = 0$$

Hence we get
$$\frac{1}{\rho g} \frac{\partial p}{\partial s} + \frac{dz}{dx} + \frac{v}{g} \frac{\partial v}{\partial s} = 0$$

i.e.
$$\frac{1}{\gamma} dp + dz + \frac{v}{g} dv = 0$$

Integrating, (with ρ and γ as constant for incompressible fluid) we get,

$$\frac{p}{\gamma} + z + \frac{v^2}{2g} = \text{constant} \quad \ldots \text{Required Bernoulli equation.}$$

Fig. 10

Assumptions:

1. Fluid is nonviscons (ideal) and incompressible.
2. Losses of energy are negligible.
3. Flow is continuous (with continuous connection between fluid particles)
4. Flow is steady (i.e. velocity, pressure, etc. do not change from time to time)

Q.5 (a) Derive the criterion for stability of a floating body.

Ans.: When a floating body is given small angular displacement say clockwise, couple of F_B and W gets formed in counter clockwise direction as center of buoyancy gets shifted. This couple brings the body back to its original (vertical) position. Thus the floating body has a stable equilibrium. Note that M lies above G. (M is the metacenter which is the point of intersection of buoyant force F_B and axis of body).

As metacentric height $\overline{GM}$ is positive, body is stable. By equating couples we get.

$$W \cdot \left(\overline{GM} \right) \tan \theta = \omega \cdot x \ldots\ldots\ldots\ldots \theta = \text{angle of tilt}.$$

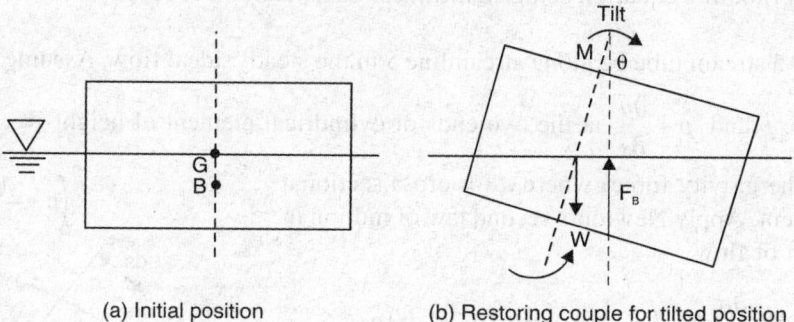

(a) Initial position   (b) Restoring couple for tilted position

Fig. 11

Where overturing moment (couple) is due to weight 'W' displaced through 'x' from axis of body and 'W' is weight of the body.

$$\therefore \qquad \overline{GM} = \frac{\omega x}{W \tan \theta}$$

(b) A bullet is fired horizontally from a rifle. The burning gases maintain a constant gauge pressure of 5 kg/cm² at the rear of the bullet. The bullet has a conical tip at the from followed by a cylindrical section 10 mm long and 5 mm in diameter. The bore of the rifle is 5.01 mm so that a uniform air gap of 0.005 mm is present all around the bullet. Assuming that viscous air resistance is present only on the cylindrical surface of the bullet, do the following ($\mu_{air} = 2 \times 10^{-5}$ Ns/m²).

 (i) Draw the free body diagram of the bullet showing all the horizontal forces acting of it.
 (ii) Set up an equation to determine the acceleration of the bullet as a function of time. Separate variables, and integrate this equation to get the velocity as a function of time.
 (iii) If the mass of the bullet is 10 gm and the bullet is expelled from the rifle after 0.1 what is the velocity of the bullet at exit?

Ans.: Note: pressure 5 kgf/cm² is not correct SI unit.

Using Newton's second Law of motion.

$$\sum F = m.a \text{ we get}$$

$$F - \tau \cdot A = m \cdot a$$

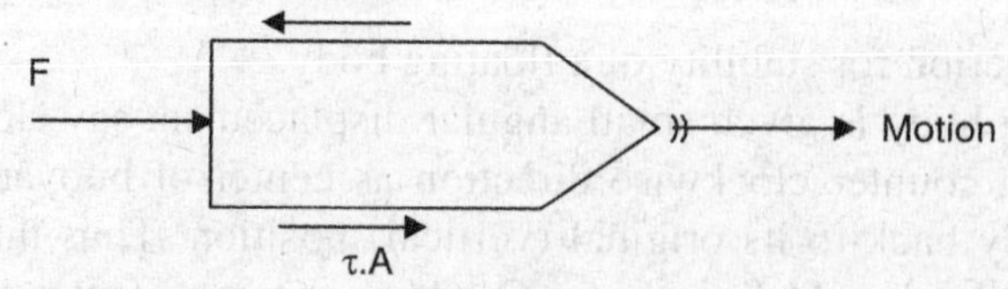

F →   → Motion

τ.A

Fig. 12

Now $\qquad \tau = \mu \dfrac{du}{dy} = \left(2 \times 10^{-5}\right)\left(\dfrac{du}{0.005 \times 10^{-3}}\right) = 4\,du$

$$F = 50 \, \text{N}/\text{cm}^2 = 5000 \, \text{N}/\text{m}^2$$

$$A = \pi L d = \pi (10 \times 10^{-3})(5 \times 10^{-3})$$

$$\therefore \qquad A = 50\pi \times 10^{-6} \, \text{m}^2$$

$\therefore$

$$(10 \times 10^{-3}) a = 5000 - 4 \, du (50\pi \times 10^{-6}) \ldots \ldots \ldots m = 10 \, \text{gm} = 10 \times 10^{-3} \, \text{kg}$$

$$\therefore \qquad a = 500 \times 10^{-3} - 0.0628 \, du$$

Using $\qquad a = \dfrac{du}{dt}$ we get

$$\dfrac{du}{dt} = 500 \times 10^{-3} - 0.0628 du. \text{ Putting } dt = 0.1s \, (\text{given})$$

$$\dfrac{du}{0.1} + 0.0628 du = 500 \times 10^{-3} \text{ or } du = 49.69 \times 10^3$$

As initial velocity is zero, $v_{\text{exit}} = 49.69 \, \text{km/s} = 13802.21 \, \text{km/h}$

(c) In the Chezy equation $V = C\sqrt{RS}$, explain the physical meaning of the terms C, R and S.

Ans.: $C = \sqrt{\dfrac{8g}{f}}$ is known as Chezy's constant or coefficient which varies inversely as square root of Durcy weisbach resistance coefficient f and is not dimensionless. C is obtained by several imperical formulae. $R = \dfrac{A}{P}$ is hydraulic radius, where A is wetted area and P is wetted perimeter. $s = \sin\theta \approx \tan\theta$. It is slope of channel bed (bottom) with θ being angle made by bed of channel with horizontal.

Q.6 (a) A tank contains fluid of variable density. The density at a depth h is given by:

$$\rho = \rho_0 (1 + \beta h^2), \ \beta = 0.05/\text{m}^2, \ \rho_0 = 900 \, \text{kg}/\text{m}^3$$

Determine the gauge pressure at a depth of 5 m if the fluid is at rest. The tank is open at the top. (Hint: Start with the differential equation for hydrostatic pressure variation). (6)

Ans.: We have $\rho = 900(1 + 0.05 \, h^2)$

Hence $\qquad dp = (\rho g) dh$ gives $dp = (900 + 45h^2) g \, dh$

$$\therefore \qquad \int_2^1 dp = g \int_0^5 (900 + 45h^2) \, dh$$

$$\therefore \qquad p_1 - p_2 = (9.81) \left[900h + 15h^3 \right]_0^5$$

or $\qquad p = 9.81 \left[4500 + 15(125 - 0) \right] \qquad \qquad \ldots \ldots p_1 = p \text{ and } p_2 = 0$

$$p = 62538.75 \text{ N/m}^2$$

or $$\boxed{p = 62.539 \text{ kPa}}.$$

(b) Derive the head *Vs* discharge relation for a *V* notch. State all the assumptions made.

Ans.: Assumptions:

1. Head '*H*' over crest is constant.
2. Losses are negligible (fluid is ideal and incompressible)
3. Flow is continuous and steady.

Let θ = Angle of the notch, H = head over notch, dH = small strip of thickness at depth '*h*'.

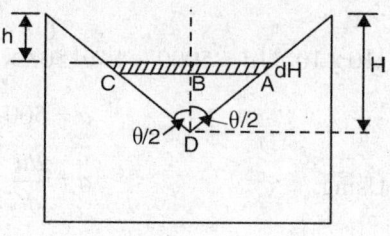

Fig. 13

From geometry, $$\tan\left(\frac{\theta}{2}\right) = \frac{CB}{BD}$$

$$CB = (H - h)\tan\left(\frac{\theta}{2}\right)$$

$$CA = 2(H - h)\tan\left(\frac{\theta}{2}\right)$$

$\therefore$ $$\text{Area of strip} = dH\left(2(H - h)\tan\left(\frac{\theta}{2}\right)\right)$$

$$Q_{th} = V_{\text{theoretical}} \times \text{Area}$$

$\therefore$ Discharge through strip $$dQ = \sqrt{2gh}\ dH\left(2(H - h)\tan\left(\frac{\theta}{2}\right)\right)$$

$\therefore$ $$Q = \int_0^H dQ \text{ and } Q_{\text{actual}}$$

$$= C_d \cdot Q_{th}$$

$\therefore$ Actual discharge $$= \int C_d \times Q_{th}$$

$$= \frac{8}{15}C_d\sqrt{2g}\ H^{5/2}.$$

(c) Given the following velocity field:

$$V = V_0(1 + \gamma t)\left[axi + \beta y^2 j\right]$$

(i) Determine the position (x_p, y_p) of a particle released at $(1, 1)$ at $t = 0$, as a function of time.

(6)

(ii) What happens to a particle released at (0, 0)?

Ans.:
$$\bar{V} = \left[V_0\alpha \times (1+\gamma t)\right]i + \left[V_0\beta y^2(1+\gamma t)\right]j$$

Integrate with respect to time $\left(\because \bar{V} = \dfrac{d\bar{r}}{dt}\right)$

$$\therefore \qquad \bar{r} = \left[V_0 a \times \left(t + \frac{\gamma t^2}{2} + C_1\right)\right]i + \left[V_0\beta y^2\left(t + \frac{\gamma t^2}{2} + C_2\right)\right]j$$

At $t = 0$, $x = 1$. Hence $V_0\alpha \times C_1 = 1$ and $V_0\beta y^2 C_2 = 1$

Assume $V_x = 0$ and $V_y = 0$ at $t = 0$

$\therefore$ $V_0\alpha \times (1-0) = 0$ and $V_0\beta y^2(1+0) = 0$

Hence $x = 0$ and $y = 0$

Note: There is some mistake/data error in the problem.

Q.7 (a) The stream function for 2D incompressible flow is given by:
$$\psi = xy^3 + x^2 y$$

   (i) Find Φ if it exists   (4)

  (ii) What is the equation of the steam line passing through (1, 1)?   (2)

Ans.:

$$\psi = xy^3 + x^2 y. \text{ Hence } \frac{\partial \psi}{\partial x} = y^3 + 2xy \text{ and } \frac{\partial \psi}{\partial y} = 3xy^2 + x^2$$

Also $\dfrac{\partial^2 \psi}{\partial x^2} = 2y$ and $\dfrac{\partial^2 \psi}{\partial y^2} = 6xy$. Hence flow is not possible.

or $u = \dfrac{\partial \psi}{\partial y}$ and $v = \dfrac{-\partial \psi}{\partial x}$ by definition of function ψ

$\therefore$ $\quad -v = y^3 + 3xy$ and $u = 3xy^2 + x^2$. Hence $\dfrac{\partial v}{\partial x} = 3x$ and $\dfrac{\partial u}{\partial y} = 6xy$.

$\therefore$ $\quad \dfrac{\partial v}{\partial y} = 3y^2 + 3x$ and $\dfrac{\partial u}{\partial x} = 3y^2 + 2x$.

As $\dfrac{\partial u}{\partial x} + \dfrac{\partial v}{\partial y} \neq 0$ the flow is not continuous (possible) ϕ does not exist as flow is rotational

i.e. $\dfrac{\partial u}{\partial y} \neq \dfrac{\partial v}{\partial x}$

(b) Compare a Venturimeter and an orifice plate, based on the following points   (4)

    (i) Cost and ease of manufacture

    (ii) Accuracy

    (iii) Energy loss

    (iv) sensitivity (output manometer deflection per unit flow rate)

Ans.:

Venturimeter	Orifice meter
(i) Costly and complicated to manufacture	(i) Cheaper and very easy to manufacture.
(ii) More accurate	(ii) Less accurate
(iii) Less loss of energy due to conical portion.	(iii) Large losses due to sudden contraction.
(iv) More sensitive	(iv) Less sensitive

(c) The velocity distribution in a boundary - layer is given by:

$$u = U_\infty \sin\left(\pi/2\, y/\delta\right) \qquad\qquad 0 \le y \le \delta$$

$$u = U_\infty \qquad\qquad\qquad\qquad y \ge \delta$$

    (i) Determine the displacement and momentum thicknesses if δ = 1 cm and U_∞ = 10m/s. $\hfill$ (8)

    (ii) What would happen to the answers of part (i) above, if the shape of the velocity profile remains the same and δ remains at 1 cm but U_∞ is doubled. $\hfill$ (2)

Ans.:

$$u = U_\infty \sin\left(\frac{\pi}{2}\cdot\frac{y}{\delta}\right). \text{ Hence } \frac{u}{U_\infty} = \sin\left(\frac{\pi}{2}\cdot\frac{y}{\delta}\right)$$

$\therefore$ Displacement thickness $\delta^* = \int\limits_0^\infty \left(1 - \frac{u}{U_\infty}\right) dy$

$\therefore$ $\delta^* = \int\limits_0^\infty \left[1 - \sin\left(\frac{\pi}{2}\cdot\frac{y}{\delta}\right)\right] dy$

$$= \delta - \frac{2\delta}{\pi}$$

$\therefore$ $\boxed{\delta^* = \delta\left(1 - \frac{2}{\pi}\right)}$

    Momentum thickness $\theta = \int\limits_0^\infty \left[\frac{u}{U_\infty}\left(1 - \frac{u}{U_\infty}\right)\right] dy$

$$= \int\limits_{0}^{\infty} \sin\left(\frac{\pi}{2} \cdot \frac{y}{\delta}\right)\left[1 - \sin\left(\frac{\pi}{2} \cdot \frac{y}{\delta}\right)\right] dy$$

$$\therefore \; \theta = \frac{2\delta}{\pi}\left(1 - \frac{\pi}{4}\right)$$

(i) Putting $\delta = 1$ cm, we get $\delta^* = 0.363$ cm and $\theta = 0.137$ cm

(ii) Since velocity profile is same and δ^* or θ does not depend on U_∞, answers would remain exactly same as in part (i) even if $U_\infty = 20$ m/s.

Q.8 (a) An aircraft is flying at a uniform speed of 1 km/s at a height of 2 km. The density and pressure at that altitude are 0.6 kg/m³ and 4×10^4 N/m² respectively.

(i) How long after it flies directly overhead will the sonic boom be heard? R for air is 287 J/kg/K and the ratio of specific heats, k is 1.4. (6)

(ii) Calculate the pressure felt at the nose of the aircraft, assuming that, in the reference frame of the plane, the air is entropically brought to rest at the nose. (4)

Ans.: $C = \sqrt{kRT} = \sqrt{\dfrac{kp}{\rho}}$. Hence $C = \sqrt{\dfrac{(1.4)(4 \times 10^4)}{(0.6)}} = 305.505$ m/s

Mach number $M_a = \dfrac{V}{C} = \dfrac{1000}{305.505}$

$\therefore \qquad\qquad M_a = 3.273$

(i) $t = \dfrac{\text{Height (altitude of plane)}}{\text{sonic velocity}\,(C)}$

$\therefore \qquad \boxed{t = 6.55 \text{ seconds}}$

Now $\quad p_s = p\left[1 + \left(\dfrac{k-1}{2}\right)(M_a)^2\right]^{\frac{k}{k-1}}$ gives the stagnation pressure.

$\therefore \qquad p_s = (4 \times 10^4)\left[1 + \left(\dfrac{0.4}{2}\right)(3.273)^2\right]^{\left(\frac{1.4}{0.4}\right)}$

$\therefore \qquad \boxed{p_s = 2.2 \times 10^6 \text{ Pascal}}$

(b) A bubble of air is released at depth of 1 m in a tank of water. If the diameter of the bubble at the time of release is 0.2 mm, calculate the gauge pressure inside the bubble (Surface tension for the air water interface is 0.073 N/m)

Ans.: $p = \dfrac{8\sigma}{d}$ gives gauge pressure in bubble in excess of outside pressure.

$$\therefore \qquad p = \frac{8(0.073)}{(0.2) \times 10^{-3}} = 2.92 \text{ kPa}$$

Now at in depth in water, $P_0 = rh$ gives $P_0 = 9.81$ kPa

Using $p_i - p_0 = \dfrac{8\sigma}{d}$ we get $p_i = p_0 + p$

or $\boxed{p_i = 12.73 \text{ kPa}}$.

(c) Compare and contrast the following:
- (i) Path line Vs streak line
- (ii) Turbulent Vs laminar flow
- (iii) 1D Vs 3D flow.

Ans.:
- (i) Path line is the locus of a single fluid particle over a period of time. Streak line is the locus of all the fluid particles at an instant of time which have passed through a reference point (or section). Smoke or dye line which can be visible show as streak line. For steady uniform flow path line coincides with stream lines.
- (ii) If fluid particles in motion (flow) move in irregular or haphazard manner, it is a turbulent flow. For this flow, inertia force dominates while viscous flow is negligibly small. This is characterised by very thorough mixing of flow, high Reynolds number and frictional loss proportional to square of velocity.
- (iii) In one dimensional (1-D) flow, flow parameters like velocity, pressure change in any one direction only, e.g. along axis of pipe having very small diameter Mathematically $v = f(x)$ or $p = f(y)$ represents 1-D flow.

Flow parameters like pressure, velocity change in all directions, e.g. flow of water in gig river or fluid flowing through large diameter pipe, etc.